Random Processes in Nuclear Reactors

Random Processes in Nuclear Reactors

M. M. R. WILLIAMS

Professor of Nuclear Engineering, Queen Mary College, University of London

PERGAMON PRESS

Oxford · New York · Toronto · Sydney

Pergamon Press Ltd, Headington Hill Hall, Oxford
Pergamon Press Inc., Maxwell House, Fairview Park, Elmsford, New York 10523
Pergamon of Canada Ltd, 207 Queen's Quay West, Toronto 1
Pergamon Press (Aust.) Pty Ltd, 19a Boundary Street, Rushcutters Bay, N.S.W. 2011, Australia

---

First edition 1974

Library of Congress Cataloging in Publication Data

Williams, Michael Maurice Rudolph.
Random processes in nuclear reactors.

Includes bibliographies.
1. Nuclear reactors—Noise. 2. Nuclear reactors—Mathematical models. 3. Stochastic processes.
I. Title.

TK9202.W53 1974 621.48'3'0184 74–4066
ISBN 0–08–017920–7

Printed at the University Printing House, Cambridge

To my Mother

Contents

Preface

RANDOM processes in one form or another have influenced scientific thinking for at least 200 years. Profound works have been written on the subject by Laplace, Bertrand, Poincaré, Borel, Markoff and more recently by Keynes and by Fisher. At the same time practical applications of the method were made in the interpretation of the physical world, the success of which had a deep influence on the old mechanistic beliefs of many classical scientists. The explanation of Brownian motion by Einstein on purely statistical grounds was of immense importance in this respect and initiated the subject of stochastic processes and the subsequent developments that have led to our present understanding of random phenomena. Having made these remarks it must now be emphasized that the present book in no way attempts to interpret or discuss any philosophical implications of the probabilistic versus deterministic viewpoints of nature. We shall accept the fact that some phenomena, whilst possibly originating from mechanistic and reversible events, can only in practice be usefully understood in terms of a random process defined by some prescribed law of probability. With this in mind, the methods of random processes are applied to the relatively new branch of engineering science known as Nuclear Engineering.

The book itself is an attempt to complement the works of Thie (1963)† and Uhrig (1970)‡ which are the only existing publications on random processes in nuclear reactors of any depth written in the English language. The aim of this book, therefore, can be summed up by the word *pragmatic*. In other words, it describes the problems that a nuclear engineer may meet which involve random fluctuations and sets out in detail how they may be interpreted in terms of various models of the reactor system. This practical approach is preceded in Chapter 1 by a brief discussion of the origins of random processes and a list of other fields of study in which random problems arise. Many of the techniques developed for the understanding of these older problems, in particular biology, are of direct interest to the nuclear engineer who can often "borrow" them with only a small change in notation and re-identification of the random source. Chapter 2 continues this line of thought and illustrates the general mathematical methods employed by applying them to simple birth and death problems such as arise in some biological and ecological situations. The application to neutrons, which can be interpreted as members of a population, then follows in an obvious manner.

It is stressed in the book that noise phenomena in nuclear reactors are divided distinctly into two classes: zero energy systems, where the branching process due to

† *Reactor Noise*, Rowman & Littlefield, Inc.
‡ *Random Noise Techniques in Nuclear Reactor Systems*, Ronald Press, N.Y.

fission is of paramount importance, and power reactors where the noise sources arise from mechanical origins rather than from any specific nuclear effect. In the case of power reactors the neutronic behaviour is indirectly influenced by feedback through the effect of the mechanical perturbations on the neutron cross-sections or geometrical arrangement of the system. In Chapter 3, therefore, we apply the general technique to zero-power problems and bring out the basic effect of fission, and fluctuations in the lifetime of neutrons, on the measured response. From this, the notion that the steady-state reactor is of little interest, from the point of view of kinetic studies, is shown to be false, since the noise superimposed on the steady background is rich in dynamic information.

Chapter 4 takes the problem further and shows how probability distributions themselves can be obtained for quite complex situations. In Chapter 5, a rather different view of random processes is put forward as promulgated by Langevin. Here the basic equations of the deterministic system are written down and then parameters are identified which could give rise to noise. In this chapter the very important formulation of the Fokker–Planck equation and some of its applications are discussed.

Chapter 6 concentrates on the interpretation of power reactor noise. From the point of view of the engineer this is an important chapter since, unless noise analysis can be accepted as a reliable technique for day-to-day analysis of commercial reactor systems, it is likely to remain a scientific curiosity of academic interest and of only peripheral practical value. This would certainly be unfortunate since it contains a wealth of information which, when interpreted correctly, would provide considerable financial savings to commercial users of power reactor systems. Indeed, paradoxically, it is the amount of information contained in noise signals that is the reason for the present caution regarding its acceptability as a diagnostic tool. This is because the information needed for a particular problem is often masked by a number of other effects. For this reason the interpretation of noise from reactor systems on the basis of simple models can be dangerous and, in Chapters 7 and 8, a detailed space-dependent study of noise is made and the limitations of many "point-model" approximations highlighted. This is particularly stressed in Chapter 8, which deals with power reactors, and it is emphasized there that a full understanding of the heat-transfer, fluid-flow and vibrational problems is required before any definitive and unequivocal statements can be extracted from noise analysis. Here it is also stressed that the noise measurements are not necessarily most effective when applied to the neutrons themselves, but that additional and sometimes superior information can be obtained by studying the fluctuations of temperature or bubble dynamics directly. Thus the last chapter, Chapter 9, deals with a number of associated problems connected with mechanical, hydraulic and thermal noise sources which must be understood before they can be incorporated into the full reactor system.

In all of the work described, full use has been made of, and credit given to, the publications of many engineers and scientists throughout the world. Whilst the books of Thie and Uhrig are referred to, probably the main assistance came from the various symposia on noise analysis that have appeared over the last ten years. Thus it is a pleasure to thank the contributors to the USAEC Proceedings on noise analysis held at the University of Florida in 1963 and again in 1966. Also, the Proceedings of the Japan–United States Seminar on Nuclear Reactor Noise Analysis in 1968 were of immense value as far as practical applications are concerned and for the personal viewpoints of those present regarding the future potential of noise in the nuclear engineering industry which, on the whole, were favourable but cautious.

With these sentiments in mind, the present monograph is written for practising nuclear engineers who wish to understand the scope of noise analysis and to apply it to their own particular problems. Research students will also benefit from the book since it is fairly self-contained whilst providing many references that will either complete any gaps or enable the reader to study some related topic in more depth. One regret is the author's unfamiliarity with the significant East European contributions to the understanding of random processes in nuclear reactors and he has therefore not been able to give due credit to much of this work. However, the author would not like the West European and North American bias of his work to be misunderstood, but simply to be regarded as a personal limitation to carry out the necessary extensive literature searches.

Finally, it is with great pleasure that the author extends his thanks to the Instituto Energia Atomica, São Paulo, Brazil. In 1973 the author was invited to give a course of lectures in São Paulo on reactor noise and it was those lectures, together with a number of constructive criticisms from the audience, that developed into the present book. The author would also like to acknowledge discussions with George Kosály of the Central Research Institute for Physics in Budapest, with Chris Greef of the Central Electricity Generating Board and with Ken Mansfield of Queen Mary College. These discussions have been of considerable value during the writing of this text. Naturally, however, any shortcomings of the book are to be attributed to the author. Ann Williams had the doubtful pleasure of typing the manuscript, for which her husband is profoundly grateful.

M. M. R. Williams

London

CHAPTER 1

Historical Survey and General Discussion

1.1. *Introduction*

The concept of statistical methods has permeated science for many years. It can be found in the work of Laplace, and the seeds of the method are evident in the Buffon needle problem (1777) which is familiar to most sixth-formers. However, in the sense of application to practical problems and recognition of the fact that a random process is involved we must begin with the kinetic theory of gases as formulated by Boltzmann (1872). It was Boltzmann's ideas on the molecular chaos assumption and the nature of irreversibility which led to the development of the subject as we know it today. However, it is not the purpose of this book to give a fundamental exposition on the nature of causal and statistical effects or, in modern parlance, deterministic and stochastic theory, but rather to apply the techniques that have been developed to the understanding of a specific subject, namely that of statistical fluctuations in nuclear reactors. Nevertheless, before dealing specifically with nuclear reactor problems it will be of value to examine the way in which the subject has developed at the practical level in its attempts to explain physical, biological, engineering and other phenomena involving some degree of randomness. We shall therefore give a short history of the subject with occasional digressions to explain the physical reasoning behind certain arguments.

Following Boltzmann's pioneering efforts to understand gas behaviour at the microscopic level, Einstein (1905) and Smoluchowsky (1906) introduced statistical methods into the study of physical phenomena in connection with the explanation of Brownian motion. Their methods were simple but effective and were developed more rigorously by Langevin (1908) whose technique had the property that it could be generalized and applied to other physical phenomena involving more general random disturbances. Using the idea of random impulses, Campbell (1909, 1910*a*, 1910*b*) was able to relate the mean square behaviour of a system to individual microscopic events. This work subsequently led to the prediction of shot noise in electronic valves. The mathematical formalism developed by these early workers in the field of random processes initiated what is today called the theory of stochastic differential and integral equations (Wax, 1954; Bharucha-Reid, 1968; Srinivasan and Vasudevan, 1971).

In the years since the explanation of Brownian motion by statistical methods many branches of science and technology have benefited from its use. It is difficult to catalogue all of the topics but the following list is fairly comprehensive:

1. Statistical mechanics (Moyal, 1949).
2. Cosmic ray showers (Arley, 1949).
3. Random telegraph signals (Rice, 1943).
4. Fatigue stresses in structural materials (Miles, 1954).
5. Turbulence (Hinze, 1959).
6. Population growth (Bartlett, 1960).
7. Bacterial colonies (Gani, 1965).
8. Insurance risks (Bharucha-Reid, 1960).
9. Economic studies (Bharucha-Reid, 1960).
10. Ecological problems (Bartlett, 1960).
11. Neurone behaviour in the brain (Rashevsky, 1948).
12. Diffusion through membranes (Breton, 1963, 1968, 1969, 1970*a*, 1970*b*).
13. Operational research and queueing theory (Bharucha-Reid, 1960).
14. Wave propagation through random media (Chernov, 1960).
15. Physical chemistry (Dainton, 1966).
16. Neutron density fluctuations (Thie, 1963).

It is, of course, the last topic which concerns us in this book. However, in many instances, either the techniques developed for one of the other subjects will be invoked or, indeed, our neutron problem may be directly linked up with, say, a stress or turbulence problem. We shall therefore continue to discuss the general development of the subject and only later specialize to reactor problems.

1.2. *The Boltzmann equation*

It is useful to discuss Boltzmann's early work on kinetic theory for several reasons but mainly because it will enable the limitations and interrelationships of later work to be more fully appreciated.

Boltzmann first published his famous equation in 1872 (Boltzmann, 1872). The prime purpose of the work was to obtain an accurate picture of the velocity distribution function of gas atoms under various conditions of non-equilibrium. The equation so derived was a particle balance for what we now call the one-particle distribution function $f(\mathbf{v},\mathbf{r},t)$, where $f(\mathbf{v},\mathbf{r},t)\,d\mathbf{v}\,d\mathbf{r}$ is the average number of atoms in the volume element $d\mathbf{r}$ about $\mathbf{r}$ with velocities in the range $(\mathbf{v},\mathbf{v}+d\mathbf{v})$ at time t.

The equation was written, for a single species gas, in the form:

$$\frac{\partial f}{\partial t}+\mathbf{v}\,.\,\nabla_{\mathbf{r}}f+\frac{\mathbf{F}}{m}\,.\,\nabla_{\mathbf{v}}f = \hat{J}(f,f), \tag{1.1}$$

where the left-hand side represents losses from phase space due to applied forces and gradients in density and the right-hand side denotes a term whose presence attempts to restore the thermal equilibrium destroyed by the loss terms. $\hat{J}(f,f)$ is in fact a collision term which is non-linear in f and assumes that the molecular chaos hypothesis is valid, i.e. all velocity correlations of atoms are destroyed after a collision.

Whilst Boltzmann was able to make very general statements about the solution of eqn. (1.1), he was unable to obtain a closed form solution. This was to be the task of Hilbert and of Chapman and Enskog. It is a sobering thought that at the beginning of this century there were still many influential critics of Boltzmann's equation, and its acceptance as a method for understanding gaseous motion was not as universal as one might have supposed. No one, today, would now doubt the validity of Boltzmann's ideas and indeed we use Boltzmann's equation to describe the statistical motion of transport for a very large number of different particles in addition to the atoms and molecules originally considered by Boltzmann.

The Boltzmann equation is, among other things, used to describe the motion of the following types of particle or pseudo-particle:

1. Neutrons.
2. Electrons and ions.
3. Radiation.
4. Phonons.
5. Motor cars.

Of paramount interest to reactor physicists are items 1 and 3, i.e. neutrons and radiation, since these arise in most aspects of reactor design.

Electron and ion behaviour are of interest to plasma physicists and designers of electronic valves, whilst phonons enter into the realms of the solid state physicist.

Motor cars constitute traffic-flow problems and Boltzmann-like equations can be derived for the flux of traffic under certain conditions (Newell, 1956).

In general, the problems that we have discussed above involve the simultaneous interaction of different particles and, for this reason, there will exist particle distribution functions $f_i(\mathbf{v},\mathbf{r},t)$ for each species i. In such cases the single Boltzmann equation is replaced by a set of coupled equations of the following form (Williams, 1971):

$$\frac{\partial f_i}{\partial t}+\mathbf{v}\,.\,\nabla_{\mathbf{r}} f_i+\frac{\mathbf{F}_i}{m_i}\,.\,\nabla_{\mathbf{v}} f_i=\sum_{j=1}^{N}\hat{J}_{ij}(f_i,f_j)+\sum_{j=1}^{N}\hat{S}_{ij}(f_i,f_j)\quad (i=1,2,\ldots N), \tag{1.2}$$

where $\hat{J}_{ij}(f_i,f_j)$ denotes collisions between particles of types i and j, and $\hat{S}_{ij}(f_i,f_j)$ is introduced to account for sources and sinks of particles arising from multiple production processes and/or removal by "capture".

This then in the classical sense is the most general Boltzmann system but it can be simplified for different physical situations. Before specializing to the case of neutrons, however, let us stress the hypothesis of molecular chaos and the consequent reduction of the equation to a consideration of *mean values* only. In a more general theory, based upon the Liouville equation for N-particles, we would have an expression for the function

$$W_N(\mathbf{r}_1,\mathbf{v}_1;\,\mathbf{r}_2,\mathbf{v}_2;\,\ldots\mathbf{r}_N,\,\mathbf{v}_N|t) \tag{1.3}$$

which defines the joint probability that, at time t, particle 1 is at $(\mathbf{r}_1,\mathbf{v}_1)$, particle 2 at $(\mathbf{r}_2,\mathbf{v}_2)$, etc. Our one particle distribution function, $f(\mathbf{r}_1,\mathbf{v}_1,t)$, is then given by

$$f(\mathbf{r}_1,\mathbf{v}_1,t) = N\int W_N(\mathbf{r}_1,\mathbf{v}_1;\,\mathbf{r}_2,\mathbf{v}_2;\,\ldots\,\mathbf{r}_N,\mathbf{v}_N|t)\,d\mathbf{r}_2 d\mathbf{v}_2\,\ldots\,d\mathbf{r}_N d\mathbf{v}_N, \tag{1.4}$$

i.e. W_N integrated over all particles except 1 (Williams, 1971). If this type of averaging is applied to the Liouville equation directly, we obtain a Boltzmann-like equation except that, on the right-hand side, $\hat{J}$ will depend not on f but on the two-particle distribution function

$$f_2(\mathbf{r}_1,\mathbf{v}_1;\,\mathbf{r}_2,\mathbf{v}_2,t) \tag{1.5}$$

which is a measure of the correlation in position and velocity between particles 1 and 2 as a result of the "memory" of their previous collisions. Boltzmann's assumption was to set

$$f_2(\mathbf{r}_1, \mathbf{v}_1; \mathbf{r}_2, \mathbf{v}_2, t) = f(\mathbf{r}_1, \mathbf{v}_1, t) f(\mathbf{r}_2, \mathbf{v}_2, t) \tag{1.6}$$

which is molecular chaos and implies no correlation.

Whilst the assumption of molecular chaos is good for dilute systems involving binary collisions, it becomes poor for dense systems. Thus high-pressure gases and, in particular, liquids must include equations for f_2 explicitly. However, we shall not discuss this point in any detail here.

1.3. *Applications in other fields*

Clearly, the application of noise methods to neutrons is a relatively recent matter and to obtain a more precise picture of how noise methods in general developed we should discuss some of the earlier applications. Thus, in 1908, Erlang (Brockmayer *et al.*, 1948) carried out work on telephone traffic in order to obtain an estimate of the optimum number of exchanges such that there was a minimum engaged period subject to the fluctuating demand of calls not exceeding some prescribed average rate. This work turned out to be a major contribution to what is now known as the theory of queues. Queueing theory is the basis of operational research and is used in such diverse activities as the loading and routing of vehicles, machine breakdown and repair, the timing of traffic lights, checkout times in supermarkets and inventory control. It is also finding use in the nuclear power field in connection with fuel reprocessing. The connection with stochastic processes follows from the fact that the source or input (e.g. customer) and the operation (e.g. service) are random variables. Thus a probability balance equation may be formulated and the conditions for prescribed probabilistic criteria evaluated.

A further very early application of random processes was in insurance for which the various risks had to be evaluated. Actuarial science is now a highly developed field of study involving the most sophisticated probabilistic methods. It is, however, a semi-empirical science since its laws have been built up by experience of actual cases and not from any basic theory of the microscopic (i.e. individual) behaviour. Abnormal situations such as epidemics are unlikely to be allowed for in actuarial tables of insurance premiums.

This discussion of insurance risks is an obvious introduction to the problems of fluctuation problems in the life sciences in general. In fact the use of mathematical methods in biology can be traced back to Galton's work on *co-relations* (i.e. the propagation of males in a family) in the last part of the nineteenth century or, if we include demographic studies as part of biology, to Graunt's *Bills of Mortality* at the beginning of the seventeenth century. Later, such famous people as Pearson, Fisher, Haldane, Lotka and McKendrick contributed to the statistical analysis of biological characteristics, agriculture, mathematical genetics, animal ecology and epidemic theory, the latter subject being of particular relevance to neutron multiplication in a fissile medium. Later work by Schlesinger, Delbruck and Luria in bacterial virology (1930s and 1940s) makes use of stochastic models for bacteriophage. The state of the art in this particular subject has been admirably surveyed by Gani (1965) from a mathematical and physical viewpoint.

The theory of epidemics, population growth and bacterial colonies are important fields in which stochastic fluctuations arise and are of paramount importance in predicting the probable outcome of a situation. Thus, for example, to study population growth, whether it be people or bacteria, it is necessary to set up a probabilistic balance equation in which the birth and death rates are included together with immigration and emigration. As well as the normal population behaviour, it is possible to consider predatory mechanisms such as wars and epidemics. From these calculations the chances of extinction of particular species can be obtained as well as the changes in the age distribution of the population with time. Knowledge of such factors is of considerable importance to sociologists and local government whose job it is to estimate future demand both in time and place. A related problem, still in the biological context, is the growth of population subject to mutation: this can apply to viruses which may subsequently become resistant to certain antibiotics and also to human populations subject to radiation hazards from both natural and artificial sources. It is important to emphasize that in all of these works average values can be very misleading and the stochastic approach is essential to study the probable (and often very large) deviations from the mean.

Another important application of stochastic processes and one which is again related to neutron behaviour in reactors is that of the theory of cascade or branching processes. For a definitive study and discussion of this subject the reader is referred to Arley (1949) who discusses basic theory and applications to cosmic radiation in the Earth's atmosphere. The problems of cascade came to the forefront of interest in the 1930s when physicists were attempting to explain the energies of electrons and photons. These particles in turn collide with the atoms of the atmosphere and cause ionization, i.e. produce further particles, are captured and also are degraded in energy. The problem was to explain the distribution at the ground or at some higher level. Basically, the radiation received consisted of the *soft-component* consisting of electrons and photons and the *hard-component* consisting of the much more penetrating mesons. Cascade theory, which is based upon a space, time and energy probability balance, enables one to calculate the mean, mean square, etc., number of particles to be expected in a shower initiated by a primary photon or electron of a given energy after a given thickness of material has been traversed. Efforts to calculate the fluctuations in cascade showers have been made by Furry (1937), Nordsieck *et al.* (1940), Scott and Uhlenbeck (1942), Heitler and Janossy (1949*a*, *b*), Janossy (1950), Janossy and Messel (1950), Bhabha and Ramakrishnan (1950), Scott (1951) and Ramakrishnan and Matthews (1954). Many of the earlier authors mentioned above have replaced the actual problem by a model which corresponds to it to a greater or lesser extent. The later authors were able to include more details of the actual energy-loss mechanisms. Solutions of the equations, which were coupled probability equations for photons and electrons, enabled the means and variances of the two types to be computed for comparison with experiment. A comprehensive summary of the cascade equations may be found in Bharucha-Reid (1960). Further applications of cascade theory have been in the development of inelastic nuclear scattering (Le Couter, 1952) in which a probability function is found for the probability that a nucleus of mass A_1 charge Z_1 with excitation energy U_1 disintegrates to leave a residual nucleus of mass A_2 charge Z_2 with excitation energy in the range (U_2, U_2+dU_2); see also Messel (1952).

Additional problems of statistical theory arise in physics and we can list the theory of radioactive decay chains in which a series of successive disintegrations occur and it is desired to find the average number and variance of a particular species after a certain time. The theory of particle counters which involve ionization cascade processes and dead time effects is a further example. There is also the theory of tracks in nuclear emulsions: here the charged particle activates some of the crystals in the gelatin as it passes through. When developed, the activated crystal grows into a grain of large size whilst the inactivated grains per unit track length leads to a random set of results which must be correlated in accordance with the statistical laws involved (Blatt, 1955).

Further applications in astronomy and astrophysics arise (Chandrasekhar, 1943), one of the most famous being an interpretation of the fluctuations in brightness of the Milky Way and another the spatial distribution of galaxies (Neyman and Scott, 1952).

Chemical reaction kinetics must also in certain situations be treated on a probabilistic basis. It is often necessary in the early stages of a reaction to examine the fluctuations in the reaction rate of certain species since this will determine the subsequent behaviour and its reproducibility: an important aspect in industrial applications. The mathematical equations are not unlike those involved in radioactive decay chains (Dainton, 1966).

Additional applications can be found in the interpretation of laser radiation. For example, Fleck (1966*a*, *b*, *c*) has examined the fluctuations of the photon populations in certain modes and excited atoms in certain states. Mandel (1958) has also examined the fluctuation of counts in a photo-electric detector by associating the incident photons with Gaussian random waves.

Let us consider now a different aspect of randomness which relies not on probability balance concepts, but rather on the "input–output" philosophy. By this terminology we mean that the conventional equation for the average value of the system is set up and *then* certain terms in it are identified as being random and prescribed statistically: these are inputs. The output is the variable whose fluctuation is of interest; the randomness of the input is communicated to the output via the response characteristics of the system. What we have described is essentially the Langevin technique and Brownian motion is its earliest application. It has been extended to more complicated systems by Uhlenbeck and Ornstein (1930) who considered a Brownian particle acted upon by harmonic forces in addition to the random effects of molecular bombardment. Coupled oscillations have also been investigated (Wang and Uhlenbeck, 1945), as have the Brownian motion of strings and rods (Van Lear and Uhlenbeck, 1931). More recent research on the random motion of continuous systems includes the work of Spiegel (1952) and Lyon (1956) on random motions of strings. Eringen (1957) has also examined random motions of bars and plates. These problems bring out very clearly the input–output nature of the problem. An interesting series of papers in which the effect of Brownian fluctuations in galvanometers has been investigated are due to Jones and McCombie (1952). These authors follow up the work of Ornstein and Van Lear which is based upon an equation for the current in the galvanometer and its associated circuit including the inductive and resistive effects. The potential generated is affected by random electromotive forces generated in the circuit and the effect of these on the deflection of the galvanometer is examined using a technique borrowed from the Langevin–Brownian theory. Further work along these lines, both experimental and theoretical, may be found in Milatz *et al.* (1953) and Milatz and Van Zolingen (1953). A general theory of thermal fluctuations in non-linear electrical systems is given by Van Kampen (1963). The comprehensive articles by Chandrasekhar (1943) and by Rice (Wax, 1954) are mandatory reading for a basic understanding of stochastic processes and their range of applications. A more fundamental study covering the theory of random functions and applications to Brownian motion, electrical current theory and quantum mechanics is to be found in Moyal (1949). Further techniques of analysis that prove useful for solving reactor problems can be found in the article on wave propagation in random media by Frisch (Bharucha-Reid, 1968). A comprehensive yet concise survey of noise phenomena may also be found in MacDonald (1962).

1.4. *Applications to nuclear reactors*

We consider now the specific application of random processes in neutron diffusion and reactor behaviour by returning to the general Boltzmann equation discussed earlier. It is from these equations that the basic equations of neutron transport can be derived and therefore an understanding of the reduction will be of importance in locating possible noise sources in reactor problems.

In this connection it is useful to regard neutrons and the nuclei with which they collide as a mixture of two gases. The fact that the moderator is actually a solid or liquid can be shown to affect only the resulting cross-sections and not the structure of the equation (Yip and Osborne, 1966).

Thus, with $f(\mathbf{v},\mathbf{r},t)$ as the neutron density and $F(\mathbf{V},\mathbf{R},t)$ as the moderator (nuclear) density, we can write eqns. (1.2) as:

$$\frac{\partial f}{\partial t}+\mathbf{v}.\nabla f = \hat{J}_{ff}(f,f)+\hat{J}_{fF}(f,F)+\hat{S}_{ff}(f,f)+\hat{S}_{fF}(f,F), \tag{1.7}$$

$$\frac{\partial F}{\partial t}+\mathbf{V}.\nabla F = \hat{J}_{FF}(F,F)+\hat{J}_{fF}(f,F)+\hat{S}_{FF}(F,F)+\hat{S}_{fF}(f,F). \tag{1.8}$$

We have omitted the force term since the only external forces acting, i.e. gravitational and magnetic, are negligible for the problems to be considered.

To obtain the basic Boltzmann equation it is important to note that the neutron particle density n in any practical situation is given by

$$n = \int f d\mathbf{v} < 10^{10}\ \text{cm}^{-3}, \tag{1.9}$$

whilst the moderator nuclei density

$$N = \int F d\mathbf{v} \simeq 10^{22}\ \text{cm}^{-3}. \tag{1.10}$$

Thus it is valid to neglect f with respect to F. For this reason, in eqn. (1.7) we can neglect $\hat{J}_{ff}(f,f)$ and $\hat{S}_{ff}(f,f)$ to arrive at

$$\frac{\partial f}{\partial t}+\mathbf{v}.\nabla f = \hat{J}_{fF}(f,F)+\hat{S}_{fF}(f,F). \tag{1.11}$$

Similarly, eqn. (1.8) becomes

$$\frac{\partial F}{\partial t}+\mathbf{V}.\nabla F = \hat{J}_{FF}(F,F)+\hat{S}_{FF}(F,F). \tag{1.12}$$

If it is further assumed that the moderator atoms are in thermal equilibrium, or that their distribution function F is known, we can accept eqn. (1.11) as a linear equation for f.

The calculation of $\hat{J}_{fF}$ and $\hat{S}_{fF}$ remains, however. Now $\hat{J}_{fF}$ denotes scattering and, in view of the rarefied nature of the neutron gas, it is reasonable to suppose that f and F are uncorrelated and that molecular chaos is valid. Moreover, it is possible to rewrite $\hat{J}_{fF}$ as the difference between two terms, viz.

$$\hat{J}_{fF}(f,F) = \int d\mathbf{v}_1' \int d\mathbf{v}_2' \int d\mathbf{v}_2 W_{fF}(\mathbf{v}\to\mathbf{v}_1';\mathbf{v}_2\to\mathbf{v}_2')F(\mathbf{v}_2')f(\mathbf{v}_1',\mathbf{r},t)$$
$$-f(\mathbf{v},\mathbf{r},t)\int d\mathbf{v}_1' \int d\mathbf{v}_2' \int d\mathbf{v}_2 W_{fF}(\mathbf{v}\to\mathbf{v}_1';\ \mathbf{v}_2\to\mathbf{v}_2')F(\mathbf{v}_2), \tag{1.13}$$

which physically represent the scattering in and scattering out of the velocity range $(\mathbf{v},\mathbf{v}+d\mathbf{v})$. $W_{fF}(\ldots)$ is the interparticle scattering function (Williams, 1971) and we find it convenient to condense eqn. (1.13) to the form

$$\hat{J}_{fF}(f,F) = \int d\mathbf{v}'\Sigma(\mathbf{v}'\to\mathbf{v})f(\mathbf{v}',\mathbf{r},t)-v\Sigma_s(v)f(\mathbf{v},\mathbf{r},t), \tag{1.14}$$

where $\Sigma(\mathbf{v}'\to\mathbf{v})$ is the neutron–nucleus scattering kernel and $\Sigma_s(v)$ is the total scattering cross-section defined by

$$v\Sigma_s(v) = \int d\mathbf{v}'\Sigma(\mathbf{v}\to\mathbf{v}'). \tag{1.15}$$

These cross-sections are assumed known in all subsequent studies of the equation.

The production term $\hat{S}_{fF}(f,F)$ is in many respects more difficult and subtle than the scattering one—at least in the context of neutronics. It is from this term, for example, that neutron noise will originate, although the simple treatment of $\hat{S}_{fF}(\ldots)$ would not suggest this. Following the simple theory we can obtain an expression for $\hat{S}_{fF}$ by assuming that neutrons are produced from an external source $Q(\mathbf{v},\mathbf{r},t)$ and from fission. Neutrons are also removed by capture. The fission source itself can be written

$$S_f = (1-\beta)\frac{\chi(v)}{4\pi}\int d\mathbf{v}\,\bar{\nu}v\Sigma_f(v)f(\mathbf{v},\mathbf{r},t)+\frac{1}{4\pi}\sum_i\lambda_i C_i(\mathbf{r},t)\chi_i(v), \tag{1.16}$$

where λ_i is the decay constant for delayed neutrons of precursor concentration $C_i(\mathbf{r},t)$. $\chi(v)$ is the fission spectrum of prompt neutrons and $\chi_i(v)$ that of the ith delayed neutron group.

C_i is obtained from the equation

$$\frac{dC_i}{dt} = -\lambda_i C_i+\beta_i\int d\mathbf{v}\,\bar{\nu}v\Sigma_f(v)f(\mathbf{v},\mathbf{r},t), \tag{1.17}$$

where

$$\beta = \sum_i\beta_i, \tag{1.18}$$

β_i being the fraction of neutrons in the ith delayed group from a fission event. $\Sigma_f(v)$ is the fission cross-section and $\bar{\nu}$ is the *mean* number of neutrons per fission.

Absorption can be written as a negative source term, viz.

$$S_a = -v\Sigma_a(v) f(\mathbf{v}, \mathbf{r}, t). \tag{1.19}$$

Thus

$$\hat{S}_{fF}(f, F) = Q + S_f + S_a. \tag{1.20}$$

We note that the moderator atom distribution is included in the definition of the macroscopic cross-sections $\Sigma_f(v)$ and $\Sigma_a(v)$ as follows:

$$v\Sigma_{f,a}(v) = \int d\mathbf{v} F(V) |\mathbf{v} - \mathbf{V}| \sigma_{f,a}(|\mathbf{v} - \mathbf{V}|). \tag{1.21}$$

This, of course, leads to Doppler broadening in the case of resonance type microscopic cross-sections $\sigma_{f,a}(v)$ (Bell and Glasstone, 1970). We have now the complete Boltzmann equation for the average particle density $f(\mathbf{v}, \mathbf{r}, t)$ and it is this equation that is used in conventional studies of reactor physics problems. However, there is little evidence of any noise or fluctuation phenomena in this equation: indeed the only obvious source of noise or *correlation* was discarded with the introduction of molecular chaos. Nevertheless, whilst molecular chaos is indeed valid for neutrons, the correlation distance being of the order of a nuclear diameter ($\sim 10^{-13}$ cm) and the correlation time of the order of a collision time ($\sim 10^{-14}$ sec), there is a new source of noise introduced which is, in general, not present in the gaseous systems discussed by Boltzmann. Such sources of noise arise from the extraneous source Q, the fission source S_f and from absorptions. These correlations are of a completely different order from those involved in molecular chaos and have time scales of milliseconds, and correlation distances of the order of a migration length. It is these phenomena, together with the additional mechanical perturbations in a power reactor, with which we associate the term *neutron noise*. The precise way in which the correlations mentioned above affect the form of the Boltzmann will be discussed in detail in the following chapters.

The detailed mathematical reasoning behind the acceptance of the molecular chaos assumption and the necessary inclusion of correlations due to sources and fission may be found in Yip and Osborne (1966).

Before attempting a full mathematical discussion on neutron noise it will be useful to talk in more general terms about its origin and the manner in which it manifests itself in a nuclear reactor.

1.5. *Reactor noise*

The concept of reactor noise is best explained by reference to the steady-state reactor. For example, if one is asked to draw a graph of the output of a reactor operating at steady power as a function of time, the answer might appear to be trivial as shown in Fig. 1.1. However, a closer examination of the output of the detector which measures this power will show that superimposed upon the steady output is an apparently random fluctuation, so that the output takes the form shown in Fig. 1.2. These fluctuations about the average are referred to as reactor noise and the purpose of reactor noise investigation is to see whether an analysis of the fluctuations will give information about the reactor behaviour. Quite clearly, since the fluctuations are functions of time the information will concern the kinetic response of the system. We have therefore the possibility of gaining dynamic information from measurements at

FIG. 1.1. Power output from ideal deterministic steady-state reactor.

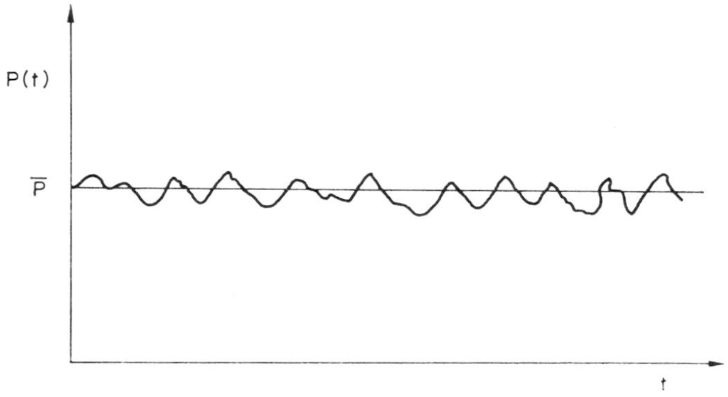

FIG. 1.2. Actual power output from a steady-state reactor.

steady state. If such an investigation is successful it would indeed be a significant advance both economically and technically since it would obviate the need for the reactor to be shut down to perform dynamic safety tests. Also it would present us with a method for continuous monitoring of the reactor behaviour with the possibility of predicting malfunctions in advance.

The problems, then, are: (1) how to extract the information from the random signal; (2) how to investigate the causes of the randomness; and (3) how to develop a theory to compare with experiment. Answers to these questions will be given in subsequent chapters.

1.6. *Classification of reactor noise*

It is appropriate at this point to classify reactor noise studies into two distinct types: (1) zero energy systems, (2) power reactors. This distinction is important since we shall find that in general very different sources of noise are prevalent in the two situations.

1.7. *Zero-energy systems*

In such systems we assume that effects due to temperature and mechanical or hydraulic effects of any type are absent. Thus the noise source arises entirely from nuclear origins and any additional independent neutron sources that may be present (de Hoffmann, 1958; Feynman, 1956).

The following noise sources are responsible for fluctuations in zero-power systems:

1. Fluctuations in the number of neutrons per fission, i.e. there is a probability distribution $p(\nu)$ that ν neutrons will be emitted per fission. In conventional reactor theory we are only concerned with the average value

$$\bar{\nu} = \sum_{\nu} \nu p(\nu).$$

In fact momentary local increases in ν can lead to corresponding local increases in the multiplication factor and hence the neutron flux.

2. Fluctuations in the time between nuclear events. In general, one thinks in terms of mean lifetimes $\bar{\tau}_a$, $\bar{\tau}_f$, $\bar{\tau}_s$ for absorption, fission and scattering, respectively. In fact, there is a probability law describing the time, τ, spent between collisions, viz.

$$P(\tau)\,d\tau = e^{-\tau/\bar{\tau}} \frac{d\tau}{\bar{\tau}}, \tag{1.22}$$

where $P(\tau)\,d\tau$ is the probability that a particular event will occur in the time interval $(\tau, \tau+d\tau)$ after the previous event of that type. This law is entirely analogous to the well-known collision probability law

$$F(x)\,dx = e^{-x/\bar{\lambda}} \frac{dx}{\bar{\lambda}}, \tag{1.23}$$

where $\bar{\lambda}$ is the mean free path and $F(x)\,dx$ is the probability that a neutron starting at $x = 0$ will suffer a collision in the region $(x, x+dx)$. The connection between $P(\tau)$ and $F(x)$ can be obtained by noting that $v = x/\tau$, where v is the neutron speed.

Bearing these laws in mind, we see the possibility of a succession of short times for fission events occurring, thereby leading to a momentary increase in the fission reaction rate. Alternatively, absorptions could predominate, leading to a decrease in neutron population. These are transient phenomena whose average fluctuation time is of the order of milliseconds. Nevertheless, they can be measured and, as we shall see later, can lead to dynamic information useful in reactor design and stability.

3. The probability that a particular event will be a fission, a scattering or an absorption. It is possible, for example, for a succession of fissions to occur, or absorptions, leading once again to momentary increases or decreases in multiplication.

1.8. *Power reactor noise*

In addition to the specifically nuclear effects discussed above, power reactors contain additional sources of noise introduced by mechanical perturbations. These can be due to vibration of mechanical parts, boiling of the coolant, fluctuations of temperature and pressure and many other phenomena peculiar to a particular reactor. The mathematical treatment of this type of noise is generally most conveniently treated by the Langevin technique and this, and the associated physical and engineering problems, will be discussed fully in Chapter 6. There is little doubt that if noise analysis in nuclear systems is to have any real future in nuclear engineering it must lead to results of practical engineering value. Two recent reviews by Seifritz and Stegemann (1971) and by Uhrig (1970) indicate that the theory of zero-power noise is fully understood and only cumbersome mathematical details remain to be solved. On the other hand, the noise analysis of power reactors is in its infancy due mainly to a lack of knowledge about the variety of noise mechanisms involved. As we have mentioned above, power reactors are strongly influenced by mechanical excitations which can arise from boiling, vibration, or temperature variations and a host of other unavoidable technological processes. In terms of the input–output concept we are not only ignorant of the nature of the input but in many cases of the system response function as well. A very clear picture of the problems facing workers in power reactor noise has been given by Kosály (1973) and many of these problems will be dealt with in depth in this book; nevertheless, it is worth quoting Kosály who says: "...the great number of noise sources and fluctuating parameters is the main stumbling-block for scientists in the field, especially for those who have become accustomed to the fascinating clearness of zero-power noise studies. At the same time it is precisely the difficulties that constitute the merit of the topic; ..." These views are held by the present author and constitute a major reason for the presentation of this book to the engineering and scientific community.

References

ARLEY, N. (1949) *On the Theory of Stochastic Processes and their Application to the Theory of Cosmic Radiation*, John Wiley, N.Y.

BARTLETT, M. S. (1960) *Stochastic Population Models*, Methuen Monograph on Applied Probability and Statistics.

BELL, G. and GLASSTONE, S. (1970) *Nuclear Reactor Theory*, Van Nostrand.

BHABHA, H. J. and RAMAKRISHNAN, A. (1950) The mean square deviation of the number of electrons and quanta in the cascade theory, *Proc. Ind. Acad. Sci.* A**32**, 141.

BHARUCHA-REID, A. T. (1960) *Elements of the Theory of Markov Processes and Their Applications*, McGraw-Hill.

BHARUCHA-REID, A. T. (1968) *Probabilistic Methods in Applied Mathematics*, Vol. 1, Academic Press, N.Y.

BLATT, J. M. (1955) Theory of tracks in nuclear research emulsions, *Australian J. Phys.* **8**, 248.

BOLTZMANN, L. (1872) *Lectures on Gas Theory*, translated by Stephen G. Brush, Univ. of California Press, 1964.

BRETON, J. P. (1963) *J. Chim. Phys.* **60**, 294.

BRETON, J. P. (1968) D.Sc. Thesis, Faculté des Sciences de Paris.

BRETON, J. P. (1969) *Phys. Fluids*, **12**, 2019.

BRETON, J. P. (1970*a*) *J. Phys. Paris*, **13**, 613.

BRETON, J. P. (1970*b*) *Physica*, **50**, 365.

BROCKMAYER, E. *et al.* (1948) *The Life and Works of A. K. Erlang*, Copenhagen Telephone Co., Copenhagen.

BUFFON, GEORGES-LOUIS LECLERC, COMTE DE (1777) *Essai d'arithmétique morale*, published in *Supplément à' l'Histoire Naturelle*, IV, pp. 46–148 (Paris).

CAMPBELL, N. R. (1909) *Proc. Camb. Phil. Soc.* **15**, 117, 310.

CAMPBELL, N. R. (1910*a*) *Proc. Camb. Phil. Soc.* **11**, 826.

CAMPBELL, N. R. (1910*b*) *Z. Phys.* **11**, 826.

CHANDRASEKHAR, S. (1943) Stochastic problems in physics and astronomy, *Rev. Mod. Phys.* **15**, 1.

CHERNOV, L. A. (1960) *Wave Propagation in Random Medium*, McGraw-Hill.

DAINTON, F. S. (1966) *Chain Reactions: an Introduction*, Methuen Monograph.

DE HOFFMANN, F. (1958) Statistical aspects of pile theory, in *The Science and Engineering of Nuclear Power*, **2**, Addison-Wesley.

EINSTEIN, A. (1905) *Ann. Phys.* **17**, 549, or see: A. EINSTEIN, *Investigations on the Theory of the Brownian Movement* (collected papers), Dover, 1956.

ERINGEN, A. C. (1957) Response of beams and plates to random loads, *J. Appl. Mech.* **24**, 46.

FEYNMAN, R. P. *et al.* (1956) Dispersion of the neutron emission in U^{235} fission, *J. Nucl. Energy*, **3**, 64.

FLECK, J. A. (1966*a*) Quantum theory of laser radiation. I. Many-atom effects, *Phys. Rev.* **149**, 309.

FLECK, J. A. (1966*b*) Quantum theory of laser radiation. II. Statistical aspects of laser light, *Phys. Rev.* **149**, 322.
FLECK, J. A. (1966*c*) Intensity fluctuations and statistics of laser radiation, *Phys. Rev.* **152**, 278.
FURRY, W. H. (1937) On fluctuation phenomena in the passage of high energy electrons through lead, *Phys. Rev.* **52**, 569.
GANI, J. (1965) *Stochastic Models for Bacteriophage*, Methuen Review Series in Applied Probability, Vol. 4.
HEITLER, W. and JANOSSY, L. (1949*a*) On the absorption of meson producing nucleons, *Proc. Phys. Soc.* (*Lond.*), Ser. A, **62**, 374.
HEITLER, W. and JANOSSY, L. (1949*b*) On the size-frequency distribution of penetrating showers, *Proc. Phys. Soc.* (*Lond.*), Ser. A, **62**, 669.
HINZE, J. O. (1959) *Turbulence, An Introduction to its Mechanism and Theory*, McGraw-Hill.
JANOSSY, L. (1950) Note on the fluctuation problem of cascades, *Proc. Phys. Soc.* (*Lond.*), Ser. A, **63**, 241.
JANOSSY, L. and MESSEL, H. (1950) Fluctuations of the electron–photon cascade moments of the distribution, *Proc. Phys. Soc.* **63A**, 1101.
JONES R. V. and MCCOMBIE, C. W. (1952) Brownian fluctuations in galvanometers and galvanometer amplifiers, *Phil. Trans. Roy. Soc.* A**244**, 205.
KOSÁLY, G. (1973) Remarks on a few problems in the theory of power reactor noise, *J. Inst. Nucl. Engr.* **14**, 67.
LANGEVIN, P. (1908) *Comptes Rend. Acad. Sci., Paris,* **146**, 530.
LE COUTER, K. J. (1952) Statistical fluctuations in nuclear evaporation, *Proc. Phys. Soc.* (*Lond.*), A**65**, 718.
LYON, R. H. (1956) Response of strings to random noise fields, *J. Acoust. Soc. Am.* **28**, 391.
MACDONALD, D. K. C. (1962) *Noise and Fluctuations: an Introduction*, Wiley.
MANDEL, L. (1958) Fluctuations of photon beams and their correlations, *Proc. Phys. Soc.* (*Lond.*), **72**, 1037.
MESSEL, H. (1952) The solution of the fluctuation problem in nuclear cascade theory: homogeneous nuclear matter, *Proc. Phys. Soc.* (*Lond.*), A**65**, 465.
MILATZ, J. M. W. *et al.* (1953) The reduction in the Brownian motion of electrometers, *Physica*, XIX, 195.
MILATZ, J. M. W. and VAN ZOLINGEN, J. J. (1953) The Brownian motion of electrometers, *Physica*, XIX, 181.
MILES, J. W. (1954) On structural fatigue under random loading, *J. Aero. Sci.* **21**, 753.
MOYAL, J. E. (1949) Stochastic processes and statistical physics, *J. Roy. Statist. Soc.*, Ser. B, **11**, 150.
NEWELL, G. F. (1956) Statistical analysis of the flow of highway traffic through a signalized intersection, *Quart. J. Appl. Math.* **13**, 353.
NEYMAN, J. and SCOTT, E. L. (1952) A theory of the spatial distribution of galaxies, *Astrophys. J.* **116**, 144.
NORDSIECK, A. *et al.* (1940) On the theory of cosmic ray showers. I. The Furry model and the fluctuation problem, *Physica*, **7**, 344.
RAMAKRISHNAN, A. and MATTHEWS, P. M. (1954) Studies on the stochastic problem of electron–photon cascades, *Prog. Theo. Phys.* **11**, 95.
RASHEVSKY, N. (1948) *Mathematical Biophysics*, Chicago Univ. Press.
RICE, S. O. (1943) Mathematical analysis of random noise, *Bell System Technical J.* **23**, 24 (also see WAX, N., 1954).
SCOTT, W. T. and UHLENBECK, G. E. (1942) On the theory of cosmic ray showers. II. Further contributions to the fluctuation problem, *Phys. Rev.* **62**, 497.
SCOTT, W. T. (1951) On the fluctuations and the general distribution problem in electron cascades, *Phys. Rev.* **82**, 893.
SEIFRITZ, W. and STEGEMANN, D. (1971) Reactor noise analysis, *Atomic Energy Rev.* (*IAEA*), **9**, 129.
SMOLUCHOWSKY, M. VON (1906) *Ann. Phys.* **21**, 756.
SPIEGEL, M. R. (1952) The random vibrations of a string, *Q. Appl. Maths.* **10**, 25.
SRINIVASAN, S. K. and VASUDEVAN, R. (1971) *Introduction to Random Differential Equations and Their Applications*, Elsevier, N.Y.
THIE, J. (1963) *Reactor Noise*, Rowman & Littlefield, Inc.
UHLENBECK, G. E. and ORNSTEIN, L. S. (1930) On the theory of the Brownian motion, *Phys. Rev.* **36**, 823.
UHRIG, R. E. (1970) *Random Noise Techniques in Nuclear Reactor Systems*, Ronald Press, N.Y.

Van Kampen, N. G. (1963) Thermal fluctuations in non-linear systems, *J. Math. Phys.* **4**, 190.
Van Lear, G. A. and Uhlenbeck, G. E. (1931) The Brownian motion of strings and elastic rods, *Phys. Rev.* **38**, 1583.
Wang, M. C. and Uhlenbeck, G. E. (1945) On the theory of the Brownian motion, II, *Rev. Mod. Phys.* **17**, 323.
Wax, N. (1954) *Selected Papers on Noise and Stochastic Processes*, Dover, N.Y.
Williams, M. M. R. (1971) *Mathematical Methods in Particle Transport Theory*, Butterworth, London.
Yip, S. and Osborne, R. K. (1966) *The Foundations of Neutron Transport Theory*, Gordon & Breach, N.Y.

CHAPTER 2

Introductory Mathematical Treatment

2.1. *Introduction*

The mathematical treatment of noise in nuclear reactors operating at zero power is based upon the construction of an equation for the probability function

$$P_{N_1,N_2}\ldots(R_1,t_1;\,R_2,t_2;\,\ldots\,|\mathbf{r},\mathbf{v},t) \tag{2.1}$$

which denotes the probability that, given one neutron at position **r**, velocity **v** at time t, there will result N_1 neutrons in a region R_1 of phase space at time t_1, N_2 neutrons in R_2 at t_2, etc.

In a sense this quantity is a stochastic Green's function from which all other probabilities can be derived. It was first introduced into reactor theory by Pál (1958) and its use has been extended by Bell (1965). Both Pál and Bell have deduced an equation from which $P_N(\ldots)$ can be obtained. This equation is rather complicated and interested readers are advised to consult the original literature for the details.

A much simpler quantity to study in order to gain basic understanding of the noise process is $P(N,t)$ which is the probability that, at time t, there are N neutrons in the system. We shall use this simple description to understand the way in which the neutron population fluctuates in the point model approximation of one-speed theory. First, however, we shall discuss two random processes, which are simpler than that of fission, to illustrate the methods employed to solve this type of problem.

2.2. *The Poisson process*

In the Poisson process it is assumed that the chance of an event being recorded in any short interval of time, Δt, is independent, not only of the previous states of the system at earlier times (i.e. the definition of a Markov process), but is also independent of the present state (Srinivasan and Vasudevan, 1971).

This process applies to radioactive decay, provided Δt is short compared to the half-life of the substance under investigation. Let us illustrate the Poisson process, which is one of the simplest problems of a random nature, by considering the probability $P(n,t)$ that a counter detects n counts in time t.

If λ is the probability per unit time of a detection, then we may write for $P(n,t)$ the following balance equation:

$$P(n,t+\Delta t) = P(n-1,t)\lambda\Delta t+P(n, t)(1-\lambda\Delta t). \tag{2.2}$$

We ignore terms $O(\Delta t^2)$ since it is assumed that the chance of two detections occurring in Δt is negligibly small.

The above equation arises by consideration of joint and mutually exclusive probabilities. Thus, in words, we can say that: the probability of n detections in time $t+\Delta t$, $[=P(n,t+\Delta t)]$, is equal to the probability that at time t there were $(n-1)$ detections, multiplied by the probability of a detection in Δt, $[=P(n-1,t)\lambda\Delta t]$, plus the probability that at time t there were already n-detections, multiplied by the probability of no detection in Δt, $[=P(n,t)(1-\lambda\Delta t)]$.

Allowing $\Delta t \to 0$ converts this balance equation to a differential–difference equation, viz.

$$\frac{dP(n,t)}{dt} = \lambda\{P(n-1,t)-P(n,t)\}. \tag{2.3}$$

If we start with the counter set to zero, the initial condition is

$$P(n,0) = \delta_{n,0} \tag{2.4}$$

and the solution of the equation is readily obtained as

$$P(n,t) = e^{-\lambda t}\frac{(\lambda t)^n}{n!} \quad (n = 0,1,2,\ldots), \tag{2.5}$$

which is the Poisson distribution.

We note that the mean value $\bar{n}$ is given by

$$\bar{n} = \sum_{n=0}^{\infty} nP(n,t) = \lambda t \tag{2.6}$$

and the variance σ^2 by $\quad \sigma^2 = \overline{n^2}-\bar{n}^2 = \bar{n} \equiv \text{mean value}. \quad (2.7)$

The Poisson distribution also arises in other fields, such as the probability of a telephone exchange receiving a call in a given time. However, in that connection, more complicated probability laws frequently arise (Bharucha-Reid, 1960).

As far as the practical utility of the probability $P(n,t)$ is concerned we could equally well define the mean time $\langle t\rangle$ and the associated variance $\langle t^2\rangle-\langle t\rangle^2$ for N particles to be detected. Thus an experiment could be performed in which, after N counts had been registered, the time t_i was recorded. Repeating the experiment a number of times would allow $\langle t\rangle$ and $\langle t^2\rangle$ to be measured.

According to the theory

$$\langle t\rangle = \frac{\int_0^{\infty} tP(N,t)\,dt}{\int_0^{\infty} P(N,t)\,dt} = \frac{N+1}{\lambda} \tag{2.8}$$

and

$$\langle t^2\rangle-\langle t\rangle^2 = \frac{N+1}{\lambda^2} = \frac{1}{\lambda}\langle t\rangle. \tag{2.9}$$

Thus, in practice, either method could be used to determine the system parameter λ.

An interesting extension of the Poisson process as applied to the detector count rate problem is to include a dead time, a, during which, due to instrumental limitations, no count can be registered even if a particle does enter the detector. The probability balance eqn. (2.2) is now modified as follows:

$$P(n,t+\Delta t) = P(n-1,t-a)\, e^{-a\lambda}\lambda\Delta t + P(n,t-a)e^{-a\lambda}(1-\lambda\Delta t). \tag{2.10}$$

The first term on the right-hand side arises in the following manner: $P(n-1,t-a)$ is the probability of $n-1$ counts at time $t-a, \exp(-a\lambda)$ from eqn. (2.5) is the probability of no counts in a time a, and $\lambda\Delta t$ is the probability of a detection in time Δt. A similar argument leads to the second term. Converting the equation to a differential difference one, we find

$$\frac{dP(n,t)}{dt} = \lambda e^{-a\lambda}\{P(n-1,t-a)-P(n,t-a)\}. \tag{2.11}$$

This equation may be solved by a technique that we shall introduce in detail later (the generating function method), namely, multiplying by x^n and summing over all n. We then find that

$$G(x,t) = \sum_{n=0}^{\infty} P(n,t)x^n \tag{2.12}$$

is given by

$$\frac{\partial G(x,t)}{\partial t} = \lambda e^{-a\lambda}(x-1)\,G(x,t-a). \tag{2.13}$$

The boundary condition is that $P(n,t) = \delta_{n,0}$ for all $t<a$, or $G(x,t) = 1$ for $t<a$. This leads to the solution

$$P(n,t) = \sum_{m=n}^{[t/a-1]} \lambda^m \frac{(-)^{m-n}}{m!} \binom{m}{n} e^{-\lambda ma}(t-ma)^m, \tag{2.14}$$

where $[t/a-1]$ is the largest integer in $t/a-1$.

The mean value of the number of detections in time t is given by $\bar{n}(t) = \lambda t \exp(-a\lambda)$, i.e. it is reduced by the dead-time factor as we expect. Interestingly, the variance $\overline{n^2}-\bar{n}^2$ remains equal to the mean as for the Poisson distribution. Thus the introduction of a dead time, a, reduces the variance by a factor $\exp(-a\lambda)$ compared with the true Poisson process.

2.3. *A simple birth and death problem*

In a great many practical situations the appearance of new particles, i.e. births, or the disappearance of existing particles, i.e. deaths, does depend to some extent on the present population size—even if it still remains independent of past events. This is particularly true of neutronic fluctuations and, in order to set the scene for that more complicated situation, we shall consider first the simple birth and death problem.

Let us define the probability of a discrete random variable taking the value n at time t by $P(n,t)$.

Let us now make the assumption that the chance of a given particle producing a new one in time Δt is $\lambda\Delta t$. Thus in time Δt the population will increase by one. Now if the chance of any given entity reproducing in time Δt is $\lambda\Delta t$, the corresponding probability that the complete population of size n will increase by one unit is $n\lambda\Delta t$.

Similarly, if $\mu\Delta t$ is the chance that an entity will disappear (i.e. die) in Δt, then the probability that the complete population of size n will decrease by one unit is $n\mu\Delta t$.

The chance of no change is

$$1-(\lambda+\mu)n\Delta t$$

and we recall that the chance of more than one of these events taking place in Δt is negligibly small.

Let us now write the balance equation for $P(n,t)$, viz.

$$\begin{aligned} P(n,t+\Delta t) = {} & \underbrace{P(n-1,t)}_{\substack{\text{Probability of } (n-1)\\ \text{entities at } t}} \times \underbrace{(n-1)\lambda\Delta t}_{\substack{\text{Probability of one}\\ \text{birth in } \Delta t}} \\ & + \underbrace{P(n+1,t)}_{\substack{\text{Probability of } (n+1)\\ \text{entities at } t}} \times \underbrace{(n+1)\mu\Delta t}_{\substack{\text{Probability of one}\\ \text{death in } \Delta t}} \\ & + \underbrace{P(n,t)}_{\substack{\text{Probability of } n\\ \text{entities at } t}} \times \underbrace{[1-(\mu+\lambda)n\Delta t]}_{\substack{\text{Probability of no}\\ \text{change in } \Delta t}}. \end{aligned} \tag{2.15}$$

Letting $\Delta t \to 0$, this reduces to the following differential–difference equation:

$$\frac{dP(n,t)}{dt} = \lambda(n-1)\,P(n-1,t)-(\lambda+\mu)nP(n,t)+\mu(n+1)P(n+1,t). \tag{2.16}$$

If we use the initial condition $\quad P(n,0) = \delta_{n,n_0}, \qquad (2.17)$

i.e. there is an initial number of particles equal to n_0, we can obtain a solution to the equation. This solution is rather complicated and not of sufficient relevance here to write down. The mean $\bar{n}$, and variance σ^2, however, can be obtained rather more simply and lead to

$$\bar{n}(t) = n_0 e^{-(\lambda-\mu)t} \tag{2.18}$$

and

$$\sigma^2(t) = \overline{n^2}-\bar{n}^2 = n_0\frac{(\lambda+\mu)}{(\lambda-\mu)}e^{(\lambda-\mu)t}\{e^{(\lambda-\mu)t}-1\}. \tag{2.19}$$

Thus, as we might expect, the mean value follows the exponential growth law whose exponent is determined by the difference between birth and death rates. The variance also increases for $\lambda > \mu$, showing that when births exceed deaths the fluctuations grow with time. Indeed, even when $\lambda = \mu$, i.e. births and deaths balance, we find that

$$\overline{n^2}-\bar{n}^2 = 2n_0\,\lambda t, \tag{2.20}$$

i.e. the variance increases linearly.

It can be shown that this simple model is identical to the problem of neutron multiplication in the point model approximation for the case when the number of neutrons per fission is two. Thus a study of this problem is not as academic as it may seem.

Before proceeding to the case of neutron multiplication, one further point is worth noting from this preliminary analysis, viz. the *extinction probability*, which we write mathematically as

$$\lim_{t\to\infty} P(0,t)$$

and which physically is the probability that, after a very long time, there will be no particles remaining in the system. This quantity is relevant in connection with reactor safety studies. The general form of the "birth and death" process value of $P(0,t)$ can be calculated exactly as

$$P(0,t) = \left\{\frac{\mu(e^{(\lambda-\mu)t}-1)}{\lambda e^{(\lambda-\mu)t}-\mu}\right\}^{n_0}. \tag{2.21}$$

For $\lambda \leqslant \mu$, clearly

$$P(0,\infty) = 1, \tag{2.22}$$

i.e. by waiting long enough the initial population, n_0, will die out—become extinct. In neutronic language we have a sub-critical system.

On the other hand, for $\lambda > \mu$, we find

$$P(0,\infty) = \left(\frac{\mu}{\lambda}\right)^{n_0}, \tag{2.23}$$

which indicates that there is a finite probability for the initial population to become extinct. The system is super-critical although, because of the probabilistic nature of the physical process, it has a finite probability of decay. Such behaviour could not be predicted by conventional average value arguments and we shall return to this problem in Chapter 4.

2.4. *Application to neutrons: prompt effect only*

We shall now apply these techniques to neutrons. For simplicity space and energy dependence will be neglected, as also will delayed neutrons. However, it should be noted that these restrictions are unnecessary and are only used to simplify the exposition.

The fluctuations in the neutron population, $P(N,t)$, due to fission differs from the simple birth and death process only in so far as we must define $p(n)$, where

$p(n)$ = probability that n prompt neutrons are produced by a single fission event.

Also we shall allow an independent source of S neutrons per second to be present: in the jargon of the birth and death problem this is known as *immigration* and it will be assumed to be independent of the present state of the system.

We also define, in analogy with λ and μ, the following parameters:

$$\tau_f = \text{mean lifetime of a neutron for fission} = \frac{1}{v\Sigma_f} = \frac{1}{\lambda_f},$$

$$\tau_c = \text{mean lifetime of a neutron for capture} = \frac{1}{v\Sigma_c} = \frac{1}{\lambda_c}.$$

To put it another way:

λ_f is the probability per unit time that a neutron induces a fission,
λ_c is the probability per unit time that a neutron is captured.

Now we set up a balance equation for $P(N,t)$ as follows:

$$P(N,t+\Delta t) \quad = \quad P(N-1,t) \quad \times \quad S\Delta t$$

Probability that there are N neutrons at time $t+\Delta t$	=	Probability that there are $(N-1)$ neutrons at time t	×	Probability that in time Δt the source emits one neutron

$$+ \quad P(N+1,t) \quad \times \quad (N+1)\lambda_c \Delta t$$

Probability that at time t there are $(N+1)$ neutrons × Probability that in time Δt one neutron is captured, given that there are $(N+1)$ neutrons in the system

$$+\sum_n P(N+1-n,t)\times p(n)\times \quad (N+1-n)\lambda_f \Delta t$$

Probability that at time t there are $(N+1-n)$ neutrons × Probability that a neutron loss gives rise to n new neutrons × Probability that in time Δt one neutron causes a fission, given that there are $(N+1-n)$ neutrons in the system

$$+ \quad P(N,t) \quad \times \{1-S\Delta t - N(\lambda_c+\lambda_f)\Delta t\}.$$

Probability that there are N neutrons at time t × Probability that no neutrons are produced in time Δt

N.B. The term involving the sum over n describes the fission branching process.

In principle, Δt should be large enough such that one neutron only can be emitted during it but small enough to make each of the events mutually exclusive, i.e. none of the events written above take place simultaneously in the same Δt interval.

If we allow $\Delta t \to 0$ and take the limit, this equation reduces to:

$$\begin{aligned}\frac{dP(N,t)}{dt} &= S[P(N-1,t)-P(N,t)] \\ &\quad +\lambda_c(N+1)P(N+1,t)-(\lambda_c+\lambda_f)NP(N,t) \\ &\quad +\lambda_f \sum_n p(n)(N+1-n)P(N+1-n,\,t). \qquad (2.24)\end{aligned}$$

(N.B. For $p(n) = \delta_{n,2}$ this becomes the birth and death problem discussed earlier.)

This is a differential–difference equation and is most conveniently dealt with by the introduction of a probability generating function, defined as

$$F(x,t) = \sum_{N=0}^{\infty} x^N P(N,t), \qquad (2.25)$$

and the auxiliary generating function

$$f(x) = \sum_{n=0}^{\infty} x^n p(n) \qquad (2.26)$$

(Bharucha-Reid, 1960; Frisch, 1959; Courant and Wallace, 1947).

Now multiplying (2.25) by x^N and summing over N we get after some rearrangement

$$\frac{\partial F(x,t)}{\partial t} = (x-1)SF(x,t)+[\lambda_c(1-x)+\lambda_f(f(x)-x)]\frac{\partial F(x,t)}{\partial x}, \qquad (2.27)$$

which is a partial differential equation for $F(x,t)$.

The importance of $F(x,t)$ lies in the fact that from it we may calculate the density moments of interest. For example, the average density $\overline{N}(t)$ is given by

$$\overline{N}(t) = \left.\frac{\partial F}{\partial x}\right|_{x=1}. \qquad (2.28)$$

Similarly, the mean square density is

$$\overline{N^2(t)} = \overline{N(t)}+\left.\frac{\partial^2 F}{\partial x^2}\right|_{x=1}. \qquad (2.29)$$

Higher moments can be obtained in a similar fashion. Thus if we can obtain F we can begin to study fluctuations about N, which is the prime purpose of noise theory, without the necessity of obtaining an explicit expression for $P(N, t)$.

A number of revealing facts can be obtained by direct differentiation of eqn. (2.27), viz.

$$\frac{d\overline{N}}{dt} = [\lambda_f(\bar{\nu}-1)-\lambda_c]\overline{N}+S, \tag{2.30}$$

$$\equiv (\bar{\nu}\lambda_f-\lambda_a)\overline{N}+S \tag{2.31}$$

($\lambda_a = \lambda_c+\lambda_f$), which is the conventional point model equation for reactor kinetics (neglecting delayed neutrons).

Similarly, by differentiating again and setting $x = 1$ we get an equation for $\overline{N^2(t)}$ in the form:

$$\frac{d}{dt}\overline{\mu_{NN}(t)} = 2\overline{\mu_{NN}(t)}[\bar{\nu}\lambda_f-\lambda_a]+\overline{\nu(\nu-1)}\lambda_f\overline{N(t)}, \tag{2.32}$$

where

$$\overline{\mu_{NN}} = \overline{N^2}-\overline{N}^2-\overline{N}, \tag{2.33}$$

$$\bar{\nu} = \sum_n np(n) \tag{2.34}$$

and

$$\overline{\nu(\nu-1)} = \sum_n n(n-1)p(n). \tag{2.35}$$

Solutions of these two equations are

$$\overline{N}(t) = \overline{N}(0)\,e^{-\rho t/\Lambda}+\frac{S\Lambda}{\rho}[1-e^{-\rho t/\Lambda}] \tag{2.36}$$

and

$$\overline{\mu_{NN}(t)} = \overline{\mu_{NN}(0)}e^{-2\rho t/\Lambda}+\overline{\nu(\nu-1)}\lambda_f\left[\left\{\overline{N}(0)-\frac{S\Lambda}{\rho}\right\}\right.$$

$$\left.\times\frac{\Lambda}{\rho}e^{-\rho t/\Lambda}+\frac{S\Lambda^2}{2\rho^2}-\left\{\left[\overline{N}(0)-\frac{S\Lambda}{\rho}\right]\frac{\Lambda}{\rho}+\frac{S\Lambda^2}{2\rho^2}\right\}e^{-2\rho t/\Lambda}\right], \tag{2.37}$$

where we have set $\rho/\Lambda = -\bar{\nu}\lambda_f+\lambda_a$, Λ being the prompt neutron generation time $1/\bar{\nu}\lambda_f$.

For a sub-critical system, we can allow the initial source transients to decay and obtain

$$\overline{N(\infty)} = \frac{S\Lambda}{\rho}, \tag{2.38}$$

which is the expected value from normal reactor kinetics (Glasstone and Edlund, 1952).

The value of $\overline{\mu_{NN}}$ becomes

$$\overline{\mu_{NN}(\infty)} = \overline{\nu(\nu-1)}\lambda_f\frac{S\Lambda^2}{2\rho^2} \tag{2.39}$$

or in terms of the variance

$$\overline{N^2}-\overline{N}^2 = \overline{N}\left\{1+\overline{\nu(\nu-1)}\frac{\lambda_f\Lambda}{2\rho}\right\}. \tag{2.40}$$

We note that the fluctuations caused by the fission process result in the variance being increased by the fraction $\overline{\nu(\nu-1)}\lambda_f\Lambda/2\rho$ above that for a pure Poisson process. From this we conclude that in a pure moderator ($p(\nu) = 0$) the neutrons have a variance given by the Poisson distribution and a probability law $P(N)$ such that

$$P(N) = \frac{1}{N!}\left(\frac{S}{\lambda_c}\right)^N. \tag{2.41}$$

For a fissile material with $\eta = 2S\tau_f/\overline{\nu(\nu-1)}$ it has been shown by Bell (1963) that the value of $P(N,t)$ is given by

$$P(N,t) = \left(\frac{\eta N(t)}{\overline{N(t)}}\right)^{\eta-1} \frac{\eta}{\overline{N(t)}(\eta-1)!} \exp\left\{-\eta \frac{N(t)}{\overline{N(t)}}\right\}. \tag{2.42}$$

For large values of η, i.e. strong sources, the distribution is peaked about $N = \overline{N}$, while for weak sources it is unlikely for N to be as large as $\overline{N}$ and the width of the distribution is much greater.

2.5. *Detector and delayed neutrons*

The theory described in the previous sections was deliberately simplified to indicate the basic technique of probability balance. The most serious omission was that of delayed neutrons. In addition, in any real system it is essential to have a measuring instrument, e.g. a counter, to record the events taking place. Thus a more appropriate probability function would be

$$P(N, C_1, \ldots, C_J, Z, t) \tag{2.43}$$

which is the probability that at time t there are N neutrons and C_1 delayed neutron precursors with decay constant λ_1, etc., and also that in the time interval $(0,t)$, neutrons have been detected by the counting device.

We can formulate a probability balance equation for this function (Harris, 1958; Pluta, 1962; Courant and Wallace, 1947) and obtain an equation for the more general probability generating function

$$F(x, y_1, \ldots, y_J, z, t) = \sum_N \sum_{C_1} \cdots \sum_{C_J} \sum_Z x^N y_1^{C_1}, \ldots, y_J^{C_J} z^Z P(\ldots, t). \tag{2.44}$$

From such an equation we may derive the steady-state point model equation of reactor kinetics for neutrons and precursors, i.e. $\overline{N}$, $\overline{C_i}$. Also, however, we can obtain equations for expressions such as $\overline{N^2}$, $\overline{NC_i}$, $\overline{Z}$, $\overline{Z^2}$, $\overline{NZ}$, etc., i.e. the covariance as well as the variance.

We sketch below the basic method and discuss some results obtained by Bell. In order to set up the probability balance, it is necessary to define some additional quantities, viz. λ_d is the detection rate per unit time and $p(n, m_1, \ldots, m_N)$ is the probability that n prompt neutrons are formed in a fission event together with m_1 precursors of type 1, m_2 of type 2, etc.

Following the rules of probability balance we can argue that

$$P(N, C_1, \ldots, C_N, Z, t+\Delta t) = P(N-1, C_1, \ldots, C_J, Z, t) S\Delta t$$

$$+ \sum_{i=1}^{J} P(N-1, C_1, \ldots, C_i+1, \ldots, C_J, Z, t)(C_i+1)\lambda_i \Delta t$$

$$+\sum_{n=0}^{\infty}\sum_{m_1=0}^{\infty}\cdots\sum_{m_J=0}^{\infty}P(N+1-n,C_1-m_1,\ldots,C_J-m_J,Z,t)(N+1-n)$$
$$\times p(n,m_1,\ldots,m_J)\lambda_f\Delta t$$
$$+P(N+1,C_1,\ldots,C_J,Z,t)(N+1)\lambda_c\Delta t$$
$$+P(N+1,C_1,\ldots,C_J,Z-1,t)(N+1)\lambda_d\Delta t$$
$$+P(N,C_1,\ldots,C_J,Z,t)\left\{1-S\Delta t-(\lambda_f+\lambda_c+\lambda_d)N\Delta t-\sum_{i=1}^{J}\lambda_i C_i\Delta t\right\}. \quad (2.45)$$

This differs from eqn. (2.24) in the addition of the delayed neutron term and the detection term involving λ_d. The basic ideas in the formulation, i.e. mutually exclusive and independent probabilities, remain the same.

By allowing $\Delta t \to 0$, we obtain the differential–difference equation for $P(\ldots,t)$. However, it is more convenient to obtain the equation for the generating function, which we do by introducing F, thereby obtaining the following equation:

$$\frac{\partial F}{\partial t} = (x-1)SF+\sum_{i=1}^{J}(x-y_i)\lambda_i\frac{\partial F}{\partial y_i}$$
$$+[\lambda_c(1-x)+\lambda_f\{f(x,y_1,\ldots,y_J)-x\}+\epsilon\lambda_f(z-x)]\frac{\partial F}{\partial x}, \quad (2.46)$$

where $\lambda_d = \epsilon\lambda_f$, ϵ being the detection efficiency, and we have introduced the new generating function

$$f(x,y_1,\ldots,y_J) = \sum_{n=0}^{\infty}\sum_{m_1=0}^{\infty}\cdots\sum_{m_J=0}^{\infty}x^n\,y_1^{m_1},\ldots,y_J^{m_J}p(n,m_1,\ldots,m_J). \quad (2.47)$$

Whilst this equation is too complicated to solve explicitly, a number of its general properties have been elucidated by Bell (1963). From the practical point of view, however, only the first few moments are of interest and these can be obtained by differentiation of the equation for the generating function as discussed earlier. Thus if we note that

$$\overline{N} = \left.\frac{\partial F}{\partial x}\right|_{\mathbf{x}=1} \quad \text{and} \quad \overline{C_i} = \left.\frac{\partial F}{\partial y_i}\right|_{\mathbf{x}=1}, \quad (2.48)$$

where $\mathbf{x} = \mathbf{1}$ indicates that all x, y_i and z are set equal to unity, we find after a single differentiation with respect to x that

$$\frac{d\overline{N}}{dt} = [\lambda_f(\bar{\nu}_0-1-\epsilon)-\lambda_c]\overline{N}+\sum_i\lambda_i\overline{C}_i+S, \quad (2.49)$$

where $\bar{\nu}_0 = \sum_{n,m_i} np(\ldots)$ is the average number of prompt neutrons produced per fission.

Defining the prompt neutron lifetime l as

$$l = \frac{1}{\lambda_f+\lambda_c'}, \quad (2.50)$$

where $\lambda_c' = \lambda_c+\epsilon\lambda_f$, we can rearrange eqn. (2.49) to the form

$$\frac{d\overline{N}}{dt} = \frac{1}{l}\left\{\frac{\lambda_f\bar{\nu}_0}{\lambda_f+\lambda_0}-1\right\}\overline{N}+\sum_i\lambda_i\overline{C}_i+S. \quad (2.51)$$

If we define
$$k(1-\beta) = \frac{\bar{\nu}_0\lambda_f}{\lambda_f+\lambda_c'}, \tag{2.52}$$

where k is the conventional multiplication constant and β the total delayed neutron fraction, we recover the normal form of the point model reactor kinetics equation.

Similarly, by differentiating with respect to y_i we obtain the following delayed neutron equations:
$$\frac{d\overline{C_i}}{dt} = \lambda_f\bar{\nu}_i\bar{N}-\lambda_i\bar{C}_i \quad (i = 1,2,\ldots,J), \tag{2.53}$$

where
$$\bar{\nu}_i = \sum_{n,\, m_1} m_i p(\ldots) \tag{2.54}$$

is the average number of precursor atoms produced per fission. Using the definition of k and introducing
$$\bar{\nu}_i = \frac{\bar{\nu}_0\beta_i}{1-\beta} \tag{2.55}$$

we find that the equation for $\bar{C}_i$ becomes
$$\frac{d\overline{C_i}}{dt} = \frac{\beta_i k}{l}\bar{N}-\lambda_i C_i \tag{2.56}$$

in accordance with normal practice.

The average detection rate is given by the equation
$$\frac{d\bar{Z}}{dt} = \epsilon\lambda_f\bar{N}, \tag{2.57}$$

which is an intuitively obvious result.

The fluctuations about these average values can be studied by taking second derivatives of F to obtain variances and covariances. Thus, if we differentiate twice with respect to x we find that the quantity $\overline{\mu_{NN}}$ is given by the equation

$$\frac{d\overline{\mu_{NN}}}{dt} = 2\bar{\mu}_{NN}[\lambda_f(\bar{\nu}_0-1-\epsilon)-\lambda_0]+\bar{N}\lambda_f\bar{\nu}_{00}+2\sum_i\lambda_i\bar{\mu}_{NC}, \tag{2.58}$$

where
$$\bar{\nu}_{00} = \sum_{n,\, m_i} n(n-1)p(\ldots) \tag{2.59}$$

and
$$\bar{\mu}_{NC_i} \equiv \overline{(N-\bar{N})(C_i-\bar{C}_i)} \equiv \overline{NC_i}-\bar{N}\bar{C}_i. \tag{2.60}$$

The only exceptions to these covariance definitions are the averages $\bar{\mu}_{NN}$ and $\bar{\mu}_{ZZ}$, which are defined as
$$\bar{\mu}_{NN} = \overline{N^2}-\bar{N}^2-\bar{N} \equiv \sigma_N^2-\bar{N} \tag{2.61a}$$

and
$$\bar{\mu}_{ZZ} = \overline{Z^2}-\bar{Z}^2-\bar{Z} \equiv \sigma_Z^2-\bar{Z} \tag{2.61b}$$

and may be referred to as *modified* variances.

The modified variance in the number of detected neutrons is readily obtained as

$$\frac{d\bar{\mu}_{ZZ}}{dt} = 2\epsilon\lambda_f\bar{\mu}_{NZ}, \tag{2.62}$$

which, like $\bar{Z}$, is an observable quantity.

Equations for $\bar{\mu}_{NZ}$, $\bar{\mu}_{ZC_i}$, $\bar{\mu}_{C_iC_j}$ can be obtained to complete the coupled set of forty-four simultaneous first-order differential equations (assuming six delayed neutron groups). Third-order equations can also be derived but these are generally of little practical interest for zero-power studies. We have in our possession, therefore, a set of equations to describe the mean values and variances of the neutron density population in the framework of one-speed point model kinetics. In the next chapter we shall develop these equations and apply them to situations of practical value.

References

BELL, G. I. (1963) Probability distribution of neutrons and precursors in a multiplying assembly, *Ann. Phys. (N.Y.)* **21**, 243.

BELL, G. I. (1965) On the stochastic theory of neutron transport, *Nucl. Sci. Engng*, **21**, 390.

BHARUCHA-REID, A. T. (1960) *Elements of the Theory of Markov Processes and Their Applications*, McGraw-Hill.

COURANT, E. D. and WALLACE, P. R. (1947) Fluctuations of the number of neutrons in a pile, *Phys. Rev.* **72**, 1038.

FRISCH, O. (1959) Fluctuations in chain processes, AECL 748.

GLASSTONE, S. and EDLUND, M. (1952) *Nuclear Reactor Theory*, Van Nostrand.

HARRIS, D. R. (1958) Stochastic fluctuations in a power reactor, USAEC, WAPD-TM-190.

PÁL, L. (1958) On the theory of stochastic processes in nuclear reactors, *Il Nuovo Cimento*, Supplemento VII, 25.

PLUTA, P. R. (1962) An analysis of fluctuations in a boiling water reactor by methods of stochastic processes, USAEC, APED-4071.

SRINIVASAN, S. K. and VASUDEVAN, R. (1971) *Introduction to Random Differential Equations and Their Applications*, Elsevier, N.Y.

CHAPTER 3

Applications of the General Theory

3.1. *Introduction*

The basic theory developed in the last chapter is capable, with some minor modifications, of explaining and predicting a wide range of experimental neutron-fluctuation problems in zero power systems. In this chapter, therefore, we will describe some of these problems, show how the theory can be used to extract parameters of interest in reactor kinetic studies and also present additional theoretical details which must be included in the analysis of certain problems.

3.2. *The variance to mean method* (*Feynman technique*)

If we assume that the reactivity and external source are constant with time, then the following conditions hold: $\dot{\bar{N}} = 0$, $\dot{\bar{C}}_i = 0$, $\dot{\bar{\mu}}_{NC_i} = 0$, $\dot{\bar{\mu}}_{NN} = 0$, $\dot{\bar{\mu}}_{C_iC_j} = 0$. In view of these quantities being zero, the quantities $\dot{\bar{Z}}$, $\dot{\bar{\mu}}_{ZZ}$, $\dot{\bar{\mu}}_{ZC_i}$ and $\dot{\bar{\mu}}_{NZ}$ describe a cumulative distribution which are, in general, non-zero. For example, $\bar{Z}(t)$, the average number of neutrons detected in a time t, increases linearly with time.

The appropriate steady values of $\bar{N}$, $\bar{C}_i$, $\bar{\mu}_{NN}$, $\bar{\mu}_{NC_i}$ and $\bar{\mu}_{C_iC_j}$ can be obtained from eqn. (2.49) and its associates in the last chapter and we readily find that

$$\bar{N} = \frac{Sl}{1-k}, \tag{3.1}$$

$$\bar{\mu}_{NN} = \frac{\bar{N}\lambda_f\bar{\nu}_{00} + 2\sum_i \lambda_i \bar{\mu}_{NC_i}}{2[\lambda_c - \lambda_f(\bar{\nu}_0 - 1 - \epsilon)]}, \tag{3.2}$$

$$\bar{\mu}_{NC_i} = \frac{\bar{\mu}_{NN}\lambda_f\bar{\nu}_i + \bar{N}\lambda_f\bar{\nu}_{0i} + \sum_i \lambda_i \bar{\mu}_{C_iC_j}}{\lambda_c + \lambda_i - \lambda_f(\bar{\nu}_0 - 1 - \epsilon)}, \tag{3.3}$$

$$\bar{\mu}_{C_iC_j} = \frac{1}{\lambda_i + \lambda_j}\{\bar{\mu}_{NC_i}\lambda_f\bar{\nu}_j + \bar{\mu}_{NC_j}\lambda_f\bar{\nu}_i + \bar{N}\bar{\nu}_{ij}\lambda_f\}. \tag{3.4}$$

These constitute a set of twenty-eight algebraic equations for the twenty-eight unknowns. When the solutions are obtained the equations for $\bar{Z}$, $\bar{\mu}_{NN}$, $\bar{\mu}_{ZZ}$, etc., can be solved and the observable ratio

$$\text{Var}(t) = \frac{\overline{Z^2} - \bar{Z}^2}{\bar{Z}} \equiv 1 + \frac{\bar{\mu}_{ZZ}}{\bar{Z}} \tag{3.5}$$

can be calculated. This is best done by use of the Laplace transform and leads after some tedious but straightforward algebra to

$$\frac{\bar{\mu}_{ZZ}}{\bar{Z}} = \sum_{i=1}^{J+1} Y_i \left\{ \frac{1-(1-e^{-\alpha_i t})}{\alpha_i t} \right\}, \tag{3.6}$$

where

$$Y_i = \frac{2\epsilon\lambda_f}{\bar{N}} \left\{ \frac{\bar{\mu}_{NN} + \bar{N} + \sum_{j=1}^{J} \frac{\lambda_j(\bar{N}\bar{\nu}_j + \bar{\mu}_{NCj})}{\lambda_j - \alpha_i}}{\alpha_i \left[1 + \lambda_f \sum_{j=1}^{J} \frac{\lambda_j \bar{\nu}_j}{(\lambda_j - \alpha_i)^2} \right]} \right\}. \tag{3.7}$$

α_i are the roots of the inhour equations, viz. $p = -\alpha_i$ where

$$\frac{k_{\text{eff}}-1}{k_{\text{eff}}} = \frac{lp}{k_{\text{eff}}} + p \sum_{i=1}^{J} \frac{\beta_i}{p+\lambda_i}. \tag{3.8}$$

If we redefine the inhour equation as

$$\rho = p\left(\Lambda + \sum_{i=1}^{J} \frac{\beta_i}{p+\lambda_i} \right) \tag{3.9}$$

where $\rho = (k_{\text{eff}}-1)/k_{\text{eff}}$ and $\Lambda = l/k_{\text{eff}}$ is the generation time, then with

$$Y_i \alpha_i \equiv 2\epsilon \frac{\overline{\nu(\nu-1)}}{(\bar{\nu})^2} A_i G(\alpha_i) \tag{3.10}$$

we can recast the variance as

$$\frac{\bar{\mu}_{ZZ}}{\bar{Z}} = 2\epsilon \frac{\overline{\nu(\nu-1)}}{(\bar{\nu})^2} \sum_{i=1}^{J+1} A_i \frac{G(\alpha_i)}{\alpha_i} \left\{ 1 - \frac{(1-e^{-\alpha_i t})}{\alpha_i t} \right\}, \tag{3.11}$$

where in conventional reactor kinetics parlance $G(p)$ is the zero-power transfer function, viz.

$$\frac{1}{G(p)} = p\left(\Lambda + \sum_{i=1}^{J} \frac{\beta_i}{\lambda_i + p} \right) - \rho, \tag{3.12}$$

and A_i are the residues of the transfer function. Thus

$$G(p) = \sum_{i=1}^{J+1} \frac{A_i}{p+\alpha_i}$$

In these expressions, α_1 = prompt decay constant and $\alpha_2, \ldots, \alpha_{J+1}$ are the delayed neutron terms. α_{J+1} becomes zero at criticality and describes the stable period for a super-critical system. However, the method does not appear to be applicable to a reactor that is not operating in the steady state (but see Section 3.5).

In the associated experiment the procedure is to record the number of counts Z_i in a given time interval t (gate length) for a large number of runs. Then because of the fluctuations inherent in the fission process, discussed in an earlier chapter, the number of counts will be a random variable. The comparison with theory is made by calculating the experimental quantity (for N runs):

$$\text{Var}(t) = \frac{\frac{1}{N}\sum_{i=1}^{N} Z_i^2 - \left(\frac{1}{N}\sum_{i=1}^{N} Z_i \right)^2}{\frac{1}{N}\sum_{i=1}^{N} Z_i} \tag{3.13}$$

for a range of values of t and then fitting to the theory to extract the parameters of interest.

Before proceeding to discuss these matters further it is interesting to note, once again, that fission is responsible for the deviation of the variance from the Poisson value, i.e. if $\lambda_f = 0$, then $\bar{\mu}_{ZZ} = 0$ and the variance is equal to the mean. For accurate experimental results therefore it is essential that the detection efficiency ϵ be as high as possible.

In practice, the gate length t is very important, since by appropriate choice of it we can simplify (3.6) and thereby deduce important reactor parameters more accurately.

Let us consider the values of α_i, which we may write as

$$\alpha_1 \gg \alpha_2 > \alpha_3 > \ldots > \alpha_7 > 0 \quad (J = 6). \tag{3.14}$$

As stated earlier, α_1 is the prompt decay and for all practical situations is large compared with the other α_i.

Now if we choose t, the counting time (gate length), such that

$$\frac{1}{\alpha_1} \ll t \ll \frac{1}{\alpha_2}, \tag{3.15}$$

we can neglect all terms in the sum in (3.6) except the first, thus

$$\mathrm{Var}(t) = 1 + Y_1, \tag{3.16}$$

which is commonly referred to as the prompt-variance method since the role of delayed neutrons is only indirect: moreover, only one counting time need be used.

The conditions (3.15) cannot be met in graphite or D_2O systems, but McCulloch (1958) has used the method in a plutonium-fuelled system surrounded by a matrix of natural uranium. Under the conditions of (3.15), Y_1 can be written for a sub-critical system as

$$Y_1 = \epsilon \frac{\overline{\nu(\nu-1)}}{(\bar{\nu})^2} \frac{1}{(\beta-\rho)^2}, \tag{3.17}$$

where $\rho = (k_{\mathrm{eff}} - 1)/k_{\mathrm{eff}}$ is the reactivity.

It should be noted in connection with this experiment that, at criticality, Y_7 diverges owing to $\alpha_7 = 0$. In practice, therefore, the experiment is performed only on sub-critical systems.

To test the apparatus, McCulloch performed some preliminary experiments with α-particles to test that a truly random distribution of pulses did indeed lead to a zero value of Y_1. His results showed that $1 + Y_1 \simeq 0{\cdot}95$ which was considered satisfactory confirmation. Experiments were then carried out in the graphite envelope of ZEPHYR to measure Y_1. Using a 50-msec gate length and 1000 gate samples it was found that for a detector efficiency ϵ of $5{\cdot}6 \times 10^{-6}$ and the Diven's factor $\overline{\nu(\nu-1)}/(\bar{\nu})^2 = 0{\cdot}80$, $\beta = (2{\cdot}83 \times 0{\cdot}36) \times 10^{-3}$. The sub-criticality ρ was derived from earlier control rod calibrations obtained in previous kinetic measurements. In general, it is more usual to assume a value of β and use this method to calculate the shutdown reactivity.

If the gate length is such that only the inequality $|\alpha_2| t \ll 1$ is satisfied, then

$$\mathrm{Var}(t) = 1 + Y_1 \left\{1 - \frac{1 - e^{-\alpha_1 t}}{\alpha_1 t}\right\}, \tag{3.18}$$

where, to a good approximation, Y_1 is given by eqn. (3.17) and

$$\alpha_1 = \frac{(\beta-\rho)}{\Lambda}, \tag{3.19}$$

where Λ is the generation time. The condition for eqn. (3.18) is accurate for a reactor near criticality if $t < 0{\cdot}1$ sec. In such a case the delayed neutron terms in the expression for $\bar{\mu}_{ZZ}$ are negligibly small. Experiments based on eqn. (3.18) will involve sampling for a range of gate times and a least-squares fit will yield Y_1 and α_1. This technique generally constitutes a convenient method of measuring neutron lifetime or shut-down reactivity.

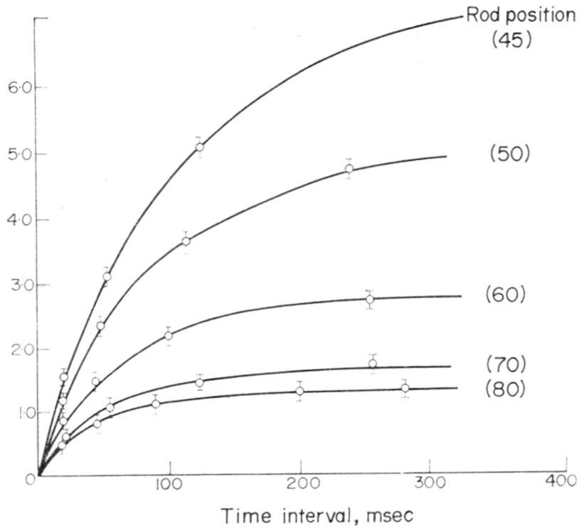

FIG. 3.1. Variance to mean $(\overline{Z^2}-\overline{Z}^2)/\overline{Z}-1{\cdot}0$ versus time interval in a D_2O moderated thermal reactor. (From Gotoh (1964) *J. Nucl. Sci. Technol.*, **1**, 11.)

Figure 3.1 shows the results of some variance to mean measurements made by Gotoh (1964) in a D_2O moderated thermal reactor. The curves are for different control rod positions and demonstrate the feasibility of this method as a rod worth calibration technique. The accuracy claimed is about 15% in neutron lifetime (e.g. in the case shown $l = 0{\cdot}621 \pm 0{\cdot}093$ msec).

A number of modifications to the basic Feynman technique have been proposed. That by Bennett (1960) is particularly interesting since it yields a theoretical expression which does not suffer from the divergence at criticality and can therefore be employed over a wide range of situations. The basic idea is to compute the variance in the form

$$\langle (Z_{k+1}-Z_k)^2 \rangle = \frac{1}{N^2}\sum_{i=1}^{N}(Z_{k+1}^{(i)}-Z_k^{(i)})^2, \tag{3.20}$$

i.e. to cross-correlate the count rates in adjacent gate intervals.

A further variation of the basic technique described here would be to fix the number of counts recorded and treat the time t_i as the random variable. The associated theory for this procedure does not appear to have been developed.

3.3. *Correlation function method*

The most common method of noise analysis is based upon the auto- and cross-correlation techniques by analogy with methods developed in electrical noise measurements.

Here we note that, in a reactor operating steadily, there will be random fluctuations about the mean value $\bar{N}$. Clearly, these fluctuations contain kinetic information despite the "average" steady behaviour. Thus it would be very useful if the information contained in these fluctuations could be extracted since it would obviate the need to disturb the reactor in any way. This would generally constitute a saving in both time and money.

To fix our ideas let us visualize the output from a detector in a steady-state reactor as shown by the fluctuating line in Fig. 3.2.

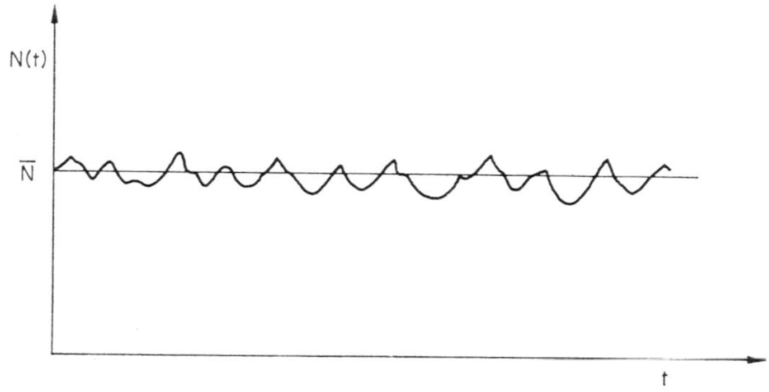

FIG. 3.2. Fluctuating neutron density in a reactor operating at steady power. $\bar{N}$ is the mean value.

Thus we can measure $N(t)$: the problem is how to analyse this information. To this end, we first define the auto-correlation function of $N(t)$, viz.

$$\phi_{NN}(\tau) = \lim_{T\to\infty} \frac{1}{2T} \int_{-T}^{T} dt\, N(t) N(t+\tau). \tag{3.21}$$

Thus we take the product of two values of N at different times and integrate over the sample length $2T$. In principle one should choose an infinite sample length but, in practice, provided T is large compared with any characteristic relaxation time of the system we can use a finite length. We can also define cross-correlation between say the neutron density N and some other noise source X and obtain the cross-correlation function

$$\phi_{NX}(\tau) = \lim_{T\to\infty} \frac{1}{2T} \int_{-T}^{T} dt\, N(t)\, X(t+\tau). \tag{3.22}$$

We shall use the idea of cross-correlation later.

3.4. *Time average and ensemble average*

At this point it will be useful to digress somewhat to discuss the relationship of the correlation functions defined above with the averages obtained from the probability functions $P(N,t)$, etc., discussed earlier. Here we meet a very important hypothesis, without which noise analysis would not be the powerful tool that it is; namely the *ergodic hypothesis*. This states that, in a stationary random system, *time averages and ensemble averages are equal.* By stationary it is meant that the basic noise sources are not changing with time, i.e. the statistics of the system, or probability laws, are constant.

With this in mind we can write for average value

$$\underset{\text{(ensemble average)}}{\bar{N} = \sum_n NP(N,t)} = \underset{\text{(time average)}}{\lim_{T\to\infty} \frac{1}{2T}\int_{-T}^{T} dt\, N(t).} \tag{3.23}$$

Thus the same average value will result whether we consider the average from one system over an infinite time, or whether we consider an infinite number of identical systems at the same time. Quite clearly from a practical point of view it is easier to measure the time average. However, in some theoretical considerations, it may be easier to calculate ensemble averages.

For the auto-correlation function $\phi_{NN}(\tau)$, we have either the ensemble average

$$\phi_{NN}(\tau) = \sum_{N_1}\sum_{N_2} N_1 N_2 P(N_1, N_2;\, t_1, t_1+\tau), \tag{3.24}$$

where $P(N_1, N_2;\, t_1, t_2)$ is the joint probability that at time t_1 there are N_1 neutrons in the system and at time t_2 there are N_2 neutrons, or we have the time average of eqn. (3.21).

We shall in all our calculations accept the ergodic hypothesis.

3.5. *Calculation of auto-correlation function*

Based upon the probability balance technique discussed earlier, it is not immediately clear how to obtain $\phi_{NN}(\tau)$. However, it has been shown by Pluta (1962) that a knowledge of the variance

$$\sigma^2(t) = \overline{Z^2(t)} - \overline{Z(t)}^2$$

enables $\phi_{NN}(\tau)$ to be obtained from the expression

$$\phi_{NN}(\tau) = \frac{1}{2}\frac{d^2\sigma(\tau)}{d\tau^2}, \tag{3.25}$$

which is an important and useful result.

Alternatively, of course, if the auto-correlation function is known we can obtain $\sigma^2(t)$ from

$$\sigma^2(t) = 2\int_0^t dt_2 \int_0^{t_2} dt_1\, \phi_{NN}(t_1). \tag{3.26}$$

Using eqn. (3.6) in (3.25) we find that

$$\phi_{NN}(\tau) = \tfrac{1}{2}\epsilon\lambda_f \bar{N} \sum_i Y_i \alpha_i e^{-\alpha_i|\tau|}. \tag{3.27}$$

To this must be added a term $\epsilon\lambda_f\bar{N}\delta(\tau)$ to account for uncorrelated neutrons which have a white noise spectrum (see Section 3.6). Thus the actual auto-correlation function may be written

$$\phi_{NN}(\tau) = \epsilon\lambda_f\bar{N}\{\delta(\tau) + \tfrac{1}{2}\sum_i Y_i\alpha_i e^{-\alpha_i|\tau|}\} \tag{3.28}$$

$$\equiv \epsilon\lambda_f\bar{N}\left\{\delta(\tau) + \epsilon\frac{\overline{\nu(\nu-1)}}{(\bar{\nu})^2}\sum_i A_i G(\alpha_i) e^{-\alpha_i|\tau|}\right\} \tag{3.29}$$

and it is this function which is to be compared with experiment.

In practice, for small values of τ, the first term in the sum is mainly due to prompt neutron effects which predominate. Thus to simplify matters we can write eqn. (3.28) as

$$\phi_{NN}(\tau) = \epsilon\lambda_f \bar{N}\{\delta(\tau)+\tfrac{1}{2}Y_1\alpha_1 e^{-\alpha_1|\tau|}\} \tag{3.30}$$

or, using the values of α_1 and Y_1 given earlier, as

$$\phi_{NN}(\tau) = \epsilon\lambda_f \bar{N}\left\{\delta(\tau)+\epsilon\,\frac{\overline{\nu(\nu-1)}}{(\bar{\nu})^2}\,\frac{e^{-\alpha_1\tau}}{2\Lambda(\beta-\rho)}\right\} \quad (\tau > 0). \tag{3.31}$$

We see therefore that from an apparently random signal we can deduce a deterministic function containing parameters of interest for reactor kinetic studies.

In the application of this technique it is sometimes necessary to consider the effect of finite bandwidth of the detection channel on the measured auto-correlation function. If, therefore, we denote by $R(\tau)$ the impulse response function of the detection system, we can write the measured random signal $C(t)$ in terms of the neutron density fluctuations as

$$C(t) = \int_{-\infty}^{\infty} R(u)N(t-u)\,du. \tag{3.32}$$

The measured auto-correlation function of the output is therefore

$$\phi_{CC}(\tau) = \left\langle \int_{-\infty}^{\infty} R(u)N(t-u)\,du\,.\int_{-\infty}^{\infty} R(u')N(t+\tau-u')\,du' \right\rangle. \tag{3.33}$$

Now $R(\tau)$ can be a random function itself, i.e. the detection system can contain its own sources of noise; however, these will certainly be uncorrelated with the fluctuations in the neutron density. Thus we may write:

$$\phi_{CC}(\tau) = \int_{-\infty}^{\infty} du \int_{-\infty}^{\infty} du'\,\phi_{RR}(u,u')\phi_{NN}(\tau-u'+u), \tag{3.34}$$

where $\phi_{RR}(u,u') = \phi_{RR}(u-u')$ if the detector noise is stationary; but it is equal to $R(u)\,R(u')$ if the response is deterministic. We will assume that the detection system has a well-defined, deterministic transfer function. If the bandwidth of this function is sufficiently wide, $R(\tau)$ will be approximately given by

$$R(\tau) \simeq R_0\delta(\tau) \tag{3.35}$$

and the measured and actual auto-correlation functions will be identical apart from a scale factor determined by the sensitivity of the detection apparatus. Details of how to correct for this may be found in Dragt (1966).

A further problem, which has been a source of embarrassment to reactor noise theory, arises when the system is critical. In that case the value of α_7 is zero and the term $Y_7\alpha_7$ in eqn. (3.28) becomes infinite for all values of τ. This would appear to preclude the application of eqn. (3.28) in critical systems. This problem, however, has been shown by Dragt to be more apparent than real. He argues as follows:

Consider the auto-correlation function of the quantity $N(t)-\bar{N}$, i.e. the fluctuation of N from its mean value, and denote it by $\phi(\tau)$. Then, by definition,

$$\phi(\tau) = \phi_{NN}(\tau)-\frac{1}{T^2}\int_0^T dt \int_0^T du\,\phi_{NN}(t-u). \tag{3.36}$$

Now $G(p)$ contains terms of the form $A/(p+\sigma)$ ($\sigma > 0$ and real), thus we may write a typical term in $\phi(\tau)$ as

$$\phi(\tau) \sim \frac{A^2}{2\sigma}\, e^{-\sigma\tau} - \frac{A^2}{\sigma^2 T}\left(1 - \frac{1-e^{-\sigma T}}{\sigma T}\right). \tag{3.37}$$

Thus, for $\sigma > 0$, the second term tends to zero when $T \to \infty$. In this case the use of eqn. (3.28) is correct provided $\sigma T \gg 1$, a condition that is fulfilled for prompt neutrons and most of the delayed neutron terms in the transfer function. For a critical reactor, the pole α_7 is zero, and instead of allowing $T \to \infty$ we let $\sigma \to 0$. Then the expression for $\phi(\tau)$ becomes

$$\phi(\tau) \sim \tfrac{1}{6}A^2 T \quad (T \gg \tau). \tag{3.38}$$

To evaluate the effect on the measured correlation function we consider a measuring time of 40 sec. Then from fundamental data typical of an ARGONAUT reactor, we calculate $A_7 = 12\ \text{sec}^{-1}$ and hence that $A_7^2 T/6 = 960\ \text{sec}^{-1}$. Similarly, the corresponding prompt neutron contribution, at $\tau = 0$, is

$$A_1 G(\alpha_1) \simeq \frac{1}{2\Lambda\beta} = 5{\cdot}23 \times 10^5\ \text{sec}^{-1}. \tag{3.39}$$

The term in the sum, which appears to cause the divergence, is therefore of order

$$\frac{960}{5{\cdot}23 \times 10^5} \simeq 0{\cdot}0018 \tag{3.40}$$

compared with the dominant prompt neutron contribution. Over most practical measuring times therefore the effect of this divergence can be ignored. Similar arguments can also be applied to the Feynman or variance to mean method discussed in Section 3.2.

The experimental implementation of the ideas described above is relatively simple. The output of the measuring device is recorded, usually on magnetic tape, and when a sufficiently long record is available it is processed by numerically computing the quantity

$$\phi(\tau) = \frac{1}{2M} \sum_{n=-M}^{M} N(n\Delta t)\, N(n\Delta t + \tau), \tag{3.41}$$

where Δt is the sampling interval. Some rough indication of α_1 is needed to decide upon the mesh size Δt and the time step τ. In the Dragt experiment $\Delta t = 2$ msec and $2M = T/\Delta T = 20{,}000$. Also it was arranged that $\tau = j\Delta t$ where in this case max $(j) = 100$.

The experiment was able to measure β/Λ, β and Λ with considerable accuracy ($<2\%$).

Figure 3.3 shows the experimental results for the auto-correlation function obtained by Dragt on a critical ARGONAUT LFR with the detector at different positions. On the basis of eqn. (3.31) it is clear that α_1 and a number of other parameters can be extracted. An interesting point is that the amplitude of the noise varied markedly with the position but the functional form of the auto-correlation function did not change appreciably, thus α_1 could be considered unique to the system. This result further indicated that, for this particular reactor, a point model space-independent analysis was adequate. However, this is not a general conclusion since in some systems space-dependent effects can be very important.

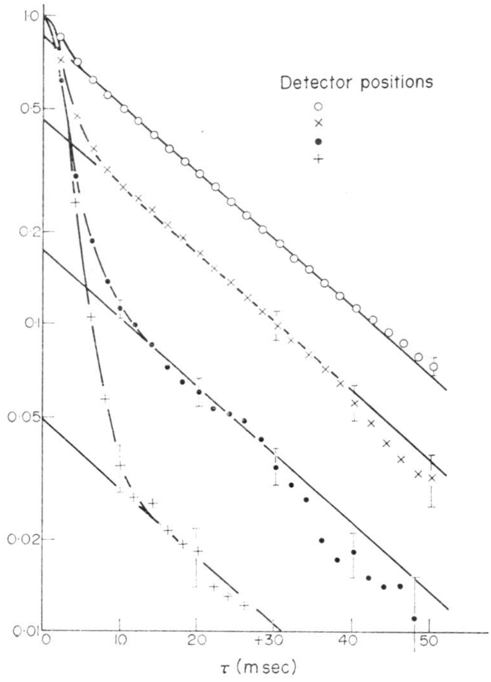

FIG. 3.3. Auto-correlation functions measured at different positions in the critical reactor. (From Jan B. Dragt (1966) *Nukleonik*, **8**, 188.)

Alternative methods of analysing the output, using Δt sufficiently small such that only one or zero counts are recorded in a channel, have been discussed by Pacilio (1970). This technique is particularly useful for direct on-line data processing via digital analysers.

3.6. *Power spectral density* (*p.s.d.*)

An alternative representation of noise phenomena can be made through the so-called power spectral density (p.s.d.) which is defined as the Fourier transform of the correlation function. Thus we can write the auto-p.s.d. as

$$\Phi_{NN}(\omega) = \int_{-\infty}^{\infty} d\tau\, e^{-i\omega\tau}\, \phi_{NN}(\tau) \tag{3.42}$$

and interpret it physically as a measure of the frequency content of the noise spectrum. For example, in deterministic problems, an harmonically oscillating system of frequency ω_0 would be characterized by a correlation function of the form

$$\phi_{NN}(\tau) = \frac{A}{\pi} \cos\omega_0\tau. \tag{3.43}$$

Thus the corresponding $\Phi_{NN}(\omega)$ would be

$$\Phi_{NN}(\omega) = A\{\delta(\omega-\omega_0)+\delta(\omega+\omega_0)\}, \tag{3.44}$$

which indicates that frequencies of $\omega = \pm\omega_0$ only are present.

In noisy systems there is in general a continuous spectrum of frequencies present and $\Phi_{NN}(\omega)\,d\omega$ is the strength of these in the range $(\omega, \omega+d\omega)$. However, any predominant frequency will show up as a peak in the p.s.d. or alternatively an oscillation in the auto-correlation function. White noise is a special but rather artificial situation in which frequencies of all values are present with equal probability in the range of $\omega(0, \infty)$. This is clearly a physical impossibility owing to such a situation corresponding to an infinite total power. However, in practice, we may consider a noise to be white provided that it is constant over the frequency range of interest in the system under investigation.

In the particular case of zero-power noise discussed above, we can insert (3.28) into (3.42) to obtain

$$\Phi_{NN}(\omega) = \epsilon\lambda_f \bar{N}\left\{1+\tfrac{1}{2}\sum_i Y_i\alpha_i \int_{-\infty}^{\infty} d\tau\, e^{-i\omega\tau-\alpha_i|\tau|}\right\} \tag{3.45}$$

$$\equiv \epsilon\lambda_f \bar{N}\left\{1+\sum_i \frac{Y_i\alpha_i^2}{\alpha_i^2+\omega^2}\right\}. \tag{3.46}$$

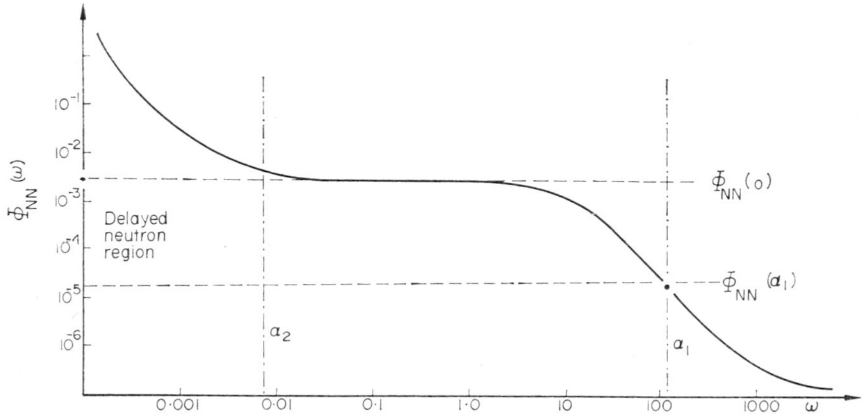

FIG. 3.4. General form of power spectral density in a zero-power assembly. The region below α_2 where delayed neutrons are dominant is not usually resolvable due to instrumental noise contamination. The dashed lines indicate the method of extracting α_1 graphically from the experimental data.

It should be noted that $\Phi_{NN}(\omega)$ is entirely real. This is because of the symmetry of $\phi_{NN}(\tau)$ with respect to τ about $\tau = 0$. However, cross-correlation functions are in general not symmetric, i.e. $\phi_{xy}(\tau) \neq \phi_{xy}(-\tau)$ and hence there will be an imaginary part to $\Phi_{xy}(\omega)$ which we may write as

$$\Phi_{xy}(\omega) = \Psi(\omega)+i\Theta(\omega). \tag{3.47}$$

It is the real part $\Psi(\omega)$ which we call the p.s.d. (in control theory jargon it is called the gain). However, the imaginary part Θ is also of interest through measurement of the phase lag θ, where

$$\theta = \tan^{-1}\left(\frac{\Theta(\omega)}{\Psi(\omega)}\right). \tag{3.48}$$

Returning to eqn. (3.46) we see that for a sub-critical system it has the form shown in Fig. 3.4.

In general the frequency range α_1 to α_2 is well separated and the functional form of $\Phi_{NN}(\omega)$ in the neighbourhood of α_1 can be written from (3.46) as

$$\Phi_{NN}(\omega) \simeq A+\frac{B}{\alpha_1^2+\omega^2}. \tag{3.49}$$

Thus a useful method for obtaining a rapid estimate of α_1 would be to either neglect or calculate approximately the background noise A and then to find the value of α_1 for which the ratio

$$\frac{\Phi_{NN}(0)-A}{\Phi_{NN}(\alpha_1)-A}$$

is equal to 2, $\Phi_{NN}(\omega)$ being given by eqn. (3.49). If A is neglected it may be seen from (3.49) that α_1 is given by the point of intersection of the line $\Phi_{NN}(\omega) = \frac{1}{2}\Phi_{NN}(0)$ with the curve $\Phi_{NN}(\omega) = B/(\alpha_1^2+\omega^2)$. A more accurate method is to employ a least-squares fitting procedure.

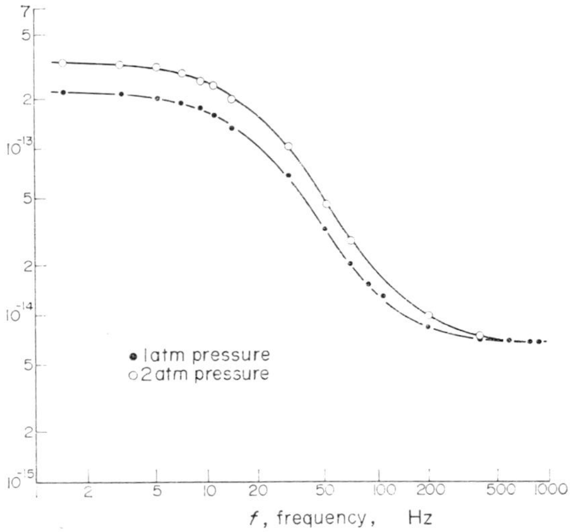

FIG. 3.5. Measured power spectral density in the BULK shielding reactor (ORNL). The effect of changing the BF_3 detector pressure is also shown. Results from Ricker *et al.* (1965), ORNL-TM-1066. (Note that the frequency is given in terms of $f = \omega/2\pi$.)

The low-frequency part of the curve, below 0·1 Hz, increases rapidly due to the critical divergence, $\alpha_7 = 0$, but also due to contamination by instrumental noise. Indeed, if we apply our Fourier transform to the equation for $\phi_{CC}(\tau)$ we find that the measured p.s.d. assumes the form

$$\Phi_{CC}(\omega) = |\bar{R}(\omega)|^2\,\Phi_{NN}(\omega), \tag{3.50}$$

where $\bar{R}(\omega)$ is the Fourier transform of the detector impulse response function $R(\tau)$. If $\bar{R}(\omega)$ is constant over the range where $\Phi_{NN}(\omega)$ varies markedly, i.e. around $\omega \simeq \alpha_1$, then it will only affect the amplitude of the measured p.s.d. which can then be compared with $\Phi_{NN}(\omega)$ to extract α_1. A typical experimental result for $\Phi_{CC}(\omega)$ for a low power reactor is shown in Fig. 3.5 (Ricker *et al.*, 1965).

Since $\alpha_1 \simeq (\beta-\rho)/\Lambda$, analysis of the noise output enables the sub-criticality to be obtained assuming prior knowledge of β and Λ or, in a critical system for which $\rho = 0$, the value of β/Λ can be obtained directly. Experiments on the Queen Mary College ARGONAUT type reactor (Mansfield, 1968) lead to $\alpha_1^{-1} \simeq 0{\cdot}02$ sec and hence to an effective neutron lifetime of about 165 μsec, a result which agrees closely with other independent measurements. This value of α_1^{-1} is to be compared with values of order 2 sec in typical power reactor situations which are to be discussed in Chapter 7.

Finally, we note that, experimentally, it is unnecessary to calculate the p.s.d. by first measuring the auto-correlation function followed by a numerical Fourier transform. By means of a band pass filter, and appropriate squaring devices, the mean square power in a frequency range $\Delta\omega$ can be obtained directly. However, modern developments in digital analysers are now so sophisticated that more accurate results are often available by the, apparently, longer route of auto-correlation and Fourier analysis (Pacilio, 1970).

3.7. *The Rossi-α technique*

A large number of other techniques exist for studying neutron density fluctuations in zero-power systems. It will not be possible to describe them all in detail here and we will therefore simply name them and give a brief indication of their purpose. First, however, we shall describe in some detail the best known of these related techniques, namely the Rossi-α method.

This method was originally developed for fast reactor systems in which the neutron lifetime is small and the nuclear chains tend not to overlap in time. It has, however, been possible with modern instrumentation to extend it to thermal systems. The basic purpose of the method is the measurement of the neutron lifetime or reactivity directly, and it relies upon a knowledge of $p_c(t_1, t_2)\Delta t_1 \Delta t_2$, which is the probability of detecting a neutron in the interval Δt_2 about t_2 given that there has been a detection in Δt_1 about t_1.

We will show below that

$$p_c(t_1, t_2)\Delta t_1 \Delta t_2 = A\Delta t_1 \Delta t_2 + B\Delta t_1 \Delta t_2 e^{-\alpha(t_2-t_1)}, \tag{3.51}$$

where A is the average detector counting rate and

$$B = \epsilon \frac{\overline{\nu(\nu-1)}}{2\alpha\tau_f^2} \tag{3.52}$$

with

$$\alpha = (\beta-\rho)/\Lambda. \tag{3.53}$$

Equation (3.51) accounts for both random events and chain correlated ones.

The method works best at low powers when individual chains do not significantly overlap, the idea being to isolate a single fission chain of related neutrons, as shown in Fig. 3.6.

By measuring the decay of a single chain we are effectively using the ability of the system to pulse itself. We note also that the theory accounts for delayed neutrons only in so far as they render the reactor sub-critical when the prompt neutrons alone are considered, i.e.

$$k_{\text{eff}} = 1 = k_{\text{eff},p} + k_{\text{eff},d} \propto \bar{\nu}(1-\beta) + \bar{\nu}\beta,$$

hence $k_{\text{eff},p} < 1$.

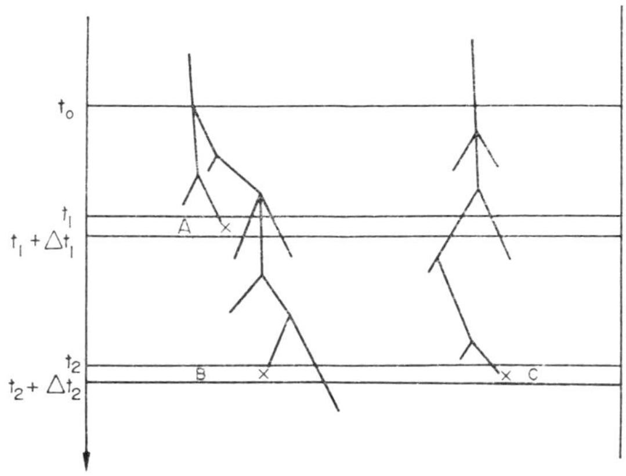

FIG. 3.6. Two typical fission chains. Detections take place in the intervals Δt_1 and Δt_2, thus (A, B) are correlated pairs and (B, C) are accidental or uncorrelated pairs.

The physical arguments used to obtain $p_c(t_1, t_2)$ are as follows:

Let the fission rate in the reactor be F_0 sec^{-1}. Consider a fission occurring at time t_0 in the interval Δt_0 and study the subsequent history of the neutrons from this event (see Fig. 3.7):

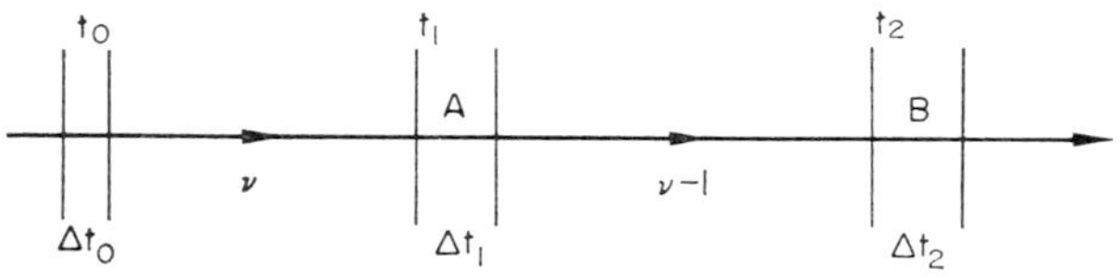

FIG. 3.7. Fission neutrons are born in Δt_0 and are detected in gates A and B.

The average number of fissions in Δt_0 is $F_0 \Delta t_0$. Consider one of the fissions in Δt_0: ν neutrons emerge from the fission but will have decayed on average by a factor $\exp[-\alpha(t_1 - t_0)]$ at time t_1, where α is the prompt neutron decay constant as given by eqn. (3.53) (or some more accurate form).

The detection probability within Δt_1 at t_1 (i.e. event A) is therefore

$$p_1 \Delta t_1 = \nu e^{-\alpha(t_1 - t_0)} \cdot \frac{\Delta t_1}{\tau_f} \cdot \epsilon. \tag{3.54}$$

↑ No. of fission neutrons reaching Δt_1 at t_1 · ↑ Probability of fission in Δt_1 · ↑ Detection efficiency

(Detection efficiency = fissions detected/total fissions occurring.)

Detection at t_1 removes one neutron, but the remaining $(\nu - 1)$ neutrons continue, and their detection probability, within Δt_2 at t_2 (i.e. event B), is therefore

$$p_2 \Delta t_2 = (\nu - 1) e^{-\alpha(t_2 - t_0)} \cdot \frac{\Delta t_2}{\tau_f} \cdot \epsilon, \tag{3.55}$$

the neutrons having decayed now, on average, by a factor $\exp[-\alpha(t_2 - t_0)]$ since t_0.

The total correlated coincidence count probability is therefore the joint probability

$$p_1 p_2 \Delta t_1 \Delta t_2$$

multiplied by the number of fissions in Δt_0, i.e. $F_0 \Delta t_0$, integrated over all possible values of t_0 up to t_1, together with a statistical average over the fluctuating yield of neutrons per fission ν, which have been emitted up to time t_1.

Thus if we denote the correlated count rate by $Q\Delta t_1 \Delta t_2$ we may write

$$Q\Delta t_1 \Delta t_2 = \left\langle \int_{-\infty}^{t_0} dt_0 F_0 p_1 p_2 \Delta t_1 \Delta t_2 \right\rangle, \tag{3.56}$$

where

$$\langle \ldots \rangle \equiv \sum_{\nu} p(\nu) \ldots,$$

$p(\nu)$ being the neutron emission probability.

Thus, inserting the expressions for p_1 and p_2, we get

$$Q\Delta t_1 \Delta t_2 = \langle \nu(\nu-1) \rangle \frac{\Delta t_1 \Delta t_2}{\tau_f^2} \epsilon^2 F_0 \int_{-\infty}^{t_0} e^{-\alpha(t_1+t_2)+2\alpha t_0} dt_0$$

$$= \Delta t_1 \Delta t_2 F_0 \frac{\overline{\nu(\nu-1)}}{2\alpha} \left(\frac{\epsilon}{\tau_f}\right)^2 e^{-\alpha(t_1+t_2)}, \tag{3.57}$$

where we have set $\langle \nu(\nu-1) \rangle \equiv \overline{\nu(\nu-1)}$.

The uncorrelated or random coincidence count rate $P\Delta t_1 \Delta t_2$ is simply the product of the probabilities of detection in the respective intervals, viz.

$$P\Delta t_1 \Delta t_2 = F_0 \Delta t_1 \epsilon . F_0 \Delta t_2 \epsilon.$$

This equation describes events like B and C of Fig. 3.4. Hence

$p_c(t_1, t_2)\Delta t_1 \Delta t_2$ = probability that, given a count at time t_1 in Δt_1, there is a second count in Δt_2 at time $t_2 - t_1 = \tau$ later

becomes

$$p_c(\tau)\Delta t_1 \Delta t_2 = F_0 \epsilon \Delta t_1 \Delta t_2 \left\{ F_0 \epsilon + \frac{\epsilon \overline{\nu(\nu-1)}}{2\alpha \tau_f^2} e^{-\alpha\tau} \right\}. \tag{3.58}$$

It is interesting to note that the average number of pairs counted in a time interval t, namely $\overline{Z(Z-1)}/2$, where Z is the number of counts during t, can be obtained by integrating eqn. (3.58) as follows:

$$\tfrac{1}{2}\overline{Z(Z-1)} = \int_0^t dt_2 \int_0^{t_2} dt_1 \, p(t_2 - t_1).$$

The result can be put in the following form:

$$\frac{\overline{Z^2} - \overline{Z}^2}{\overline{Z}} = 1 + \frac{\epsilon \overline{\nu(\nu-1)}}{\alpha^2 \tau_f^2} \left\{ 1 - \frac{(1 - e^{-\alpha t})}{\alpha t} \right\}, \tag{3.58a}$$

where $\overline{Z} = F_0 \epsilon t$ is the total count in a time t.

Equation (3.58a) is identical with eqn. (3.18) for the variance to mean ratio and indeed the method outlined above is an entirely equivalent way of computing this ratio. Experiments using this technique have been performed by Gotoh (1964) to measure the neutron lifetime and sub-criticality in a D_2O thermal reactor. The method proved particularly useful for calibrating control rods and some typical results have been shown already in Fig. 3.1.

3.8. *Experimental method for Rossi-α technique*

It is first assumed that the correlated coincidence rate is significant in magnitude compared with the uncorrelated count rate (i.e. $>10\%$). Now the procedure is to gate a detector into a number of distinct time channels, sufficiently narrow as to make the reading either 0 or 1 when in operation. Then, for example, detection in gate 3 triggers the time-analyser and counts are recorded in gate 8. The frequency of counts in two channels is now determined as a function of channel separation. If the data are recorded on magnetic tape, the analysis can be repeated conveniently for successive initial channels as well as for a different channel spacing to give the maximum data from the observations.

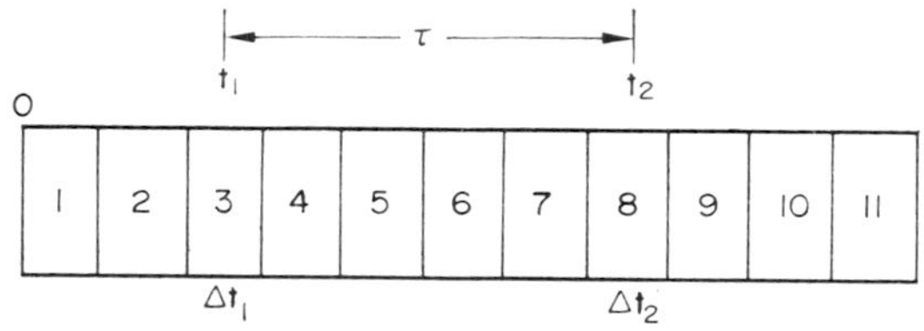

FIG. 3.8. Schematic diagram of multi-channel time analyser with gates of width Δt_i.

The uncorrelated component is subtracted from the value in each channel, i.e. the mean counting rate $F_0\epsilon\Delta t_i$, and the residual is plotted in the form

$$\log\left\{\frac{p_c(\tau)-(F_0\epsilon)^2}{(F_0\epsilon)^2}\right\} = \log\left(\frac{\epsilon\,\overline{\nu(\nu-1)}}{(\bar{\nu})^2}\right)-\alpha\tau. \tag{3.59}$$

Orndorff (1957) has applied this technique to the fast reactor, GODIVA I, which consists of U^{235} with a lifetime $l \sim 10^{-7}$ sec. Values of α in the range 10^5 to 10^6 sec^{-1} were expected so that a time resolution Δt of about $0{\cdot}25\times10^{-6}$ sec was required: this was accomplished using a ten-channel time analyser with delay lines between channels. A fission counter was used as a detector. Results of these experiments are shown in Fig. 3.9 for various reactivities starting at criticality and going to a value of reactivity ρ of 86·6 ₵ above delayed critical. Thus the experiment is actually performed on a super-critical system although the prompt decay constant, α, which is given by

$$\alpha = \frac{\beta-\rho}{\Lambda} = \frac{1-k(1-\beta)}{l}, \tag{3.60}$$

remains positive since $\rho < \beta$.

The value of β/l for the assembly was found by writing eqn. (3.60) as

$$\alpha = \frac{\beta}{l}\cdot\frac{1-k(1-\beta)}{\beta}. \tag{3.61}$$

The factor $\{1-k(1-\beta)\}/\beta$ is found from the reactor period and the inhour equation, and β/l is the constant of proportionality. In this case a value of $\beta/l = 1{\cdot}03\times10^6$ sec^{-1} is obtained, accurate to a "few" per cent.

The Rossi-α technique suffers from a number of limitations which can be summarized, with comment, as follows:

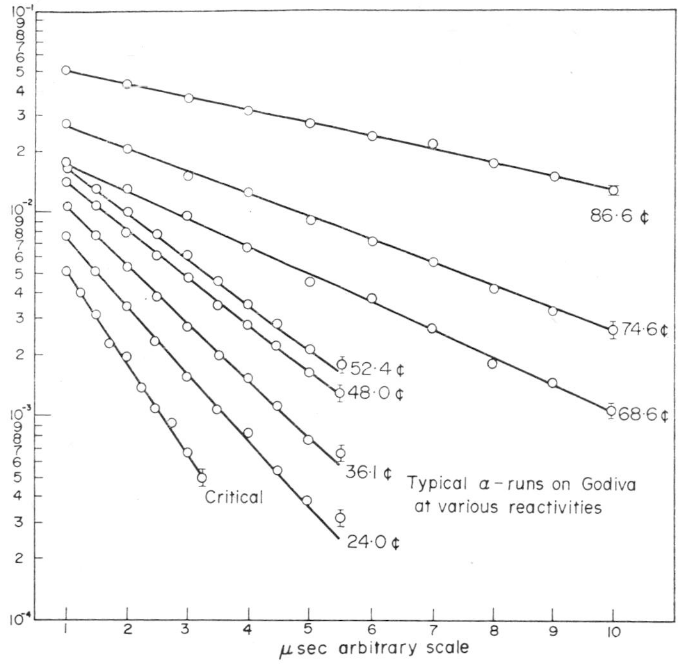

FIG. 3.9. Typical curves obtained in the Rossi-α experiment on GODIVA at various reactivities. (From Orndorff (1957) *Nucl. Sci. Engng*, **2**, 450.)

(*a*) Although the method can be applied to sub-critical systems it has to be noted that any method of defining reactivity for heavily sub-critical systems runs into difficulty over the flux shape, which is no longer in the fundamental mode. Thus a detailed space-dependent analysis of the Rossi-α theory would be required to accurately interpret for sub-critical systems, i.e. the lifetime will be space-dependent so that α will appear to depend on detector position: this is particularly important for fast reflected systems. This problem is dealt with in Section 7.12.

(*b*) If the detector is a fission chamber, spurious counts may arise from neutrons produced by the first count which had been assumed to terminate on detection. Nevertheless, because of its high efficiency the fission counter is very attractive and methods can be devised for correcting the spurious counts (Orndorff, 1957).

(*c*) An extension to the method of times longer than a neutron lifetime can be made to explore delayed neutron behaviour. This needs a more detailed examination of the theory, as given by Bennett (1960), and more sophisticated detection techniques. A discussion of the difficulties encountered when applying the Rossi-α method to thermal reactors has been given by Tachon (1962).

3.9. *Related techniques*

The Rossi-α technique was the forerunner of a host of methods based directly on calculating the probability, $P_k(\Delta t)$, of counting k pulses in a time Δt. Such methods have become feasible by virtue of recent improvements in electronics and the application of these to coincidence and high-speed counting. Detailed accounts of these methods may be found in Pál (1962), Pacilio (1970) and Uhrig (1970).

3.10. *Sign correlation techniques*

The polarity correlation method is based upon the auto- or cross-correlation function of the output of one or two detectors where the output is simply $+1$ or -1, depending upon whether the fluctuation is above or below the mean value at the instant of measurement.

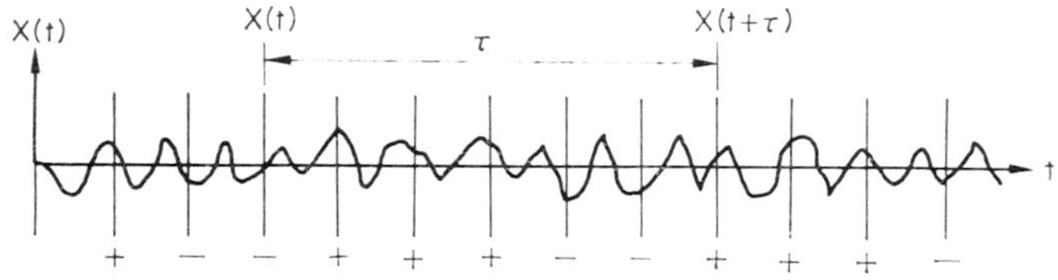

FIG. 3.10. Sample of fluctuating output used for calculation of polarity correlation function.

Thus if the output is as shown in Fig. 3.10 we can define a polarity correlation function as

$$\mathrm{Pol}(\tau) = \langle \mathrm{sgn}\, X(t)\, \mathrm{sgn}\, X(t+\tau) \rangle, \tag{3.62}$$

where $X = N - \bar{N}$, and is the difference between the fluctuating output of the detector and the mean value. The output is staggered by a time τ and the instant polarity is measured at a succession of times and averaged. Since there are only four permutations of the product possible, viz. $++$, $+-$, $-+$, $--$, we have a considerable reduction in computing time. In practice we have to calculate the quantity

$$\mathrm{Pol}(\tau) = \frac{1}{2M} \sum_{n=-M}^{M} \mathrm{sgn}\, X(n\Delta t)\, \mathrm{sgn}\, X(n\Delta t + \tau), \tag{3.63}$$

where $2T = 2M\Delta t$ = length of output analysed.

On the assumption that the amplitude distribution of the X's are Gaussian (a reasonable and tested assumption) it may be shown that $\mathrm{Pol}(\tau)$ is related to the conventional correlation function $\phi_{NN}(\tau)$ as follows:

$$R(\tau) = \frac{\phi_{NN}(\tau)}{\phi_{NN}(0)} = \sin\left(\tfrac{1}{2}\pi\, \mathrm{Pol}(\tau)\right). \tag{3.64}$$

This method is particularly useful since the nature of output required blends conveniently with digital on-line data processing techniques.

A detailed experiment using polarity correlation has been performed by Dragt (1966*a*) who based his work on the calculations of Veltman and Kwakernaak (1961). The experiment was performed on the LFR at Petten (Dragt, 1966*b*) using digital equipment. The main idea of the project was to compare the conventional correlation method with the polarity method with a view to assessing the viability of the latter technique. The experiment was performed for eight noise samples measured under identical circumstances, each containing 10^4 data points. In this way both $R(\tau)$ and

Pol(τ) were obtained. Tests were also made to establish that $X(t)$ obeyed Gaussian statistics. The results of these tests showed that deviations from the Gaussian were minimal.

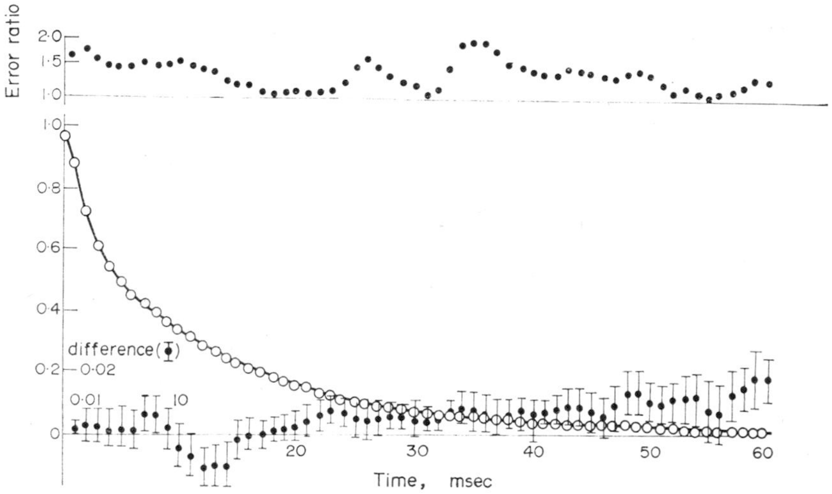

FIG. 3.11. Comparison between auto-correlation function and polarity correlation function. Open circles show $R(\tau)$, black circles show the difference between $R(\tau)$ and $\sin(\frac{1}{2}\pi \operatorname{Pol}(\tau))$. The upper curve shows the ratio of the experimental errors in the two methods as specified by their standard deviations. (From Dragt (1966) *Nukleonik*, **8**, 225.)

Dragt's results are summarized in Fig. 3.11, where the open circles show $R(\tau)$ and, on a 10 times expanded scale, the difference

$$R(\tau) - \sin(\tfrac{1}{2}\pi \operatorname{Pol}(\tau))$$

is plotted in black circles together with the associated errors. The difference is not significant. At the top of the figure the ratio of the standard deviations in $\sin(\frac{1}{2}\pi \operatorname{Pol}(\tau))$ and in $R(\tau)$ are shown. Dragt makes the very important statement: "It is very surprising that the polarity correlation function is only 1·5 times as inaccurate as the direct correlation function." This is of some significance since determining correlation functions from sgn $X(t)$ rather than $X(t)$ can be regarded as measuring $X(t)$ with an accuracy of only 1 bit. The implications for direct on-line data processing have already been mentioned above. Further work on the polarity method may be found in Szatmary and Valko (1969).

A refinement of this technique, known as *integrated polarity sampling*, averages the output over a time Δt and correlates the average polarity rather than the instantaneous value, as shown in Fig. 3.12.

Further details may be found in Pacilio's book (1970).

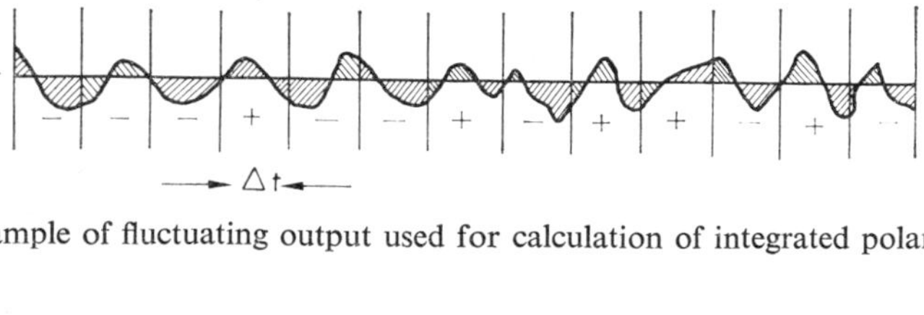

FIG. 3.12. Sample of fluctuating output used for calculation of integrated polarity sampling.

3.11. *The method of zero crossings*

The method is based upon a measurement of the mean frequency at which a fluctuating signal crosses its mean value. It is, however, a special case of a much more general problem in which it is desired to know the expected number of peaks E_1 occurring in unit time in the range $X_A < X < X_B$ where $X(t)$ is the random deviation from the mean value at time t. Figure 3.13 explains the problem, with the lines $X = X_A$ and $X = X_B$ defining the limits of interest.

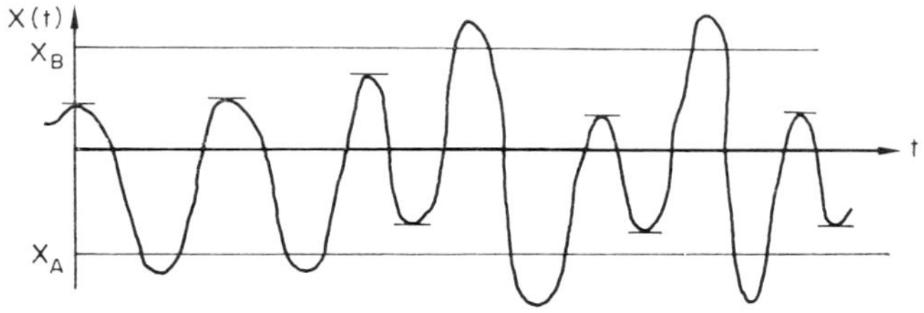

FIG. 3.13. Statistical distribution of the random signal about its mean value. The short horizontal lines indicate those peaks that lie in the range $X_B < X < X_A$.

If the fluctuations are Gaussian it may be shown with relative ease (Robson, 1963) that $E_1[X_A < X < X_B]$ is given by

$$E_1[X_A < X < X_B] = \int_{X_A}^{X_B} dP\,[\text{Peak in } dX], \tag{3.65}$$

where
$$dP[\ldots] = \frac{dX(-R'')^{\frac{1}{2}}}{2(2\pi R)^{\frac{1}{2}}} \frac{X}{(2R)^{\frac{1}{2}}} \exp\left(-\frac{X^2}{2R}\right) \times\left[1+\operatorname{erf}\left(\frac{kX}{(2R)^{\frac{1}{2}}}\right)+\frac{1}{\sqrt{\pi}}\frac{(2R)^{\frac{1}{2}}}{kX}\exp\left(-\frac{k^2X^2}{2R}\right)\right], \tag{3.66}$$

$$k^2 = \frac{(R'')^2}{RR^{\text{iv}}-(R'')^2} \tag{3.67}$$

with
$$R(\tau) = \langle X(t)\,X(t+\tau)\rangle$$

and
$$R = R(0) = \langle X^2(t)\rangle \equiv \phi_{NN}(0),$$
$$R'' = R''(0) = \phi''_{NN}(0),$$
$$R^{\text{iv}} = R^{\text{iv}}(0),$$

primes indicating differentiation with respect to τ.

As a special case we could ask for the expectation of peaks in unit time which are greater than n times the r.m.s. value $\sqrt{R}$. Then, with a number of simplifying assumptions, the integration over X may be performed leading to

$$E_1 = \frac{1}{2\pi}\left(-\frac{R''}{R}\right)^{\frac{1}{2}} \exp(-\tfrac{1}{2}n^2). \tag{3.68}$$

Thus the frequency of occurrence depends upon the random process through R and the quantity $\exp(-\frac{1}{2}n^2)$, which decreases rapidly as the value of n increases. This type of analysis is used in connection with the calculation of peak stress and failure of system components due to fatigue. Its use in understanding neutronic behaviour has not been studied to any great extent. One application, however, has been to the average number of crossings about the mean value per unit time. This can be obtained as a special case of eqn. (3.68) with $n = 0$; then E_1 is one-half of the number of zero crossings, $\mathcal{N}$, per unit time. Thus we may write for $\mathcal{N}$,

$$\mathcal{N} = 2E_1 = \frac{1}{\pi}\left(-\frac{R''}{R}\right)^{\frac{1}{2}}, \tag{3.69}$$

$$\equiv \frac{1}{\pi}\left(-\frac{\phi''_{NN}(0)}{\phi_{NN}(0)}\right)^{\frac{1}{2}}. \tag{3.70}$$

An alternative way of writing the quantity under the square root is obtained by referring back to the definition of power spectral density. Thus we find that

$$-\frac{\phi''_{NN}(0)}{\phi_{NN}(0)} = \frac{\int_0^\infty \omega^2\Phi(\omega)\,d\omega}{\int_0^\infty \Phi(\omega)\,d\omega}. \tag{3.71}$$

From measurements of $\mathcal{N}$ we can therefore obtain information regarding the moments of the p.s.d. In practice, however, it should be noted that p.s.d. of the zero power type led to divergence of the numerator in eqn. (3.71). For this reason it is necessary to pass the signal through a low pass filter which itself has a spectral density $|H(\omega)|^2$ with a suitable break frequency ω_0.

Very little work has been reported on the application of this technique to reactor analysis although Seifritz and Stegemann (1971), in a review article, mention that it has been employed on the ANNA and STARK reactors to study the dependency of $\mathcal{N}$ on reactivity. Similarly, Thie (1963) quotes results for the relaxation time $\tau^* = (\pi\mathcal{N})^{-1}$ measured on a variety of boiling-water reactors of the BORAX family. Typical values of τ^* range from 0·064 up to 0·5 sec, depending on the reactor power and operating pressure. Further, by calculating higher moments of the spectral density using knowledge of the distribution of maxima per unit time, it was possible to construct an auto-correlation function for the BORAX IV reactor.

The advantage of the technique is similar to that of the polarity method, namely the effort involved in computation is much lower than for auto-correlation or p.s.d. methods. However, unlike the polarity method, the accuracy attainable is not very high and more development is needed before the method can be considered as a reliable measure of kinetic parameters.

3.12. *Space and energy effects*

At this point it is worth noting that the above theories and techniques have been extended by several authors to cover the following more complicated situations:

1. space-dependent effects,
2. detector shape,
3. energy-dependent effects.

Thus, in principle, a more precise correlation function, including these effects, would take the form

$$\phi_{NN}(\mathbf{r}_1, \mathbf{r}_2, E_1, E_2, \tau) \tag{3.72}$$

which measures correlations between different space points $(\mathbf{r}_1, \mathbf{r}_2)$ and different energies (E_1, E_2). The work of Sheff and Albrecht (1966), Williams (1967*a*) and Cassell and Williams (1969) has shown that the effect of items 1, 2 and 3 can be summarized as follows:

(*a*) The point reactor model, as used in previous sections, is equivalent to an *infinite detector* placed in an *infinite medium*. The *point detector* in an infinite medium gives a different result. However, in practice, over the most important frequency ranges the functional form of the p.s.d. remains the same as in eqn. (3.64) with A and B different and α given by

$$\alpha_1 \simeq \frac{\beta - \rho}{\Lambda_{\text{eff}}}, \tag{3.73}$$

where $$\rho = \frac{k_{\text{eff}} - 1}{k_{\text{eff}}} \quad \text{and} \quad \Lambda_{\text{eff}} = \Lambda / P(B^2),$$

B^2 being the geometric buckling of the system and $P(B^2)$ the fast non-leakage probability.

In other words, the parameters are modified by the appropriate non-leakage probabilities.

(*b*) Cross-correlations in space indicate that the correlation distance is dependent on the migration length and the neutron lifetime. For such studies a detailed space dependent analysis has been shown by Natelson *et al.* (1966) to be essential. Similarly, Cassell and Williams (1969) have shown the importance of transport theory in the calculation of cross-correlation functions between different detectors at a distance less than about two mean free paths. In this connection a correlation length can be defined which is a measure of the distance between two points A and B such that a noise signal from A has been depleted by a factor e^{-1} by the time it reaches B. This length is of order of a wavelength as defined by the neutron wave experiment (Weinberg and Schweinler, 1948), hence it will be about a migration length in magnitude for low frequencies and become shorter for higher frequencies. Thus high-frequency noise signals are less strongly correlated as far as distance is concerned: to put it another way, separation acts as a low pass filter for noise signals. This effect is even more important in finite systems than the previous statement might imply since at low frequencies, $\omega\bar{\tau} \ll 1$, where $\bar{\tau}$ is the mean lifetime of delayed neutron precursors, the reactor intensity fluctuates as a whole with the fundamental mode predominating and the neutron flux at any point being in phase with that at any other point. In this case a point model description of noise would be acceptable. On the other hand, for $\omega\bar{\tau} \gg 1$, or if the detector and noise source are close together, the fundamental mode no longer predominates and the noise source excites different harmonics which all have different phases. It is this latter case which is more associated with the damped wave effect described earlier. This effect manifests itself mathematically in that, in a finite medium, the p.s.d. function between two space points is complex with the phase angle increasing as the frequency increases.

(*c*) Whilst conclusion (*a*) above is valid for the asymptotic region of a reactor, there are severe distortions of the p.s.d. when the detector is placed near a boundary.

Special calculations must be made for this case. Generally, boundary noise relaxes back to asymptotic reactor theory noise as $\exp(-X/L(\omega))$, where $L(\omega)$ is the attenuation length as defined by the neutron wave problem and X is the distance from the boundary (Sheff and Albrecht, 1966).

(*d*) The shape of the detector affects the functional form of the p.s.d. but tends towards the infinite medium result as the detector becomes larger. An example of the changes to be expected is illustrated by comparing eqn. (3.49), viz.

$$A+\frac{B}{\alpha^2+\omega^2} \tag{3.74}$$

for an infinite detector, with that for a point detector, viz.

$$A'+\frac{B'}{\{\alpha+[\alpha^2+\omega^2]^{\frac{1}{2}}\}^{\frac{1}{2}}}, \tag{3.75}$$

and for a plate detector: $$A''+\frac{B''}{(\omega^2+\alpha^2)^{\frac{1}{2}}\{\alpha+[\alpha^2+\omega^2]^{\frac{1}{2}}\}^{\frac{1}{2}}}, \tag{3.76}$$

and finally for a long thin cylindrical detector:

$$A'''+B'''\frac{1}{\omega}\tan^{-1}\left(\frac{\omega}{\alpha}\right). \tag{3.77}$$

In practice, detectors are made sufficiently large to obtain a high count rate and the net effect is to make the infinite medium expression (3.74) an acceptable approximation.

(*e*) In most spatial noise theories developed to account for detector shape effects, the actual depression of the surrounding flux due to the presence of the detector has been neglected. Using a technique based upon the source-sink or Feinberg–Galanin method (Galanin, 1960) the effect of this perturbation has been studied by Williams (1967*b*) for a plate detector. If the absorption rate of the detector is characterized by Galanin's constant γ, then eqn. (3.76) above is modified, due to perturbation of the flux, to the following expression:

$$A_0+\frac{B_0}{a_0\{[\gamma+2a_0(D_0)^{\frac{1}{2}}]^2+4D_0b_0^2\}}\cdot\left\{1+\frac{\gamma}{2(D_0)^{\frac{1}{2}}[(\omega_0)^{\frac{1}{2}}+a_0]}\right\}, \tag{3.78}$$

where $L^2 = D_0/v\Sigma_a$, L being the medium diffusion length, $\omega_0 = v\Sigma_a[1-k_\infty(1-\beta)]$ and

$$a_0 = \frac{1}{\sqrt{2}}\{(\omega^2+\omega_0^2)^{\frac{1}{2}}+\omega_0\}^{\frac{1}{2}},$$

$$b_0 = \frac{1}{\sqrt{2}}\{(\omega^2+\omega_0^2)^{\frac{1}{2}}-\omega_0\}^{\frac{1}{2}}.$$

For a "black" detector, $\gamma \gg (D_0\omega)^{\frac{1}{2}}$, eqn. (3.78) further changes to

$$A_0'+\frac{B_0'}{2a_0[(\omega_0)^{\frac{1}{2}}+2a_0]}. \tag{3.79}$$

Details of this calculation may be found in the paper by Williams cited above, but it is noted that the effect of detector perturbation may affect the interpretation of noise spectra in water moderated systems. Moreover, experimental evidence for the importance of detector perturbations has been found by Szatmary *et al.* (1965) in the Rossi-α experiment. They noticed that the Rossi-α value is profoundly influenced by the detector, which introduces into the system a large amount of absorbing material, thereby altering the life history of the neutrons in its neighbourhood. The detector tends to preferentially sample from shorter chains, simply because there are relatively more of these in its neighbourhood. We see, therefore, a further correction to the Rossi-α method which might profitably be investigated.

(*f*) Finally, we note that most spatial corrections to noise formulae are based upon homogeneous reactor theory. That is, the reactor behaviour and neutron density distribution have been averaged over any heterogeneities present. Although this averaging procedure is usually a good one for dealing with most reactor kinetics problems, it does not necessarily follow that it can be used for the analysis of fluctuation experiments. For example, in a heterogeneous system, the source of non-Poisson noise originates in the fuel elements and is zero in the moderator. On the other hand, in the homogeneous equivalent of the reactor, the fuel is assumed to be smeared uniformly over the core and hence leads to a uniform noise source.

The errors associated with the homogenization procedure have been assessed by Williams (1967) for a reactor consisting of fuel elements in the form of finite plates. As for the case of the detector mentioned in section (*e*), the Feinberg–Galanin method is used and an exact expression obtained for the cross-p.s.d. between any two fuel plates. For the case of auto-correlation, it is shown that the homogeneous approximation is valid provided the following criterion is obeyed:

$$a/L(\omega) \ll 1,$$

where $L(\omega)$, the neutron wave attenuation length, is given by

$$L(\omega) = L\left[\frac{2}{1+(\omega^2 l^2+1)^{\frac{1}{2}}}\right]^{\frac{1}{2}}, \tag{3.80}$$

l being the neutron lifetime in the moderator, L the diffusion length in the moderator and a the plate spacing.

In other words, instead of simply the plate spacing being very much less than a diffusion length, as required for steady-state problems, we now have the added restriction that the plate spacing must be very much less than the attenuation length of a neutron wave in the moderator. Thus there is a high frequency limit to the use of homogeneous reactor theory in the analysis of noise phenomena in heterogeneous systems.

Specific details regarding the calculation of space and energy effects in zero-power systems will be considered in Chapter 7.

3.13. *A note on divergence at criticality*

The divergence of the auto-correlation function, the variance to mean ratio and the Rossi-α formula, when the reactor is critical, is a constant source of irritation in the interpretation of noise experiments. As will have been noted, the experiments are usually performed on sub-critical systems to avoid this problem; however, it is manifestly obvious from measurements on critical systems that the divergences do not affect results in any serious manner. Several explanations have been advanced for the absence of this "critical catastrophe", the most popular being that a reactor is never operated exactly at critical because there is always a background source of some kind. This explanation, however, is not overly convincing and it is more likely that either inherent and/or an operator induced feedback mechanism exists which is of low frequency and is sufficient to prevent any divergences at criticality.

References

BENNETT, E. F. (1960) The Rice formulation of pile noise, *Nucl. Sci. Engng*, **8**, 53.

CASSELL, J. S. and WILLIAMS, M. M. R. (1969) Space dependent noise spectra via the transport equation, *J. Nuclear Energy*, **23**, 755.

DRAGT, JAN B. (1966*a*) Accurate reactor noise measurements in a low power critical reactor, *Nukleonik*, **8**, 188.

DRAGT, JAN B. (1966*b*) Reactor noise analysis by means of polarity correlation, *Nukleonik*, **8**, 225.

GALANIN, A. D. (1960) *Thermal Reactor Theory*, Pergamon Press, London.

GOTOH, Y. (1964) Measurement of neutron life in a D_2O-system by neutron fluctuation, *J. Nucl. Sci. Techn.* **1**, 11.

MANSFIELD, W. K. (1968) Private communication, Queen Mary College, London.

McCULLOCH, D. B. (1958) An absolute measurement of the effective delayed neutron fraction in the fast reactor Zephyr, AERE, R/M 176.

NATELSON, M. *et al.* (1966) Space and energy effects in reactor fluctuation experiments, *J. Nucl. Energy*, **20**, 557.

ORNDORFF, J. D. (1957) Prompt neutron periods of metal critical assemblies, *Nucl. Sci. Engng*, **2**, 450.

PACILIO, N. (1970) *Reactor Noise Analysis in the Time Domain*, AEC Critical Review Series.

PÁL, L. (1962) Statistical theory of the chain reaction in nuclear reactors, UKAEA (translation) HAP 31981.

PLUTA, P. R. (1962) An analysis of fluctuations in a boiling water reactor by methods of stochastic processes, APED-4071.

RICKER, C. W. *et al.* (1965) Measurement of reactor fluctuation spectra and subcritical reactivity, USAEC, ORNL-TM-1066.

ROBSON, J. D. (1963) *Random Vibrations*, Edinburgh Univ. Press.

SEIFRITZ, W. and STEGEMANN, D. (1971) Reactor noise analysis, *Atomic Energy Rev.* **9**, No. 1, 129.

SHEFF, J. R. and ALBRECHT, R. W. (1966) The space dependence of reactor noise. I. Theory, *Nucl. Sci. Engng*, **24**, 246.

SZATMARY, Z. *et al.* (1965) Rossi-α measurements on several EKI lattices, Central Res. Inst. for Physics, Budapest, Conference on Reactor Physics and Engineering, paper M-2.

SZATMARY, Z. and VALKO, J. (1969) An on-line spectral analysis of reactor noise, *Nukleonik*, **12**, 208.

TACHON, J. (1962) Neutron study of a plutonium thermal neutron reactor "Prosperine", correlations between neutrons in a chain reactor, UKAEA, TRG 17(D).

THIE, J. (1963) *Reactor Noise*, Rowman & Littlefield, Inc.

UHRIG, R. E. (1970) *Random Noise Techniques in Nuclear Reactor Systems*, Ronald Press, N.Y.

VELTMAN, B. B. TH. and KWAKERNAAK, H. (1961) *Regelungstechnik*, **9**, 357.

WEINBERG, A. M. and SCHWEINLER, H. C. (1948) Theory of oscillating absorber in a chain reactor, *Phys. Rev.* **74**, 851.

WILLIAMS, M. M. R. (1967*a*) An application of slowing down kernels to thermal neutron density fluctuations in nuclear reactors, *J. Nucl. Energy*, **21**, 321.

WILLIAMS, M. M. R. (1967*b*) Reactor noise in heterogeneous systems: 1. Plate type elements, *Nucl. Sci. Engng*, **30**, 188.

CHAPTER 4

Practical Applications of the Probability Distribution

4.1. *Introduction*

In the previous chapters our attention has been focused on mean values and variances; the actual structure or behaviour of the basic probability distribution has not been explicitly used. In this chapter, however, we wish to discuss some problems of considerable practical importance which require a more detailed knowledge of individual fluctuations.

Investigations of this type are certainly not new since they can be traced back to the work of Frankel (1944) and of Feynman (1946). However, it is only fairly recently that serious study has been given to the problem by Hansen (1960) and Hurwitz *et al.* (1963). A far-sighted paper by Schrödinger was also published on the subject in 1945 (Schrödinger, 1945) which will be discussed below, as also will the work of Pál (1958). It is also important to note that the basic ideas involved in dealing with neutron population fluctuations were developed not by neutron physicists but rather by biologists (Bailey, 1964) and naturalists (Galton and Watson, 1874; Galton, 1889; Haldane, 1949).

As far as neutrons are concerned there are a number of practical matters of concern; for example, it is necessary to be able to calculate the probability of obtaining various neutron density populations in systems brought to super-criticality in a number of different ways. It is this type of problem that forms the basis of the present chapter.

4.2. *Weak source start-up*

When a reactor is started up with an extremely weak source, there will be an initial period of time during which the power level is so low that statistical fluctuations are important. Eventually, the power level will rise to a higher level where these fluctuations are negligible. However, the influence of the fluctuations in the initial start-up phase will persist through to the high power level due to their effect on the initial conditions. Thus it does not follow that identical start-up procedures on the same reactor will, every time, lead to the same reactor power at a given time after start-up.

This situation can be demonstrated by reference to some experiments made on the GODIVA fast critical assembly (Wimett *et al.*, 1960). GODIVA is a small bare sphere of U^{235} which can be made super-critical by the rapid assembly of its component parts.

An experiment was performed in which the system was brought rapidly to a configuration a few cents above prompt critical. No artificial source was present and multiplication was initiated by the very low spontaneous fission and cosmic background radiation sources. Eventually the fission rate built up to a maximum and the system shut itself down due to a temperature feedback effect.

The experiment was repeated several times under identical initial conditions and results similar to those shown in Fig. 4.1 were obtained.

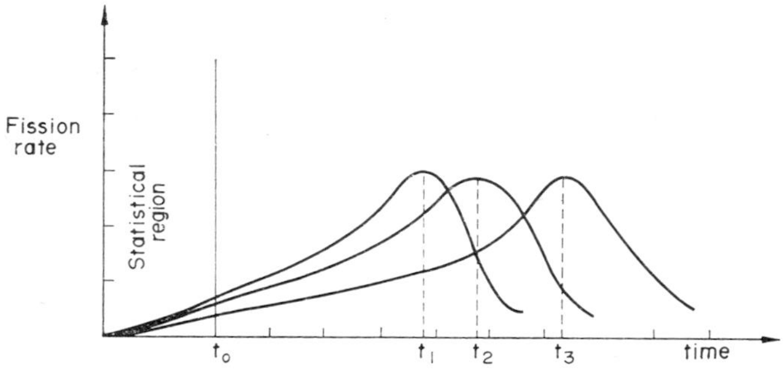

FIG. 4.1. Fission rates following three identical ramp reactivity insertions in a small, fast reactor. For $t < t_0$ the behaviour is essentially statistical, for $t > t_0$ the conventional reactor kinetics equations describe the situation.

The maximum fission rate occurred at different times for each experiment (average time ~3 sec). On the basis of the normal reactor kinetics equations such behaviour is inexplicable. However, when we consider the possible fluctuations that can arise and use the probability balance formalism, then it is clear that the different times of the peak fission rates are due to the different initial conditions arising from the random nature of the triggering mechanism. Figure 4.1 shows schematically the situation. In the early stages of multiplication the fluctuations are large; after a time t_0, however, the fission rate is sufficiently large for the average value equations to describe the situation. Nevertheless, the effect of the statistical stage remains. Physically, we can explain this by noting that on some occasions a burst of fissions will occur very quickly after core assembly, in which case the power will build up rapidly: in other cases, there may be a substantial time lag before a fission event occurs, thereby delaying the multiplication.

This effect is clearly important in certain types of reactor which have low sources and which, for some technological or economic reason, cannot have large artificial sources inserted into them. Nuclear submarine reactors probably come into this class although their precise construction is classified.

Thus an important characteristic of a safe start-up of such a system is that *the power must rise to a detectable level before the reactivity has risen to an undesirably high value.* A primary objective of the statistical calculation is, therefore, to demonstrate that the probability of the power remaining low, while the reactivity is *appreciably above* delayed critical, is vanishingly small for the range of start-up procedures which may occur in practice.

A further example of the random nature of the fission process is provided by the results of another experiment performed on GODIVA. In Table 4.1 we record the range of times for the fission rate to build up to a prescribed level (viz. $2{\cdot}7 \times 10^{11}$ fissions sec^{-1}) following a step increase in reactivity of $k-1 = 0{\cdot}0047$, i.e. $\Delta k = 0{\cdot}7\beta$. The twenty-two runs shown were made under identical initial conditions.

TABLE 4.1

Run no.	Time (sec)	Run no.	Time (sec)
1	33·1	12	25·8
2	30·5	13	34·3
3	30·1	14	35·8
4	32·4	15	43·9
5	34·2	16	28·6
6	31·9	17	25·4
7	31·4	18	39·3
8	37·8	19	26·7
9	31·2	20	31·5
10	33·8	21	27·0
11	26·3	22	27·8

We note that the times vary from 25·4 sec to 43·9 sec, i.e. they differ by almost a factor of 2.

The average time $\bar{t} = 31{\cdot}8$ sec and the r.m.s. deviation is

$$\sigma = \sqrt{(\overline{t^2}-\bar{t}^2)} = 4{\cdot}6 \text{ sec.}$$

In a comparable theoretical study using a one-speed point model approximation and assuming a source strength of 90 neutrons sec^{-1} it was predicted that

$$\bar{t} = 32{\cdot}5 \text{ sec}$$

and

$$\sigma = 3{\cdot}3 \text{ sec,}$$

which indicates that the theory used underestimated considerably the erratic nature of the fission chain growth: a more detailed theoretical investigation is clearly needed, possibly accounting for space- and energy-dependent effects.

4.3. *Theory of weak source fluctuations*

In Chapter 2, Section 2.4, we mentioned briefly the form of the probability distribution $P(N,t)$ arising from the case of the no delayed neutrons. Clearly, any detailed study must be more general than that; therefore in this section we will sketch, without too much detail, the basic mathematical technique for obtaining the information necessary to provide the data for safety assessments under low source start-up conditions.

Before writing down the basic fluctuation equations it is useful to consider the procedure which is to be adopted. As we have stated, it is during the initial period of start-up with a weak source that fluctuations are important. Eventually, the power rises to a sufficiently high level for the statistical fluctuations to have a negligible effect. Nevertheless, the influence of those early fluctuations persists in the sense that they determine the initial conditions from which the conventional reactor kinetics equations "take over".

The basic objective of the statistical calculation is to convince the operator that the probability of the reactor power remaining low, while the reactivity is appreciably above delayed critical, is below some preset value.† A procedure for doing this, therefore, is to choose a fission rate, F_0, at which statistical fluctuations are certainly unimportant and then to determine the earliest time, t_0, at which the probability that the actual fission rate, F, exceeds this chosen value will arise, i.e. $P(F > F_0; t_0)$. The value of the probability itself is chosen to be equal to the maximum uncertainty that one is willing to accept and this clearly depends on the situation under consideration: a value of 10^{-5} might be typical for $1-P$. If the subsequent behaviour of the reactor is calculated by the conventional reactor kinetics equations with initial conditions $F = F_0$ at $t = t_0$, then the probability 10^{-5} tells us the chance of such a transient arising in practice.

The mathematical apparatus for calculating the necessary probabilities has already been formulated in Section 2.5 and with minor modification can be employed for the present problem. First, let us note that, in most practical cases, the power will rise above the statistical region before prompt criticality is attained. It is therefore necessary to pay particular attention to the region in the neighbourhood of delayed critical.

We shall restrict consideration to the case of one group of delayed neutrons and ignore the presence of the detector. In this case we may set $\epsilon = 0$ in eqn. (2.46) and also

$$p(n,m_1, \ldots, m_J) = p(n)\{\delta_{m_1,0}, \ldots, \delta_{m_J,0}(1-\beta\bar{\nu}_0)+\delta_{m_1,1}, \delta_{m_2,0}, \ldots, \delta_{m_J,0}\beta\bar{\nu}_0\}, \quad (4.1)$$

where β is the ratio of delayed neutrons to prompt neutrons. The auxiliary generating function $f(x, \ldots)$ now becomes

$$f(x,y_1, \ldots, y_J) = f(x,y) = g(x)\{1-\beta\bar{\nu}_0+\beta\bar{\nu}_0 y\}, \quad (4.2)$$

where

$$g(x) = \sum_{n=0}^{\infty} x^n p(n). \quad (4.3)$$

Also we set

$$F(x,y_1, \ldots, y_J,t) = G(x,y,t),$$

where

$$G(x,y,t) = \sum_{n=0}^{\infty} \sum_{m=0}^{\infty} x^n y^m P(n,m,t) \quad (4.4)$$

to obtain

$$\frac{\partial G}{\partial t} = \lambda(x-y)\frac{\partial G}{\partial y} - S(1-x)G + [\lambda_c(1-x)+\lambda_f\{g(x)(1-\beta\bar{\nu}_0+\beta\bar{\nu}_0 y)-x\}]\frac{\partial G}{\partial x}, \quad (4.5)$$

where λ is the average decay constant of the precursors.

† An interesting biological analogue to this problem concerns the spread of an epidemic in which there is a delay time for incubation of the infectious carrier. This occurs with scabies, which can be passed by physical contact to a number of people before the first contact is made aware of its presence. By that time a very large number of people in the chain may have been infected and an epidemic can start before preventive measures can be taken. A branch of such a chain only terminates when one of the hosts is immune; however, other branches of the chain can continue and lead to a serious health problem. There are many other biological analogues with the fission process and the reader is referred to Bailey (1964) and to Gani (1965).

A further approximation is permissible if we restrict consideration to situations where the reactivity does not greatly exceed delayed critical. In this case the insertion of a single source or delayed neutron will cause a chain of prompt fissions which may persist for about one hundred generation times. Since the prompt generation time is of order 10^{-3} sec and the mean delayed neutron lifetime is about 10 sec, it is clear that over the period of the fluctuation (i.e. 100×10^{-3} sec) the reactivity and the delayed neutron precursor concentration will change very little. This fact enables us to use the zero prompt generation time approximation in which terms involving neutron emission from precursors are neglected (Meghreblian and Holmes, 1960). Of course, precursor *formation* is not neglected since these are produced in the course of a prompt fission chain. As far as eqn. (4.5) is concerned, these remarks imply that the delayed neutrons have an infinite lifetime and therefore λ can be set equal to zero. Moreover, since we are interested in the effect of a single prompt neutron chain, we shall set $S = 0$ and replace it by the initial condition

$$P(n,m,0) = \delta_{n,1}\delta_{m,0} \tag{4.6}$$

or equivalently

$$G(x,y,0) = x. \tag{4.7}$$

$P(n,m,t)$ is now the probability that at time t after the introduction of a single neutron there are n neutrons present and m precursors have been produced.

Redefining some parameters in eqn. (4.5) as follows:

$$l = \frac{1}{\lambda_c+\lambda_f}, \quad k_p = \bar{\nu}_0\lambda_f l$$

we can write our equation for $G(x,y,t)$ as

$$\frac{\partial G(x,y,t)}{\partial t} = -\frac{1}{l}\, h(x,y)\,\frac{\partial G(x,y,t)}{\partial x}, \tag{4.8}$$

where

$$h(x,y) = 1-\frac{k_p}{\bar{\nu}_0} - x + \frac{k_p}{\bar{\nu}_0}\, g(x)\{1-\beta\bar{\nu}_0+\beta\bar{\nu}_0 y\}. \tag{4.9}$$

The correct solution of this equation subject to the stated boundary condition is found by the method of characteristics to be (Smith, 1967):

$$\int_{s=x}^{s=G(x,y,t)} \frac{ds}{h(s,y)} = -\frac{t}{l}. \tag{4.10}$$

If l is very short, this solution can be simplified. If we consider times t such that $t/l \to \infty$ (very large) but still short enough such that there is no significant progenitor decay or reactivity change, then close inspection of $h(s,y)$ shows that in order that the integral becomes large and negative, its upper limit must approach that value of s given by $h(s,y) = 0$. In addition, $P(n,m,t)$ will be very small at such times, except for $n = 0$, so that the situation of physical interest is given by $G(0,y,t)$.

The solution of the problem is therefore given by the root of the equation

$$h(G(0,y,\infty),y) = 0. \tag{4.11}$$

$G(0,y,\infty)$ is related to the probability distribution $P(0,m,\infty)$ by the relation

$$G(0,y,\infty) = \sum_{m=0}^{\infty} P(0,m,\infty)\, y^m = \mathcal{G}(y), \tag{4.12}$$

where $P(0, m, \infty)$ is the probability distribution for emission of precursors from a single prompt fission chain.

A quantity of some interest is the probability, $P(0, 0, \infty)$, that no delayed neutron precursors are produced in the chain of fissions. This is equal to $\mathscr{G}(0)$ which, for $\mathscr{G}$ near unity, can be evaluated approximately from eqn. (4.11) as

$$\mathscr{G}(0) \simeq 1 - \frac{([1-k_p]^2 + 2\beta k_p^2 \overline{\nu(\nu-1)}/\bar{\nu}_0)^{\frac{1}{2}}}{k_p \overline{\nu(\nu-1)}/\bar{\nu}_0} \tag{4.13}$$

$$\simeq 1 - \left(\frac{2\beta\bar{\nu}_0}{\overline{\nu(\nu-1)}}\right)^{\frac{1}{2}}. \tag{4.14}$$

For the constants of U^{235} near delayed critical, we find $\mathscr{G}(0) \simeq 1-\beta^{\frac{1}{2}} = 0{\cdot}92$. Hence there is a 92% probability that no precursors are produced in a prompt chain. The mean number of precursors produced is

$$\bar{m} = \left.\frac{\partial G}{\partial y}\right|_{y=1} = \frac{k_p \beta}{1-k_p}. \tag{4.15}$$

In this case $\bar{m} \simeq 1$ and indicates that very large fluctuations in precursor production are to be expected. These fluctuations are made even more evident if we note that $\overline{m^2} - \bar{m}^2 \simeq 300$ for the case of $\bar{m} = 1$.

Whilst the distribution of precursor emission is of intrinsic physical interest it is also of practical value in calculating the safety factors discussed at the beginning of this section. What we have done thus far is to assume that a prompt fission chain is counted as an instantaneous event. Its role in developing the neutron and precursor levels during operation is described completely in terms of the number of precursors produced before its decay to negligible levels. This number is given by $P(0, m, \infty)$, which we now call $p_d(m)$, and is the probability distribution that m precursors are produced by the chain. In principle we should write this probability as $p_d(m,t)$ since it depends on time through the time dependence of the reactivity. Because the fission chain decayed sufficiently rapidly, p_d was calculated on the basis of constant reactivity. Now, however, we shall use p_d to predict the precursor level $P_d(m,t)$ over a much larger time scale which includes many prompt chains. In a sense, therefore, $p_d(m)$ can be regarded as analogous to the $p(\nu)$ used for prompt neutron fluctuation studies: to put it otherwise, the effect of multiplication is completely described in terms of the residual precursors left by an instantaneous fission chain.

The balance equation for $P_d(m,t)$ can be written as follows for a system containing a source of strength S neutrons per second:

$$\begin{aligned} P_d(m, t+\Delta t) &= [1-(S+\lambda m)\Delta t]P_d(m,t) \\ &+ S\Delta t \sum_{n=0}^{m} p_d(n,t) P_d(m-n,t) \\ &+ \lambda \Delta t \sum_{n=0}^{m} (m+1-n) p_d(n,t) P_d(m+1-n,t). \end{aligned} \tag{4.16}$$

Physically, the terms on the right-hand side arise as follows: the first assumes that there are m precursors at time t and that no new neutrons are produced by random decay of a precursor or by source emission. The second term assumes that, in Δt, the probability for neutron emission is $S\Delta t$; if this leads to the production of n precursors, then the contribution to $P_d(m, t+\Delta t)$ is as shown. Similarly, the third term denotes the additional precursors produced by the neutron emission from precursor decay, remembering of course that one out of the existing $(m+1-n)$ precursors disappears.

Converting eqn. (4.16) to differential form by allowing $\Delta t \to 0$ and then using the generating function technique, we find

$$\frac{\partial G_d(v,t)}{\partial v} = \lambda[g_d(v,t)-v]\frac{\partial G_d(v,t)}{\partial t}+S[g_d(v,t)-1]G_d(v,t), \tag{4.17}$$

where

$$G_d(v,t) = \sum_{m=0}^{\infty} v^m P_d(m,t) \tag{4.18}$$

and

$$g_d(v,t) = \sum_{m=0}^{\infty} v^m p_d(m,t). \tag{4.19}$$

Setting aside the solution of the above equation for the moment, let us investigate how it may be used in practical situations. As we have mentioned earlier, a system may be judged safe if the probability is sufficiently small that a power level be anomalously low in the presence of a high reactivity and correspondingly short period. If $m_1(t)$ is a prescribed precursor population, then

$$Q(m_1,t) = \sum_{m=0}^{m_1=1} P_d(m,t) \tag{4.20}$$

is the probability that, no matter what value the population may be, it is less than m_1.†

Calculations show (Hurwitz *et al.*, 1963) that use of the method of steepest descents gives an accurate value of Q over the range of practical interest, viz.

$$Q \simeq \frac{1}{\sqrt{(2\pi\sigma_0)}}\,\frac{v_0^{-m_1}G_d(v_0,t)}{1-v_0}, \tag{4.21}$$

where v_0 is given by

$$\frac{m_1}{v_0} = \frac{1}{1-v_0}+\frac{\partial}{\partial v}\ln G_d\bigg|_{v=v_0} \tag{4.22}$$

and

$$\sigma_0 = \frac{\partial^2}{\partial v^2}\ln\left(\frac{v^{-m_1}G_d}{1-v}\right)\bigg|_{v=v_0}. \tag{4.23}$$

To evaluate G_d itself, numerical methods have been applied to eqn. (4.17) for various forms of reactivity variation $k_p(t)$.

To consider a practical example, let us suppose that a non-statistical (i.e. based on average values) calculation indicates that S_m and R_m are a *just-safe* source strength and reactivity rate combination for reactor start-up. This calculation will take into account the fact that if a certain reactivity rate is required for a particular operational

† Neutron number, n, and precursor number, m, are in practice strongly correlated so that the probability $Q(m_1,t)$ is an accurate measure of the power excursions.

procedure, then the higher the rate the stronger must the source be. The information from this calculation is needed in order to set safety scram rods which are tripped at some predetermined power level: the power excursion following the trip should not damage the system. For example, suppose that the maximum available reactivity insertion rate is accidentally employed in each of the two systems, which are identical except for different source strengths. Then the reactor with the smaller source will reach the scram level later, in the presence of higher reactivity and a shorter period, and will experience a more severe post-scram power transient. Calculations can be made which give combined values of S_m and R_m which should be used together to ensure that there is no danger of a serious power excursion.

The arguments above neglect entirely the fluctuations. Thus there is the possibility that the combination (S_m, R_m) is not safe because of statistical fluctuations and one must contemplate the use of a larger source, $S > S_m$. The question to be answered is: what degree of certainty exists that an excessively large power transient will be avoided? To answer this question we fix the value of m_1 such that it corresponds to the safe-source value S_m. Then with prescribed R_m (i.e. $k_p(t)$) the equation for G_d is solved and the quantities Q and the mean value

$$\overline{m} = \left.\frac{\partial G_d}{\partial v}\right|_{v=1} \tag{4.24}$$

are calculated. It may also be shown on the basis of analytical considerations that for sufficiently large $\overline{m}$ and m_1 the ratio $\overline{m}/m_1$ is constant and equal to S/S_m.

Curves of $\overline{m}/m_1$ versus $-\log_{10} Q$ are plotted for various reactivity rates and source strengths. Figure 4.2 shows such a curve for a ramp reactivity insertion such that $R/\lambda = 0{\cdot}01$ \$ (R = constant rate of reactivity insertion) and for various values of S/λ, S being the source strength. To illustrate the use of the graph let us assume that preliminary, non-statistical calculations indicate that a constant reactivity insertion rate of 0·1 ₵ sec^{-1} and a source of strength 20 neutrons sec^{-1} are *just-safe*. We now ask: will the use of a stronger source of 100 neutrons sec^{-1} give a firm assurance that the power level during start-up will not fall below that known to be marginal for safety? Using $\lambda = 0{\cdot}1$ sec^{-1}, we have $S/\lambda = 1000$. Thus from Fig. 4.2 for $S/S_m = \overline{m}/m_1 = 100/20 = 5$, we find $Q = 10^{-3}$. Increasing the source strength to 200 neutrons sec^{-1} gives $Q = 10^{-6}$, thus the probability Q is very sensitive to source strength. Since Q is a measure of avoiding plant damage its value can only be decided upon by prevailing conditions at the site. However, numerical studies with the present model have shown that for $R/\lambda \lesssim \$\,0{\cdot}2$, and a source of 10^4 neutrons sec^{-1}, there is about one chance in 10^{10} that values of $\overline{m}/m_1$ greater than *two* will be obtained. This suggests that for sources of strength greater than 10^4 neutrons sec^{-1}, the probability of reaching a given power level with a significantly higher reactivity than that predicted by the average kinetics equation is negligibly small.

Finally we note that the discussion presented here is restricted to reactivities well below prompt critical and for reactivity insertion rates less than 2 ₵ sec^{-1}. Improvements to the theory which enable it to be used in the region of and beyond prompt criticality are given by D. R. Harris (1964), Hurwitz *et al.* (1963) and Bell (1963); the latter author used his results to explain the GODIVA experiments. Extensions to include space and energy effects are also implicit in the work of Pál (1958) and Bell (1965).

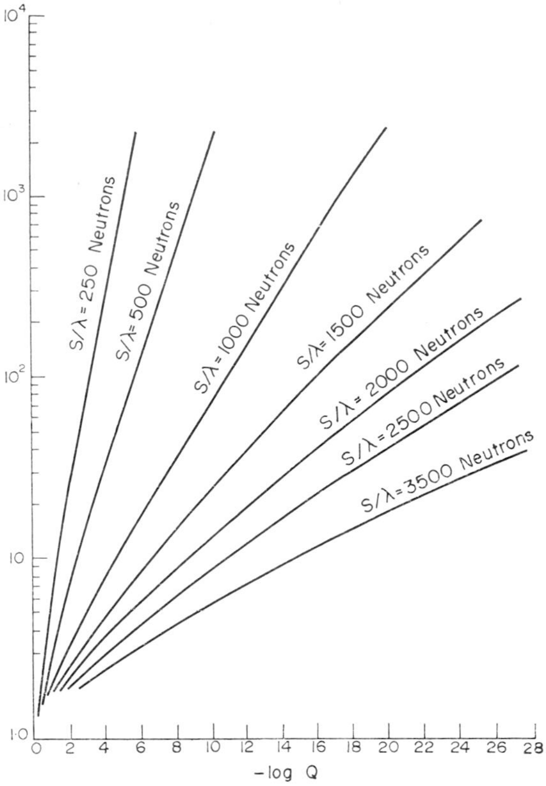

FIG. 4.2. Relative precursor population $\overline{m}/m_1$ versus log of probability (to base 10). From USAEC Report KAPL-2000 (McMillan *et al.*, 1960).

4.4. *The extinction probability*

We have encountered the term *extinction probability* in Section 2.3. It was defined there, quite generally, as the probability that given an initial burst of particles, after a very long time there will be no particles remaining in the system. That such a situation is possible can only be discovered by means of a statistical analysis of the problem. Physically, it is possible to understand the situation since, even in a multiplying medium, there is a chance that a succession of deaths will follow one another, leading to complete removal of all particles. As soon as this situation rises, the system remains empty for all time. Indeed we showed in Section 2.3 that for the simple birth and death problem the extinction probability for $\lambda > \mu$ is equal to $(\mu/\lambda)^{n_0}$, where n_0 is the number of particles in the initial burst.

The history of the extinction problem is exceedingly interesting since it was first formulated by the naturalists Galton and Watson (1874) in their studies on the probability of extinction of family names. Thus, because the surname is carried only by the male child, any family which produces girls results in the family name becoming extinct; similarly, a bachelor would lead to extinction of the family name. Galton and Watson did field trials using church records, and also live field surveys, to calculate the extinction probability. At the same time, they developed a statistical theory which is called the *Galton–Watson process* and is one of the basic techniques of the modern theory of branching processes (T. E. Harris, 1964).

The equivalent problem in neutron transport can be posed as follows: given a lump of fissile material of super-critical configuration, what is the probability that a neutron of velocity $\mathbf{v}$, injected at a point $\mathbf{r}$, will lead to a family of fission chains that eventually die out? We assume here that spontaneous fission and background sources are absent. The method of solution of this problem is implicit in the work of Pál (1958) and of Bell (1965). However, one of the earliest and most penetrating studies of the problem was published by Schrödinger (1945) in a paper entitled "Probability problems in nuclear chemistry". Schrödinger begins with a point model approximation which is fairly straightforward; his main contribution lies in the generalization to space dependence. Thus he poses the problem (in the one-speed approximation): what is the extinction probability for the fission chain initiated by a neutron isotropically injected into a lump of fissile material at position $\mathbf{r}$? The answer to such a question has importance in certain problems of nuclear safety and in the design of nuclear explosives.

Schrödinger's argument to solve this problem involved the use of the transport kernel to find the probability that a neutron born at $\mathbf{r}$ would cause a fission in $d\mathbf{r}'$ at $\mathbf{r}'$. Assuming that a neutrons were emitted per fission he then calculates the probability that S of them continue to produce further chains. Summing over all possible values of S from 0 to a leads to a non-linear integral equation for the extinction probability $\tau(\mathbf{r})$ as a function of position. The equation is solved approximately for the case when τ is close to unity.

Rather than develop Schrödinger's method in any detail it will be more profitable to derive the equation from the modern formalism of Bell (1965). The extinction probability can be calculated as a special case of the equation for the generating function $G(Z; \mathbf{r}, \mathbf{\Omega}, t)$ for the probability $p_n(R, t_f; \mathbf{r}, \mathbf{\Omega}, t)$, i.e.

$$G(Z; \mathbf{r}, \mathbf{\Omega}, t) = \sum_{n=0}^{\infty} Z^n p_n(R, t_f; \mathbf{r}, \mathbf{\Omega}, t), \tag{4.25a}$$

where $p_n(\ldots)$ is the probability that a neutron which is born at position $\mathbf{r}$, travelling in the direction denoted by the unit vector $\mathbf{\Omega}$ at time t, will lead to exactly n neutrons in the sub-region R at a later time t_f.

Defining c_j as the probability that j neutrons emerge instantaneously from a collision, Bell arrives at the following result for G:

$$-\mathbf{\Omega}.\nabla G - \frac{\partial G}{\partial t} + \Sigma G = \sum_{j=0}^{I} c_j \left[\frac{1}{4\pi}\int d\mathbf{\Omega}' G(Z; \mathbf{r}, \mathbf{\Omega}', t)\right]^j. \tag{4.25b}$$

This equation neglects delayed neutrons, assumes one-speed theory and isotropic scattering. However, Bell has shown how the approximations may be relaxed if necessary. The equation is also supplemented by the boundary condition on G for $\mathbf{\Omega}.\mathbf{n} > 0$, $\mathbf{n}$ being an outward directed normal, and by the final condition

$$\begin{aligned} G(Z; \mathbf{r}, \mathbf{\Omega}, t_f) &= Z; \quad (\mathbf{r}, \mathbf{\Omega}) \in R \\ &= 1; \quad (\mathbf{r}, \mathbf{\Omega}) \notin R. \end{aligned} \tag{4.26}$$

For a bare system, the boundary condition is

$$G(Z; \mathbf{r}_s,\mathbf{\Omega},t) = 1 \quad (\mathbf{\Omega}.\mathbf{n} > 0). \tag{4.27}$$

This equation and its modifications can be used to understand most of the zero-power fluctuation phenomena in nuclear reactors, and we refer the reader to Bell (1965) and Pál (1958, 1964) for applications. In the present case we are interested in its application to the extinction probability and we therefore note from eqn. (4.25) that this is given by

$$p_0(R,t_f; \mathbf{r},\mathbf{\Omega}, -\infty) = G(0; \mathbf{r},\mathbf{\Omega}, -\infty),$$

i.e. the probability that a neutron injected into the system in the far distant past at $(\mathbf{r},\mathbf{\Omega})$ will produce no neutrons in R at some time t_f.

We now set $G = 1-\mathscr{G}$ in eqn. (4.25*b*) and define

$$\bar{c} = \sum_{j=0}^{I} jc_j \equiv (\Sigma_s+\bar{\nu}\Sigma_f)/\Sigma \tag{4.28}$$

as the mean number of neutrons emerging from a collision and

$$\chi_i = \sum_{j=i}^{I} j(j-1)\ldots(j-i+1)c_j, \tag{4.29}$$

$\chi_2/2!$ being the mean number of pairs emerging, etc., where

$$\chi_2 = \frac{\Sigma_f}{\Sigma}\overline{\nu(\nu-1)}. \tag{4.30}$$

With these definitions, eqn. (4.25) can be rewritten for $\mathscr{G}$ as

$$-\mathbf{\Omega}.\nabla\mathscr{G}-\frac{\partial}{\partial t}\mathscr{G}+\Sigma\mathscr{G} = \frac{(\Sigma_s+\bar{\nu}\Sigma_f)}{4\pi}\int d\mathbf{\Omega}'\mathscr{G}(Z; \mathbf{r},\mathbf{\Omega},t)$$

$$-\sum_{j=2}^{I}(-)^j\frac{\chi_j}{j!}\Sigma\left[\frac{1}{4\pi}\int d\mathbf{\Omega}'\mathscr{G}(Z; \mathbf{r},\mathbf{\Omega}',t)\right]^j. \tag{4.31}$$

Since the extinction probability is a time-independent quantity, we can find it by considering the steady-state solution of eqn. (4.31). Using the boundary condition for a bare convex body we can convert this equation to integral form by the usual technique (Davison, 1957) and obtain for

$$\mathscr{G}_0(\mathbf{r}) = \int d\mathbf{\Omega}\,\mathscr{G}(0; \mathbf{r},\mathbf{\Omega}, -\infty) \tag{4.32}$$

the following:

$$\mathscr{G}_0(\mathbf{r}) = \int_V \frac{e^{-\Sigma|\mathbf{r}-\mathbf{r}'|}}{4\pi|\mathbf{r}-\mathbf{r}'|^2}\left\{(\Sigma_s+\bar{\nu}\Sigma_f)\mathscr{G}_0(\mathbf{r}')-4\pi\sum_{j=2}^{I}\frac{(-)^j}{j!}\chi_j\Sigma\left[\frac{1}{4\pi}\mathscr{G}_0(\mathbf{r}')\right]^j\right\}. \tag{4.33}$$

$\{1-\mathscr{G}/4\pi\}$ is the angle integrated extinction probability and is the extinction probability that would be obtained by averaging a large number of angle-dependent values resulting from isotropically distributed trigger neutrons.

If the system is only slightly super-critical then the extinction probability will be close to unity and hence $Q(\mathbf{r}) = \mathscr{G}_0(\mathbf{r})/4\pi$ is small. Using this fact we may obtain an approximate solution to eqn. (4.33) as follows.

Neglecting terms in Q higher than the second power, the non-linear integral equation becomes

$$Q(\mathbf{r}) = \int_V K(\mathbf{r},\mathbf{r}')\{(\Sigma_s+\bar{\nu}\Sigma_f)\,Q(\mathbf{r}')-\tfrac{1}{2}\Sigma_f\overline{\nu(\nu-1)}\,Q^2(\mathbf{r}')\}\,d\mathbf{r}', \tag{4.34}$$

where $K(\mathbf{r},\mathbf{r}')$ is used to denote the symmetric transport kernel.

The critical equation for this system can also be written as

$$\phi(\mathbf{r}) = [\Sigma_s+(\bar{\nu}\Sigma_f)_0]\int_V K(\mathbf{r},\mathbf{r}')\,\phi(\mathbf{r}')\,d\mathbf{r}', \tag{4.35}$$

where $(\nu\Sigma_f)_0$ is the value for criticality.

If the system is just super-critical, we can write

$$\bar{\nu}\Sigma_f = (\bar{\nu}\Sigma_f)_0(1+\delta) \quad (\delta \ll 1) \tag{4.36}$$

and

$$Q(\mathbf{r}) = \alpha\phi(\mathbf{r})+\Delta(\mathbf{r}), \tag{4.37}$$

where α is a constant and Δ is the deviation from the critical flux shape.

Inserting (4.36) and (4.37) into the equation for Q, neglecting second-order terms and using (4.35) leads to

$$\Delta(\mathbf{r}) = [\Sigma_s+(\bar{\nu}\Sigma_f)_0]\hat{K}\Delta(\mathbf{r})+\alpha\hat{K}\{(\bar{\nu}\Sigma_f)_0\,\delta\phi(\mathbf{r})-\tfrac{1}{2}\Sigma_f\overline{\nu(\nu-1)}\,\alpha\phi^2(\mathbf{r})\}, \tag{4.38}$$

where $\hat{K}$ denotes the integral operator.

This is an inhomogeneous integral equation and for it to have a solution the inhomogeneous term must be orthogonal to the solution of the homogeneous equation, i.e. $\phi(\mathbf{r})$, thus:

$$\int_V d\mathbf{r}\,\phi(\mathbf{r})\hat{K}\{(\bar{\nu}\Sigma_f)_0\,\delta\phi(\mathbf{r})-\tfrac{1}{2}\Sigma_f\overline{\nu(\nu-1)}\,\alpha\phi^2(\mathbf{r})\} = 0. \tag{4.39}$$

Using the identity

$$\int_V \phi(\mathbf{r})\hat{K}F(\mathbf{r})\,d\mathbf{r} \equiv \frac{1}{[\Sigma_s+(\bar{\nu}\Sigma_f)_0]}\int_V F(\mathbf{r})\,\phi(\mathbf{r})\,d\mathbf{r} \tag{4.40}$$

we can find α and hence, to first order, the extinction probability, viz.

$$\tau(\mathbf{r}) = 1-Q(\mathbf{r}) = 1-\frac{2\bar{\nu}\delta}{\overline{\nu(\nu-1)}}\,\frac{\int_V \phi^2(\mathbf{r})\,d\mathbf{r}}{\int_V \phi^3(\mathbf{r})\,d\mathbf{r}}\cdot\phi(\mathbf{r}), \tag{4.41}$$

where

$$\delta = (\Sigma_f-\Sigma_{f0})/\Sigma_{f0}.$$

Equation (4.41) is a generalization of the original equation of Schrödinger. It shows that the importance of a neutron as regards the runaway probability $(1-\tau)$ is directly proportional to the flux, a result which is not expected in one-speed theory. Unpublished extensions of this work by the author which include energy dependence indicate that $1-\tau(\mathbf{r},E)$ is proportional to a solution of the adjoint Boltzmann equation, i.e. the importance function: again not a surprising conclusion.

To obtain a rough estimate of $\tau(\mathbf{r})$ for a spherical system of radius R for pure U^{235}, let us write $\phi(\mathbf{r}) \simeq A(R^2-r^2)$, whence with $\overline{\nu(\nu-1)}/(\bar{\nu})^2 = 0{\cdot}8$ and $\bar{\nu} = 2{\cdot}5$, we get

$$\tau(r) = 1-1{\cdot}5\delta\left(1-\frac{r^2}{R^2}\right). \tag{4.42}$$

For a 10% change in fission cross-section $\delta = 0{\cdot}10$, and we find that for neutrons injected at the centre of the sphere, $\tau(0) = 0{\cdot}85$. Indeed $\tau(0)$ is completely independent of the radius of the sphere, a fact which arises from the form of $\phi(\mathbf{r})$, which is valid only for systems of size less than about one mean free path. If an asymptotic transport theory value of $\phi(\mathbf{r})$ is used, e.g. $\sin Br/r$, then the dependence of $\tau(0)$ on R remains. The present approximation indicates that the neutron chain has an 85% chance of extinction. In practice, of course, there will be spontaneous fission background sources and these will tend to reduce τ to well below its "single-neutron" value. The problem, then, becomes similar to the situation of low source start-up as discussed in the previous section. However, eqn. (4.31) provides a technique for introducing space and energy dependence which were neglected in Section 4.3. Indeed, eqn. (4.25) constitutes the basic equation from which the mean value transport equation for the neutron density, viz. eqn. (1.11), is obtained and also, by appropriate differentiation of the generating function, the mean square and higher moments (Pál, 1958, 1964; Bell, 1965; Otsuko and Saito, 1965*a*, *b*, 1966).

References

BAILEY, N. T. J. (1964) *The Elements of Stochastic Processes with Applications to the Natural Sciences*, Wiley & Sons, London.

BELL, G. (1963) Probability distribution of neutrons and precursors in a multiplying assembly, *Ann. Phys. (N.Y.)*, **21**, 243.

BELL, G. (1965) On the stochastic theory of neutron transport, *Nucl. Sci. Engng*, **21**, 390.

DAVISON, B. (1957) *Neutron Transport Theory*, Oxford University Press.

FEYNMAN, R. P. (1946) Statistical behaviour of neutron chains, Los Alamos Report, LA-591.

FRANKEL, S. P. (1944) The statistics of the hypocritical gadget, USAEC Rep. LAMS-36.

GALTON, F. (1889) *Natural Inheritance*, London.

GALTON, F. and WATSON, H. W. (1874) On the probability of the extinction of families, *J. Anthrop. Inst.* **4**, 138.

GANI, J. (1965) *Stochastic Models for Bacteriophage*, Methuen Review Series in Applied Probability, Vol. 4.

HALDANE, J. B. S. (1949) Some statistical problems arising in genetics, *J. Roy. Statist. Soc.* B, **11**, 1.

HANSEN, G. E. (1960) Assembly of weak fissionable material in the presence of a weak neutron source, *Nucl. Sci. Engng*, **8**, 709.

HARRIS, D. R. (1964) *Naval Reactors Physics Handbook*, Vol. I, edited by A. RADKOWSKY (USAEC).

HARRIS, T. E. (1964) *The Theory of Branching Processes*, Prentice-Hall, N.Y.

HURWITZ, H. *et al.* (1963) Kinetics of low source reactor start-ups, *Nucl. Sci. Engng*, **15**, 166.

MCMILLAN, D. B. *et al.* (1960) Effects of neutron population fluctuations on reactor design, Reactor Technology Report, No. 15-Physics, USAEC, KAPL-2000-12.

MEGHREBLIAN, R. V. and HOLMES, D. K. (1960) *Reactor Analysis*, McGraw-Hill, N.Y.

OTSUKO, M. and SAITO, K. (1965*a*) Theory of statistical fluctuations in neutron distributions, *J. Nucl. Sci. Tech.* **2**, 304.

OTSUKO, M. and SAITO, K. (1965*b*) Space–time correlations in neutron distributions in a multiplying medium, *J. Nucl. Sci. Tech.* **2**, 191.

OTSUKO, M. and SAITO, K. (1966) Transfer function and power spectral density in a zero power reactor, *J. Nucl. Sci. Tech.* **3**, 45.

PÁL, L. (1958) On the theory of stochastic processes in nuclear reactors, *Il Nuovo Cimento*, Supplemento VII, 25.

PÁL, L. (1964) Statistical theory of neutron chain reactors, *Proc. Int. Conf. Geneva*, Vol. 2, p. 218.

SCHRÖDINGER, E. (1945) Probability problems in nuclear chemistry, *Proc. Roy. Irish Acad.* LI, 1.

SMITH, M. G. (1967) *Introduction to the Theory of Partial Differential Equations*, Van Nostrand.

WIMETT, T. F. *et al.* (1960) Godiva II—an unmoderated pulse-irradiation reactor, *Nucl. Sci. Engng*, **8**, 691.

CHAPTER 5

The Langevin Technique

5.1. *Introduction*

An interesting and important technique for dealing with noise problems, originally conceived by Langevin to understand Brownian motion, is based upon the *average* equations for the system. Thus we assume that any system (nuclear or otherwise) can be represented by the equation

$$\hat{L}y(\mathbf{x}) = f(\mathbf{x}), \tag{5.1}$$

where $\hat{L}$ is an operator containing information about the system and $f(\mathbf{x})$ is a "driving" term.

Now if $\hat{L}$ contains parameters that are themselves random, and also if $f(\mathbf{x})$ is random, then the equation for $y(\mathbf{x})$ becomes itself random and will yield correspondingly random solutions. We have what is known as a stochastic equation and the problem is: given the probability laws of the various parameters in $\hat{L}$, say $\hat{L}(a_1, a_2, \ldots, \mathbf{x})$, calculate the probability law for $y(\mathbf{x})$ (Adomian, 1963, 1965; Caughey and Dienes, 1962).

5.2. *Brownian motion*

A simple problem of the type described above can be demonstrated by the case of Brownian motion which we may pose as follows: a particle of mass m is subjected to a random driving force $F(t)$ and a corresponding frictional force proportional to velocity. The equation of motion may therefore be written as

$$m\frac{du(t)}{dt} = -fu(t) + F(t), \tag{5.2}$$

where f is a constant friction factor and $u(t)$ the velocity at time t (Wax, 1954).

Now because $F(t)$ is random, then so will $u(t)$ be random. To completely specify the problem we should know all of the statistical details of $F(t)$; we can then solve eqn. (5.2), viz.

$$u(t) = u_0 e^{-\beta t} + e^{-\beta t}\int_0^t e^{\beta\xi} A(\xi)\, d\xi, \tag{5.3}$$

where u_0 is the initial velocity, $\beta = f/m$ and $A = F/m$.

Knowing the statistics of $A(t)$ it is easy to find those of $u(t)$. However, it is not always easy to calculate the statistics of the driving term $A(t)$ and therefore certain physical assumptions have to be introduced. In the case of Brownian motion, the following are considered reasonable:

(i) Assuming all particles start at $t = 0$ with non-random velocity u_0, then the mean value of the fluctuating force $A(t)$ is zero, i.e.

$$\langle A(t)\rangle = 0, \tag{5.4}$$

where $\langle \ldots \rangle$ denotes an ensemble average.

(ii) The values of $A(t)$ at two different times t_1 and t_2 are not correlated except for $|t_1 - t_2|$ small. More precisely

$$\langle A(t_1)A(t_2)\rangle = \phi(|t_1-t_2|), \tag{5.5}$$

where $\phi(x)$ is peaked at $x = 0$.

(iii) $A(t)$ is Gaussian so that all odd correlations $\langle A(t_1)A(t_2)\ldots A(t_{2n+1})\rangle = 0$ and all even ones can be related to $\phi(t)$. That is, the probability distribution of $A(t)$ is governed entirely by its mean value and variance.

Taking the mean value of eqn. (5.3) we get

$$\langle u\rangle = u_0e^{-\beta t}+e^{-\beta t}\int_0^t e^{\beta\xi}\langle A(\xi)\rangle\, d\xi \tag{5.6}$$

$$= u_0e^{-\beta t}. \tag{5.7}$$

Similarly the mean square value can be written

$$\langle u^2(t)\rangle = u_0^2e^{-2\beta t}+2u_0e^{-2\beta t}\int_0^t e^{\beta\xi}\langle A(\xi)\rangle\, d\xi+e^{-2\beta t}\int_0^t dt_1\int_0 dt_2\, e^{\beta(t_1+t_2)}\langle A(t_1)A(t_2)\rangle. \tag{5.8}$$

Now with $\langle A(t_1)A(t_2)\rangle = \phi(|t_1-t_2|)$ we can write (5.8) as

$$\langle u^2(t)\rangle = u_0^2e^{-2\beta t}+e^{-2\beta t}\int_0^t dt_1\int_0^t dt_2\, e^{\beta(t_1+t_2)}\phi(|t_1-t_2|). \tag{5.9}$$

Setting $t_1+t_2 = v$ and $t_1-t_2 = w$ we can reduce (5.9) to

$$\langle u^2(t)\rangle = u_0^2e^{-2\beta t}+\tfrac{1}{2}e^{-2\beta t}\int_0^{2t} dv\int_{-t}^t dw\, e^{\beta v}\phi(|w|), \tag{5.10}$$

where we have used the fact that $0 \leqslant v \leqslant 2t$, $-t \leqslant w \leqslant t$ and $dt_1, dt_2 = \frac{1}{2}dv/dw$.

Since $\phi(|w|)$ is peaked at $w = 0$ we can set

$$\int_{-t}^t dw\, \phi(|w|) \simeq \int_{-\infty}^{\infty} dw\, \phi(|w|) = \tau, \tag{5.11}$$

which is equivalent to assuming a *white noise source*, i.e. all frequencies equally probable and τ is an unknown relaxation time characterized by the particle mass and properties of the surrounding medium.

We now find that

$$\langle u^2(t)\rangle = u_0^2e^{-2\beta t}+\frac{\tau}{2}\, e^{-2\beta t}\int_0^{2t} dv\, e^{\beta v} \tag{5.12}$$

$$= u_0^2e^{-2\beta t}+\frac{\tau}{2\beta}(1-e^{-2\beta t}), \tag{5.13}$$

which tells us how the mean square velocity, or energy of the Brownian particle, relaxes to equilibrium. Thus, as $t \to \infty$,

$$\langle u^2(\infty)\rangle = \frac{\tau}{2\beta}. \tag{5.14}$$

At this point we introduce some physical information; namely equipartition of energy. Thus, in thermal equilibrium,

$$\tfrac{1}{2}m\langle u^2(\infty)\rangle = \tfrac{1}{2}kT \tag{5.15}$$

or

$$\frac{\tau}{2\beta} = \frac{kT}{m}$$

and

$$\tau = \frac{2\beta kT}{m}. \tag{5.16}$$

Thus we can write

$$\langle u^2(t)\rangle = u_0^2 e^{-2\beta t} + \frac{kT}{m}(1-e^{-2\beta t}). \tag{5.17}$$

This is an example of how the Langevin technique relies on certain physical conditions to fix the unknown constants which arise. Similar techniques are useful in the study of neutron noise as will be demonstrated later. The Langevin technique is particularly useful in power reactor systems where, generally, very little is known about the nature of the noise.

Calculations similar to those described above can also be used to obtain auto-correlation functions $\langle u(t_1)u(t_2)\rangle$ and, moreover, it is not necessary for the noise sources to be white.

5.3. *The Gaussian nature of noise*

An important theorem exists, known as the central limit theorem, which is concerned with the tendency of the distribution of a sum of N independent random variables to become normal (i.e. Gaussian) as $N \to \infty$.

We can illustrate this procedure by reference to the impact of molecules on the surface of a Brownian particle to show that the statistics of the forces may be regarded accurately as Gaussian. Let $p(F)\,dF$ be the probability that the force on the particle, due to the bombardment of a single molecule, takes a value between F and $F+dF$. Then the probability that it takes a value between F and $F+dF$ when two molecules are acting is

$$P_2(F)\,dF = dF\int dF_1\,p(F_1)\,p(F-F_1). \tag{5.18}$$

Similarly, when three molecules are acting

$$P_3(F)\,dF = dF\int dF_1\,dF_2\,p(F-F_1)\,p(F_1-F_2)\,p(F_2) \tag{5.19}$$

and when N particles act

$$P_N(F)\,dF = dF\int dF_1,\ldots,dF_N\,p(F-F_1)\,p(F_1-F_2),\ldots p(F_{n-2}-F_{n-1})\,p(F_{n-1}). \tag{5.20}$$

Now, according to the central limit theorem (Bartlett, 1962),

$$\lim_{N\to\infty} P_N(F) = \frac{1}{(2\pi\sigma)^{\frac{1}{2}}}\exp\left\{-\frac{(F-\bar{F})^2}{2\sigma^2}\right\} \tag{5.21}$$

irrespective of the nature of $p(F)$.

In this Gaussian, the mean and variance are defined by

$$\bar{F} = \int Fp(F)\,dF, \tag{5.22}$$

$$\sigma^2 = \int F^2 p(F)\,dF - \bar{F}^2. \tag{5.23}$$

Since the number of molecules hitting the particle in a time interval is short compared with the macroscopic relaxation time of the particle, we can assume that the Gaussian is an accurate description of the force law distribution function.

In general, the Gaussian law covers a wide range of physical phenomena including many in reactor dynamics. It should not be presupposed, however, that the corresponding power spectral density is white: it is possible to have Gaussian noise sources with quite arbitrary values of $\Phi(\omega)$ or, what is the equivalent thing, $\phi(\tau)$.

5.4. *Random source perturbation of a nuclear reactor*

In order to explain the Langevin technique further we shall apply it to the problem of a sub-critical nuclear reactor which is excited by an external, randomly pulsed source. For simplicity we assume the point model approximation and one-speed theory although, in principle, the method may be applied to the full space- and energy-dependent equations. We will also allow one group of delayed neutrons, thus our system is described by the equations

$$\frac{dN}{dt} = -\frac{(\rho+\beta)}{\Lambda}N + \lambda C - S(t), \tag{5.24}$$

$$\frac{dC}{dt} = \frac{\beta}{\Lambda}N - \lambda C. \tag{5.25}$$

In these equations, ρ is the sub-critical reactivity and Λ is the mean neutron generation time. $S(t)$ is the random source.

Let us now separate $S(t)$ into its mean value plus a random component, viz.

$$S(t) = \langle S\rangle - \mathscr{S}(t), \tag{5.26}$$

where by definition $\langle \mathscr{S}\rangle \equiv 0$.

Similarly, let us write

$$N = \langle N\rangle + n(t), \tag{5.27}$$

$$C = \langle C\rangle + c(t). \tag{5.28}$$

Inserting (5.26)–(5.28) into (5.24) and (5.25) and using the steady-state conditions for the average values, viz.

$$0 = -\frac{(\rho-\beta)}{\Lambda}\langle N\rangle + \lambda\langle C\rangle + \langle S\rangle, \tag{5.29}$$

$$0 = \frac{\beta}{\Lambda}\langle N\rangle - \lambda\langle C\rangle, \tag{5.30}$$

we obtain for the fluctuating components the following equations:

$$\frac{dn}{dt} = -\frac{(\rho+\beta)}{\Lambda}n + \lambda c + \mathscr{S}, \tag{5.31}$$

$$\frac{dc}{dt} = \frac{\beta}{\Lambda}n - \lambda c. \tag{5.32}$$

If we assume that the source was switched on in the distant past so that all initial transients have decayed, we can, after Laplace transformation, write:

$$n(t) = \int_{-\infty}^{t} dt' \, G(t-t') \mathscr{S}(t') \equiv \int_{0}^{\infty} dt_0 \, G(t_0) \mathscr{S}(t-t_0), \tag{5.33}$$

where

$$G(t) = \frac{(\omega_1+\lambda)}{\omega_1-\omega_2} e^{\omega_1 t} + \frac{(\omega_2+\lambda)}{\omega_2-\omega_1} e^{\omega_2 t}, \tag{5.34}$$

ω_1 and ω_2 being the roots of

$$p^2 + \left(\frac{\rho+\beta}{\Lambda}+\lambda\right)p + \frac{\lambda\rho}{\Lambda} = 0, \tag{5.35}$$

which is basically the "inhour equation" for one delayed neutron group. $G(t)$ is referred to as the Green's function of the system.

Let us note from eqn. (5.33) that $\langle n \rangle = 0$ as we expect. To find the auto-correlation function we calculate the ensemble average

$$\phi_{nn}(t_1-t_2) = \langle n(t_1) n(t_2) \rangle$$

$$= \left\langle \int_0^\infty dt_0 \, G(t_0) \mathscr{S}(t_1-t_0) \int_0^\infty dt_0' \, G(t_0') \mathscr{S}(t_2-t_0') \right\rangle \tag{5.36}$$

$$= \int_0^\infty dt_0 \int_0^\infty dt_0' \, G(t_0) G(t_0') \langle \mathscr{S}(t_1-t_0) \mathscr{S}(t_2-t_0') \rangle. \tag{5.37}$$

We thus have the auto-correlation function of n in terms of that of $\mathscr{S}$, which we assume to be known. Furthermore, if our noise source is stationary, we may write

$$\langle \mathscr{S}(t_1-t_0) \mathscr{S}(t_2-t_0') \rangle \equiv \phi_{\mathscr{S}\mathscr{S}}(t_2-t_1+t_0-t_0') \tag{5.38}$$

and with $\tau = t_2-t_1$, eqn. (5.37) becomes

$$\phi_{nn}(\tau) = \int_0^\infty dt_0 \int_0^\infty dt_0' \, G(t_0) G(t_0') \phi_{\mathscr{S}\mathscr{S}}(\tau+t_0-t_0'). \tag{5.39}$$

ϕ_{nn} is therefore stationary: a not unexpected result since n and $\mathscr{S}$ are linearly related.

We note from this result that knowledge of the input auto-correlation and a measurement of the output auto-correlation function should enable us to calculate $G(t)$, the Green's function, and hence obtain the system parameters ρ, β, l, etc.

In terms of the p.s.d. we can take the Fourier transform of (5.39) and get

$$\Phi_{nn}(\omega) = \Phi_{\mathscr{S}\mathscr{S}}(\omega) |\bar{G}(i\omega)|^2, \tag{5.40}$$

where $\Phi_{\mathscr{S}\mathscr{S}}(\omega)$ is the p.s.d. of the input and $\bar{G}(i\omega)$ is the Laplace transform of the Green's function with $p = i\omega$. Using (5.34) we therefore find that

$$\Phi_{nn}(\omega) = \Phi_{\mathscr{S}\mathscr{S}}(\omega) \frac{(\lambda^2+\omega^2)}{(\omega^2+\omega_1^2)(\omega^2+\omega_2^2)}. \tag{5.41}$$

If now we choose $\Phi_{\mathscr{S}\mathscr{S}}(\omega)$ to have a flat response over the range where $|\bar{G}(i\omega)|$ varies most, we can write

$$\Phi_{nn}(\omega) \simeq \sigma^2 \frac{(\lambda^2+\omega^2)}{(\omega^2+\omega_1^2)(\omega^2+\omega_2^2)}. \tag{5.42}$$

Such a situation can be achieved by using a white noise source of the pseudo-random type (see Section 5.11).

As an example of the practical utility of this method, consider the case for small ρ, when

$$\omega_1 \simeq -(\rho+\beta)/\Lambda, \tag{5.43}$$

$$\omega_2 \simeq -\lambda\rho/(\rho+\beta). \tag{5.44}$$

We see that the p.s.d. takes the form shown in Fig. 5.1.

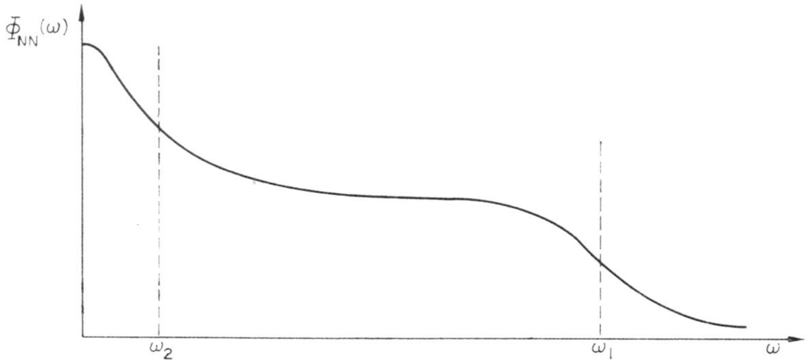

FIG. 5.1. Typical p.s.d. arising from artificially excited random noise in sub-critical, zero-power system. The scale of the figure is log-log.

In the region of ω_1, Φ_{nn} can be written approximately as

$$\Phi_{nn}(\omega) \simeq \frac{A}{\omega^2+\omega_1^2}, \tag{5.45}$$

thus a least squares fit yields ω_1 and hence $(\rho+\beta)/\Lambda$. In this way we can measure the sub-criticality of the system. This result is basically the same as that obtained from direct fission-induced noise, except that we have the advantage of choosing the type of source function and can vary its position and amplitude with some precision. Of course, in many reactors, insertion of a perturbing source is not practicable and therefore the direct fission noise source is used instead.

A further advantage of the artificial noise source generator is that it can be used to lead *directly* to the Green's function in the following manner:

Consider the cross-correlation

$$\langle n(t_1)\mathscr{S}(t_2)\rangle = \int_0^\infty dt_0\, G(t_0)\langle \mathscr{S}(t_1-t_0)\mathscr{S}(t_2)\rangle, \tag{5.46}$$

$$= \int_0^\infty dt_0\, G(t_0)\,\phi_{\mathscr{S}\mathscr{S}}(t_1-t_2-t_0) \tag{5.47}$$

or

$$\phi_{n\mathscr{S}}(\tau) = \int_0^\infty dt_0\, G(t_0)\,\phi_{\mathscr{S}\mathscr{S}}(\tau-t_0). \tag{5.48}$$

Now if we choose $\mathscr{S}(t)$ to be white noise, i.e.

$$\phi_{\mathscr{S}\mathscr{S}}(\tau) = \sigma^2\delta(\tau) \tag{5.49}$$

(white noise corresponds to a constant p.s.d. and the Fourier transform of a constant is a delta function), then

$$\phi_{n\mathscr{S}}(\tau) = \sigma^2 G(\tau). \tag{5.50}$$

Thus the cross-correlation function is directly proportional to the system Green's function.

5.5. *Random reactivity perturbation*

An equally likely and in many ways more convenient method of stochastically driving a sub-critical system is to vary its reactivity $\rho(t)$. Thus we return to eqns. (5.24) and (5.25), assume that the source term $S(t) = S_0$ is steady, and write the reactivity $\rho(t)$ as

$$\rho(t) = \langle\rho\rangle + \Delta(t). \tag{5.51}$$

$\langle\rho\rangle$ is the average value and $\Delta(t)$ is the randomly fluctuating part whose mean value is zero and which can be produced by, for example, the random motion of a control rod.

Writing N and C as in eqns. (5.27) and (5.28) we get

$$\frac{dn}{dt} = -\frac{(\langle\rho\rangle+\beta-\Delta)}{\Lambda}(\langle N\rangle+n)+\lambda(\langle C\rangle+c)+S_0, \tag{5.52}$$

$$\frac{dc}{dt} = \frac{\beta}{\Lambda}(\langle N\rangle+n)-\lambda(\langle C\rangle+c) \tag{5.53}$$

or in more detail,

$$\frac{dn}{dt} = -\frac{(\langle\rho\rangle+\beta)}{\Lambda}\langle N\rangle-\frac{(\langle\rho\rangle+\beta)}{\Lambda}n+\frac{\Delta}{\Lambda}\langle N\rangle+\frac{\Delta}{\Lambda}n+\lambda(\langle C\rangle+c)+S_0. \tag{5.54}$$

To get the average equations we apply the ensemble average $\langle\ldots\rangle$ to each term in the equation, viz.

$$0 = -\frac{(\langle\rho\rangle+\beta)}{\Lambda}\langle N\rangle+\frac{1}{\Lambda}\langle\Delta n\rangle+\lambda\langle C\rangle+S_0, \tag{5.55}$$

$$0 = \frac{\beta}{\Lambda}\langle N\rangle-\lambda\langle C\rangle. \tag{5.56}$$

Note that we have introduced a term $\langle\Delta n\rangle$ whose value we cannot find by straight solution. This problem is typical of those where the stochastic excitation arises *parametrically*, i.e. as a factor multiplying the dependent variable, rather than as an inhomogeneous driving term. It introduces the problem of *closure*. That is, to obtain $\langle\Delta n\rangle$, we must multiply the basic equation by Δ and again average: however, this in turn introduces a new correlation function $\langle\Delta\Delta n\rangle$ which again is unknown. Thus at some point it is necessary either to neglect a correlation function or to relate it to lower-order moments; only in this way can the set of equations be made equal to the number of unknowns. This is called the problem of closure and arises in the theory of turbulence and a number of other practical problems.

In our case, we will assume that the reactivity perturbation is small and neglect entirely the second-order term $\langle\Delta n\rangle$ compared with $\langle\rho\rangle\langle N\rangle$ in eqn. (5.55). However, it should be observed that the validity of such an approximation has not been fully studied and may well lead to erroneous results for some values of τ (usually large values which are of little physical interest).

With this approximation, the equations now become

$$\frac{dn}{dt} = -\frac{(\langle\rho\rangle+\beta)}{\Lambda}n+\lambda c+\frac{\Delta}{\Lambda}\langle N\rangle \tag{5.57}$$

and

$$\frac{dc}{dt} = \frac{\beta}{\Lambda}n-\lambda c. \tag{5.58}$$

These equations are identical to those with source excitation with the exception of the nature of the fluctuating term which, instead of $\mathscr{S}(t)$, is now

$$\Delta(t)\frac{\langle N\rangle}{\Lambda}. \tag{5.59}$$

All previous results are therefore true provided

$$\phi_{\mathscr{S}\mathscr{S}}(\tau) \to \frac{\langle N\rangle^2}{\Lambda^2}\phi_{\Delta\Delta}(\tau), \tag{5.60}$$

where $\phi_{\Delta\Delta}(\tau)$ is the auto-correlation function of the reactivity noise source.

In more detailed spatial studies of random reactivity perturbation and random source perturbation this simple connection is not necessarily valid.

Experiments are usually carried out *in practice* by the random reactivity method rather than by source excitation and workers at a number of laboratories have successfully extracted useful reactor parameters by this method. The reactivity perturbation is usually made by a randomly operated absorbing plate, the nature of which we shall examine in Section 5.10.†

5.6. *The Fokker–Planck equation*

The employment of the Langevin technique in Section 5.5 enabled mean values and correlation functions to be obtained for random variables. A question which now arises is whether it is possible to construct from the Langevin theory the complete probability distribution. We shall discuss this problem via the one-dimensional Brownian motion considered above. First, however, we write the following balance equation:

$$W(u,t+\Delta t) = \int W(u-\Delta u,t)\psi(u-\Delta u;\,\Delta u)\,d(\Delta u), \tag{5.61}$$

where $W(u,t)$ is the probability distribution function for the random variable u at time t. Moreover, $\psi(u;\Delta u)$ is the so-called *transition probability* that u suffers an increment Δu in a time Δt: this function contains the physics of the problem.

Equation (5.61) is the Chapman–Kolmogoroff equation for a Markoff process, i.e. a process in which the changes occurring in a system at time t depend only on the state of the system at time t. If $\psi(u;\Delta u)$ can be found from other considerations (Klahr, 1958) then (5.61) constitutes an integral equation for the distribution function. More often, however, only certain moments of ψ are easily obtained and it is therefore of interest to see whether (5.61) can be modified to make use of this reduced amount of basic information and yet still give results of practical value. We ask, therefore, if there exists a time interval Δt during which the macroscopic parameters such as

† A discussion of the Langevin technique and its application to inherent fission fluctuations is given in the Appendix.

position, velocity, density, etc., change by infinitesimal amounts while in the same time there occur a very large number of the fluctuations which characterize the transition probability, i.e. the noise source. In the case of Brownian motion this would correspond to the rapid impacts of the molecules on the surface of the particle, whilst in neutron multiplication it would necessitate the noise source fluctuations being much more rapid than the corresponding fluctuations in neutron density. If the input noise source is white, then quite clearly these assumptions will be valid.

To incorporate these physical considerations into our equation we expand W and ψ in a Taylor series as follows:

$$W(u,t+\Delta t) = W(u,t)+\Delta t\frac{\partial W}{\partial t}+O(\Delta t^2), \tag{5.62}$$

$$W(u-\Delta u,t) = W(u,t)-\Delta u\frac{\partial W}{\partial u}+\tfrac{1}{2}(\Delta u)^2\frac{\partial^2 W}{\partial u^2}+O(\Delta u^3), \tag{5.63}$$

$$\psi(u-\Delta u;\Delta u) = \psi(u;\Delta u)-\Delta u\frac{\partial \psi}{\partial u}+\tfrac{1}{2}(\Delta u)^2\frac{\partial^2 \psi}{\partial u^2}+O(\Delta u^3). \tag{5.64}$$

Inserting these expressions into the Chapman–Kolmogoroff equation leads to

$$\frac{\partial W}{\partial t}\Delta t+O(\Delta t^2) = -\frac{\partial W}{\partial u}\langle\Delta u\rangle+\frac{1}{2}\frac{\partial^2 W}{\partial u^2}\langle(\Delta u)^2\rangle-W\frac{\partial}{\partial u}\langle\Delta u\rangle+\frac{\partial W}{\partial u}$$
$$\times\frac{\partial}{\partial u}\langle(\Delta u)^2\rangle+\tfrac{1}{2}W\frac{\partial^2}{\partial u^2}\langle(\Delta u)^2\rangle+O((\Delta u)^3), \tag{5.65}$$

where the averages are given by

$$\langle\Delta u\rangle = \int\Delta u\psi(u;\Delta u)d(\Delta u), \tag{5.66}$$

$$\langle(\Delta u)^2\rangle = \int(\Delta u)^2\psi(u;\Delta u)d(\Delta u). \tag{5.67}$$

We now divide by Δt and allow $\Delta t\to 0$, when the equation becomes

$$\frac{\partial W}{\partial t} = \tfrac{1}{2}B(u)\frac{\partial^2 W}{\partial u^2}+\left(\frac{\partial B(u)}{\partial u}-\mathscr{A}(u)\right)\frac{\partial W}{\partial u}+\left(\frac{1}{2}\frac{\partial^2 B(u)}{\partial u^2}-\frac{\partial\mathscr{A}(u)}{\partial u}\right)W$$
$$\equiv -\frac{\partial}{\partial u}(\mathscr{A}(u)W)+\frac{1}{2}\frac{\partial^2}{\partial u^2}(B(u)W) \tag{5.68}$$

with the definitions

$$\mathscr{A}(u) = \lim_{\Delta t\to 0}\frac{\langle\Delta u\rangle}{\Delta t}, \tag{5.69}$$

$$B(u) = \lim_{\Delta t\to 0}\frac{\langle(\Delta u)^2\rangle}{\Delta t}, \tag{5.70}$$

$$\lim_{\Delta t\to 0}\frac{\langle(\Delta u)^k\rangle}{\Delta t} = 0 \quad (k\geqslant 3). \tag{5.71}$$

The last condition is certainly true for a transition probability of the form

$$\psi(u;\Delta u) = \frac{1}{(4\pi q\Delta t)^{\frac{3}{2}}}\exp\{-(\Delta u+\beta u\Delta t)^2/4q\Delta t\}, \tag{5.72}$$

which is characteristic of Brownian motion. We see therefore that we now have a second-order partial differential equation for $W(u,t)$ with undetermined coefficients $\mathscr{A}(u), B(u)$. These are found via the Langevin equations. Consider the case of Brownian motion for which the equation of motion is given by (5.2). Integrating this equation over the time interval $(t,\ t+\Delta t)$, we find

$$\Delta u = -\beta u \Delta t + \int_t^{t+\Delta t} A(\xi)\,d\xi. \tag{5.73}$$

Performing a statistical average over this equation and using the fact that $\langle A \rangle = 0$, we find

$$\mathscr{A}(u) = -\beta u \tag{5.74}$$

and similarly

$$B(u) = \tau_1 = \frac{2\beta kT}{m}. \tag{5.75}$$

All higher powers of Δu are proportional to powers of Δt higher than the first. Using (5.74) and (5.75) in (5.68) gives, therefore,

$$\frac{\partial W(u,t)}{\partial t} = -\beta \frac{\partial}{\partial u}(uW(u,t)) + \tfrac{1}{2}\tau_1 \frac{\partial^2 W(u,t)}{\partial u^2} \tag{5.76}$$

for the probability distribution function.

Subject to $W(u,0) = \delta(u-u_0)$, we can solve eqn. (5.76) to obtain

$$W(u,t) = \left(\frac{m}{2\pi kT(1-e^{-2\beta t})}\right)^{\frac{1}{2}} \exp\left\{-\frac{m}{2kT}\frac{(u-u_0e^{-\beta t})^2}{(1-e^{-2\beta t})}\right\}, \tag{5.77}$$

which shows that, as $t \to \infty$, the system relaxes to Maxwellian equilibrium distribution.

Not all physical situations are as simple as that of Brownian motion even if the basic noise sources are white. For example, a more common situation is to have a system described by a set of coupled equations for a number of different variables. Thus we may have the n-dimensional Markoff process described by the equations

$$\dot{y}_i = f(y_1, y_2, \ldots, y_n; \Delta_1, \Delta_2, \ldots, \Delta_m) \quad (i = 1, 2, \ldots, n), \tag{5.78}$$

where Δ_j are noise sources. The problem is then to calculate the conditional probability function

$$P(\bar{y}_1, \bar{y}_2, \ldots, \bar{y}_n, t_0 | y_1, y_2, \ldots, y_n, t), \tag{5.79}$$

which is the probability that the system will be in the state $y_1, y_2, \ldots, y_n$ at time t, provided that it was in the state $\bar{y}_1, \bar{y}_2, \ldots, \bar{y}_n$ at an earlier time t_0.

A more general form of the Fokker–Planck equation can then be derived (Middleton, 1960) in the following form:

$$\frac{\partial P}{\partial t} = -\sum_{i=1}^{n} \frac{\partial (A_i P)}{\partial y_i} + \tfrac{1}{2}\sum_{ij} \frac{\partial^2 (B_{ij} P)}{\partial y_i \partial y_j}, \tag{5.80}$$

where the coefficients A_i and B_{ij} are given by

$$A_i = \lim_{\Delta t \to 0} \frac{\langle \Delta y_i \rangle}{\Delta t}, \tag{5.81}$$

$$B_{ij} = \lim_{\Delta t \to 0} \frac{\langle \Delta y_i \Delta y_j \rangle}{\Delta t}, \tag{5.82}$$

with

$$\Delta y = y(t+\Delta t) - y(t). \tag{5.83}$$

Now for white noise processes a rather subtle argument is required to evaluate A_i. The reason is that white noise processes can be discrete or continuous; for example, let us consider a system governed by the equation

$$\frac{dy(t)}{dt}+My(t)+\alpha_1(t)y(t) = \alpha_0(t), \tag{5.84}$$

where M is a constant and $\alpha_1(t)$ and $\alpha_0(t)$ are Gaussian, physical white noise sources.

Physically, we might regard eqn. (5.84) as describing Brownian motion in a fluid with a randomly time-varying viscosity (although as we shall see below this is not strictly correct).

Yet another form of stochastic equation can be written for the transition Δy in the form

$$\Delta y = -My\Delta t-\Delta\beta_1 y+\Delta\beta_0. \tag{5.85}$$

Here we can view $\Delta\beta_1$ and $\Delta\beta_0$ as being Wiener processes, i.e. jump processes, such that they are related to α_1 and α_0 via

$$\alpha_i = \frac{\Delta\beta_i}{\Delta t}. \tag{5.86}$$

Equation (5.85) may be interpreted in a discrete sense by assuming that, at the start of each time interval, Δt, the system receives a random impulse that sends it from state P_1 to P_2 instantaneously due to the action of the term $-\Delta\beta_1 y+\Delta\beta_0$. From this stage the system moves according to the term $-My\Delta t$ until the end of the time interval when it is in state P_3. The process is then repeated indefinitely with impulses followed by steady motion in successive time intervals Δt: this is characteristic of Brownian motion or shot noise.

The other situation arises when the system receives a continuous random disturbance so that in each interval of time Δt the change of state Δy is given by integrating eqn. (5.84) in the classical sense, viz.

$$\Delta y = -M\int_t^{t+\Delta t} y(t')\,dt' - \int_t^{t+\Delta t} \{\alpha_1(t')y(t')-\alpha_0(t')\}\,dt'. \tag{5.87}$$

Since $y(t)$ varies infinitesimally over Δt we can rewrite this as

$$\Delta y = -My\Delta t - \int_t^{t+\Delta t} \{\alpha_1(t')y(t')-\alpha_0(t')\}\,dt'. \tag{5.88}$$

It is now assumed that $\Delta\beta_1$ and $\Delta\beta_0$ are continuous processes, i.e. not the "jump-like" Wiener process but rather that $\Delta\beta_i/\Delta t$ is a mathematical approximation for a Gaussian process with a very short correlation time. Thus we may assume that

$$\langle\alpha_1\rangle = 0, \quad \langle\alpha_0\rangle = 0 \tag{5.89}$$

and also that

$$\langle\alpha_1(t_1)\alpha_0(t_2)\rangle = 2\sigma_{10}^2\,\delta(t_1-t_2), \tag{5.90}$$

$$\langle\alpha_1(t_1)\alpha_1(t_2)\rangle = 2\sigma_{11}^2\,\delta(t_1-t_2), \tag{5.91}$$

$$\langle\alpha_0(t_1)\alpha_0(t_2)\rangle = 2\sigma_{00}^2\,\delta(t_1-t_2). \tag{5.92}$$

We can also formally relate the quantity β to α by the stochastic integral

$$\beta(t) = \int_0^t \alpha(t')\,dt'. \tag{5.93}$$

Now when we come to perform the statistical average $\langle \Delta y \rangle$, we obtain

$$\langle \Delta y \rangle = -My\Delta t - \int_t^{t+\Delta t} \langle \alpha_1(t')y(t') - \alpha_0(t') \rangle \, dt'. \tag{5.94}$$

In the case of the jump or discrete process, α_1 and y are completely uncorrelated and $\langle \alpha_0(t') \rangle = 0$, thus

$$\langle \Delta y \rangle = -My\Delta t. \tag{5.95}$$

This is the result given by many authors. However, the case of the continuous white noise source has given rise to some controversy (Gray and Caughey, 1965; Srinivasan and Vasudevan, 1971) and it has been shown that in this case a more careful examination of the integral in eqn. (5.94) leads, via eqns. (5.90)–(5.93), to the following result:

$$\langle \Delta y \rangle = -My\Delta t + \sigma_{11}^2 y \Delta t + \sigma_{10}^2 \Delta t. \tag{5.96}$$

Thus in one case the Fokker–Planck equation has the value $\mathscr{A} = -My$ and in the other it is $(\sigma_{11}^2 - M)y - \sigma_{10}^2$. Fortunately, the value of B is completely unambiguous and is given by $2\sigma_{11}^2 y^2 - 4\sigma_{10}^2 y + 2\sigma_{00}^2$ in both cases.

The choice of the value of $\mathscr{A}$ to use for a particular problem can be rather difficult. For example, Brownian motion clearly requires the discrete process value. However, problems in electrical circuit theory where the value of a particular component, say the capacitance, may be undergoing random changes might well involve the continuous process. It is important to note, however, that no matter which process is used, the distribution function P is continuous and cannot exhibit any discrete fluctuations such as we find in eqn. (2.24).

5.7. *Applications of Fokker–Planck equation to neutron noise*

As an example of this technique in reactor physics consider the point model equations of reactor kinetics with one group of delayed neutrons, viz.

$$\frac{dN(t)}{dt} = \frac{1}{\Lambda}(\rho(t) - \beta)N(t) + \lambda C(t) + S(t), \tag{5.97}$$

$$\frac{dC(t)}{dt} = \frac{\beta}{\Lambda} N(t) - \lambda C(t). \tag{5.98}$$

Let us assume that $\rho(t)$ and $S(t)$ are random functions of time such that

$$\rho(t) = \rho_0 + \Delta(t) \tag{5.99}$$

and

$$S(t) = S_0 + \mathscr{S}(t), \tag{5.100}$$

where $\langle \Delta \rangle = \langle \mathscr{S} \rangle = 0$ and $\Delta(t), \mathscr{S}(t)$ are white noise sources in the sense that

$$\Delta(t) = \frac{\Delta Z}{\Delta t}, \quad \mathscr{S}(t) = \frac{\Delta V}{\Delta t}, \tag{5.101}$$

where

$$\langle (\Delta Z)^2 \rangle = 2\sigma_{11}^2 \Delta t, \quad \langle (\Delta V)^2 \rangle = 2\sigma_{22}^2 \Delta t, \quad \langle \Delta V \Delta Z \rangle = 2\sigma_{12}^2 \Delta t. \tag{5.102}$$

Now if the reactivity fluctuations are due to the fission process, it would seem reasonable to employ the discrete process for calculating $\mathscr{A}$. In order to avoid mixing our processes we shall also assume that $\mathscr{S}(t)$ is a discrete process.

Using the general theory of eqn. (5.80), we find that

$$A_1 = \lim_{\Delta t \to 0} \frac{\langle \Delta N \rangle}{\Delta t} = \frac{1}{\Lambda}(\rho_0 - \beta)N + \lambda C + S_0, \tag{5.103}$$

$$A_2 = \lim_{\Delta t \to 0} \frac{\langle \Delta C \rangle}{\Delta t} = \frac{\beta}{\Lambda} N - \lambda C, \tag{5.104}$$

$$B_{12} = \lim_{\Delta t \to 0} \frac{\langle \Delta N \Delta C \rangle}{\Delta t} = 0, \tag{5.105}$$

$$B_{11} = \lim_{\Delta t \to 0} \frac{\langle (\Delta N)^2 \rangle}{\Delta t} = \frac{2}{\Lambda^2} \sigma_{11}^2 N^2 + \frac{4}{\Lambda} \sigma_{12}^2 N + 2\sigma_{22}^2, \tag{5.106}$$

$$B_{22} = \lim_{\Delta t \to 0} \frac{\langle (\Delta C)^2 \rangle}{\Delta t} = 0, \tag{5.107}$$

where we have set $y_1 = N$ and $y_2 = C$.

The Fokker–Planck equation for $P(N, C, t)$ now becomes

$$\frac{\partial P}{\partial t} = -\frac{\partial}{\partial N}\left\{\left(\frac{1}{\Lambda}(\rho_0 - \beta)N + \lambda C + S_0\right)P\right\} - \frac{\partial}{\partial C}\left\{\left(\frac{\beta}{\Lambda}N - \lambda C\right)P\right\}$$
$$+ \frac{\partial^2}{\partial N^2}\left\{\left(\frac{\sigma_{11}^2}{\Lambda^2}N^2 + \frac{2}{\Lambda}\sigma_{12}^2 N + \sigma_{22}^2\right)P\right\}. \tag{5.108}$$

We stress again that N and C are here considered as continuous random variables as opposed to the discrete values actually occurring in practice. This will be accurate for large population sizes.

We note also that if the continuous white noise process had been employed to obtain A_1, we should have instead the value:

$$A_1 = \left\{\frac{\rho_0 - \beta}{\Lambda} - \frac{\sigma_{11}^2}{\Lambda^2}\right\} N + \lambda C + S_0 + \frac{\sigma_{12}^2}{\Lambda}. \tag{5.109}$$

We choose to retain the "jump-noise" value of A_1 because the average equations predicted by it are equivalent to those for ordinary reactor kinetics. That is, if eqn. (5.108) is multiplied by N and C in turn and averaged, the conventional average equations, viz.

$$\frac{d\langle N \rangle}{dt} = \frac{1}{\Lambda}(\rho_0 - \beta)\langle N \rangle + \lambda\langle C \rangle + S_0, \tag{5.110}$$

$$\frac{d\langle C \rangle}{dt} = \frac{\beta}{\Lambda}\langle N \rangle - \lambda\langle C \rangle \tag{5.111}$$

are obtained and also, as we shall see below, the correct auto-correlation function is obtained. The method would seem therefore to be sound for zero-power noise simulation. However, other sources of reactivity excitation due, for example, to bubble formation and temperature fluctuations may be more accurately described by the continuous process and we shall say more about this below.

Returning to eqn. (5.108), we note that a general solution does not seem readily available. However, a number of moments of interest can be obtained, viz. $\langle N^2 \rangle$, $\langle C^2 \rangle$, etc., and also the auto-correlation function $\phi_{NN}(\tau)$. The method of obtaining this is lengthy and described in detail by Williams (1969) so that we shall only quote the result, namely:

$$\phi_{NN}(\tau) = A\{\alpha_1 e^{\omega_1\tau} - \alpha_2 e^{\omega_2\tau}\}, \tag{5.112}$$

where ω_1 and ω_2 are roots of

$$\omega^2 - \left\{\frac{1}{\Lambda}(\rho_0-\beta)-\lambda\right\}\omega - \frac{\lambda\rho_0}{\Lambda} = 0. \tag{5.113}$$

$$\alpha_1 = (\lambda\Lambda-\rho_0)(\lambda+\omega_1)+\beta\lambda \equiv \frac{\Lambda}{\lambda}\,\omega_2(\omega_1^2-\lambda^2), \tag{5.114}$$

$$\alpha_2 = (\lambda\Lambda-\rho_0)(\lambda+\omega_2)+\beta\lambda \equiv \frac{\Lambda}{\lambda}\,\omega_1(\omega_2^2-\lambda^2) \tag{5.115}$$

and

$$A = \Lambda(\omega_1-\omega_2)^{-1}\left[\rho_0(\rho_0-\beta-\lambda\Lambda)+(\rho_0-\lambda\Lambda)\frac{\sigma_{11}^2}{\Lambda}\right]^{-1}\cdot\left[\frac{S_0^2\sigma_{11}^2}{\rho_0^2}-\frac{2}{\rho_0}\,\sigma_{12}^2 S_0+\sigma_{22}^2\right]. \tag{5.116}$$

Thus we have obtained the auto-correlation function of a system with white noise parametric and source excitation. In Section 5.4 we calculated the auto-correlation for these two effects separately and the results are given by eqns. (5.39) and (5.60). In obtaining $\phi_{NN}(\tau)$ for reactivity perturbation it was necessary to assume a small perturbation; however, in the above calculation we did not resort to this approximation: nevertheless, we did use a white noise source, whereas this was left quite general in the earlier case.

It is useful to assess the error in the closure technique used for reactivity perturbation and we can do this by comparing the values of $\phi_{NN}(\tau)$ so obtained, with the exact value above. Returning to eqn. (5.57) and adding the source term $\mathscr{S}(t)$, so that the total random excitation is $\mathscr{S}(t)+\langle N\rangle\Delta(t)/\Lambda$, we find that for $\mathscr{S}(t)$ and $\Delta(t)$ white, the corresponding auto-correlation function is exactly the same as in eqn. (5.112) except that

$$A \to A' = \frac{\Lambda\sigma_{33}^2}{(\omega_1-\omega_2)\rho_0(\rho_0-\beta-\lambda\Lambda)}, \tag{5.117}$$

where

$$\left\langle\left(\mathscr{S}(t_1)+\frac{1}{\Lambda}\langle N\rangle\Delta(t_1)\right)\left(\mathscr{S}(t_2)+\frac{1}{\Lambda}\langle N\rangle\Delta(t_2)\right)\right\rangle = 2\sigma_{33}^2\delta(t_1-t_2) \tag{5.118}$$

with

$$\sigma_{33}^2 = \frac{S_0^2\sigma_{11}^2}{\rho_0^2}-\frac{2\sigma_{12}^2}{\rho_0}\,S_0+\sigma_{22}^2. \tag{5.119}$$

We see, therefore, that perturbation theory is valid provided that

$$\left|\frac{(\rho_0-\lambda\Lambda)\,\sigma_{11}^2/\Lambda}{\rho_0(\rho_0-\beta-\lambda\Lambda)}\right| \ll 1. \tag{5.120}$$

Now, physically, $\sigma_{11}^2 = \frac{1}{2}\Delta t\langle(\Delta\rho)^2\rangle$, where $\langle(\Delta\rho)^2\rangle$ is the mean square reactivity fluctuation and Δt a characteristic time scale of the fluctuation. Δt is in fact of the order of the inverse break frequency of the stochastic perturber. For the present theory to be valid, with typical values of λ, Λ, β and the system not too near criticality, we find that the condition (5.120) becomes

$$\frac{\Delta t\langle(\Delta\rho)^2\rangle}{2\beta\Lambda} \ll 1, \tag{5.121}$$

a condition which can usually be realized.

Our calculations thus far have been concerned with means, variances and auto-correlation functions; the explicit calculation of the probability density function $P(N, C, t)$ has been avoided. However, there are situations where a knowledge of P would be useful particularly in connection with power reactors and their associated temperature fluctuations as we shall see in a later chapter. Thus the solution of equations of the Fokker–Planck type is of importance. Whilst we can see no analytic way to solve eqn. (5.108) it is useful to consider two limiting cases. In the first, we assume that only the source is stochastic and therefore that $\sigma_{11}^2 = 0$ and $\sigma_{12}^2 = 0$. We also assume that delayed neutrons are absent. The solution can now be written down directly as follows:

$$P(N,t) = \left(\frac{1}{2\Sigma^2\pi(1-e^{-2\alpha t})}\right)^{\frac{1}{2}} \exp\left\{-\frac{\{N-\langle N\rangle-[N_0-\langle N\rangle]e^{-\alpha t}\}^2}{2\Sigma^2(1-e^{-2\alpha t})}\right\}, \qquad (5.122)$$

where $\langle N\rangle$ is the mean value given by $\Lambda S_0/(-\rho_0)$, $\alpha = -\rho_0/\Lambda$ and Σ^2 is the variance, equal to $\sigma_{22}^2\Lambda/(-2\rho_0)$. The solution is subject to the initial condition $P(N,0) = \delta(N-N_0)$. The solution is, therefore, Gaussian and has an equilibrium value as $t \to \infty$ given by

$$P(N) = \left(\frac{1}{2\pi\Sigma^2}\right)^{\frac{1}{2}} \exp\left\{-\frac{(N-\langle N\rangle)^2}{2\Sigma^2}\right\}. \qquad (5.123)$$

The other limiting case, again for no delayed neutrons, arises when we consider parametric excitation only and no source term. Thus we have $\sigma_{22}^2 = 0$, $\sigma_{12}^2 = 0$ and $S_0 = 0$ in eqn. (5.108). The resulting solution is

$$P(N,t) = \frac{1}{N}\left(\frac{\Lambda^2}{4\pi\sigma_{11}^2 t}\right)^{\frac{1}{2}} \exp\left[\frac{-\{\log(N/N_0)-\rho_0 t/\Lambda\}^2}{(4\sigma_{11}^2 t/\Lambda^2)}\right] \qquad (5.124)$$

which is markedly non-Gaussian. Moreover, it has the interesting feature of not having an equilibrium value as $t \to \infty$. The reason for this curious behaviour can be understood if we examine the moments of $P(N,t)$, viz.

$$\begin{aligned}\langle N^p\rangle &= \int_0^\infty N^p P(N,t)\,dN \\ &= N_0^p \exp\left\{p\left(\rho_0 + p\frac{\sigma_{11}^2}{\Lambda}\right)\frac{t}{\Lambda}\right\}. \end{aligned} \qquad (5.125)$$

For $p = 1$, we find the mean value to be

$$\langle N\rangle = N_0 \exp\left\{\left(\rho_0 + \frac{\sigma_{11}^2}{\Lambda}\right)\frac{t}{\Lambda}\right\}.$$

Thus even if the mean value of the reactivity, ρ_0, is zero, the neutron density will on the average continue to increase with time. Thus a random reactivity insertion *with zero mean* has a resultant effect equivalent to a *positive* reactivity addition. Even if the mean reactivity ρ_0 is adjusted such that $\rho_0 + \sigma_{11}^2/\Lambda = 0$, thereby making $\langle N\rangle$ constant, we find that the second moment

$$\langle N^2\rangle = N_0^2 \exp\left\{\frac{2\sigma_{11}^2}{\Lambda^2}t\right\}$$

increases with time, i.e. the fluctuations about the mean increase in amplitude. Thus we have the interesting result that the variance increases with time, yet the probability of N attaining large values as $t \to \infty$ becomes vanishingly small (this can be seen by studying $P(N,t)$). The reason for this effect is difficult to understand but is probably due to the fact that the excitation appears parametrically and therefore has a non-linear effect on the system response. This phenomenon has been noted in electrical systems where it can have unexpected effects on stability; however, its importance in reactor dynamic behaviour has yet to be examined.

5.8. *Generalization of Fokker–Planck equation to an nth order system*

It has been shown by Leibowitz (1963) and by Ariaratnam and Graefe (1965) that a system governed by the coupled set of linear random equations

$$\frac{dy_i(t)}{dt} = -\sum_{j=1}^{n} a_{ij}y_j(t) - \sum_{j=1}^{n} \alpha_{ij}(t)y_j(t) + \alpha_i^{(0)}(t) \quad (i = 1,2,\ldots,n), \tag{5.126}$$

where a_{ij} are constants and α_{ij} and $\alpha_i^{(0)}$ are random Gaussian white noise sources, can be represented in terms of probability by a Fokker–Planck equation if A_i and B_{ij} are defined as follows:

$$A_i = -\sum_{r=1}^{n}\left(a_{ir}y_r + D_{ir,r}^{(0)} - \sum_{s=1}^{n} D_{ir,rs}\,y_s\right), \tag{5.127}$$

$$B_{ij} = 2\left\{D_{ij}^{(0)} - \sum_{r=1}^{n}(D_{jr,i}^{(0)} + D_{ir,j}^{(0)})y_r + \sum_{r=1}^{n}\sum_{s=1}^{n} D_{ir,js}\,y_r y_s\right\}. \tag{5.128}$$

The coefficients D are defined by the following cross-correlation functions:

$$\langle \alpha_{ij}(t_1)\alpha_{rs}(t_2)\rangle = 2D_{ij,rs}\,\delta(t_1 - t_2), \tag{5.129}$$

$$\langle \alpha_i^{(0)}(t_1)\alpha_r^{(0)}(t_2)\rangle = 2D_{ir}^{(0)}\,\delta(t_1 - t_2), \tag{5.130}$$

$$\langle \alpha_{ij}(t_1)\alpha_r^{(0)}(t_2)\rangle = 2D_{ij,r}^{(0)}\,\delta(t_1 - t_2). \tag{5.131}$$

It should be mentioned, however, that Srinivasan and Vasudevan (1971) have expressed doubts about the validity of the above Fokker–Planck equations. Nevertheless, experiments have been performed by Graefe (1966) who obtained the power spectral density of the random variables $y_1(t)$ and $y_2(t)$ described by a pair of coupled differential equations by using a white noise generator to parametrically excite these equations. The corresponding random outputs $y_1(t)$ and $y_2(t)$ were processed using a digital computer and a digital to analogue converter to obtain an on-line analysis of the spectral density. Good agreement between the numerical simulation (experiment) and the results of theory based upon the above equations was obtained.

A further experiment which provides convincing evidence of the existence of mean square instability was performed by Samuels and Eringen (1959) who set up the equation

$$\ddot{y}(t) - \beta\dot{y}(t) + [\omega_0^2 + \alpha(t)]y(t) = 0 \tag{5.132}$$

on an analogue computer. $\alpha(t)$ is broad band noise, β and ω_0^2 are constants. The response $y(t)$, when noise of a proper level is injected into the system, shows clearly the tendency for the fluctuations in $y(t)$ to increase without bound as t increases. In these experiments only qualitative agreement was sought so that no conclusions could be drawn regarding agreement with theory.

An illustration of the effect of random parametric excitation on the stability of mean values can be obtained from the equation

$$\ddot{y}+[c_1+g_1(t)]\dot{y}+[c_0+g_0(t)]y = 0, \tag{5.133}$$

where $\langle g_1 \rangle = 0$ and $\langle g_0 \rangle = 0$. Now we write $y = y_1$ and $\dot{y} = y_2$ to reduce eqn. (5.133) to a pair of coupled first-order equations. Then using the general theory we can find the mean value $\langle y \rangle$ and show that it satisfies the equation

$$\langle \ddot{y} \rangle + [c_1 - \tfrac{1}{2}\sigma_1^2]\langle \dot{y} \rangle + c_0 \langle y \rangle = 0, \tag{5.134}$$

where

$$\langle g_1(t_1) g_1(t_2) \rangle = \sigma_1^2 \delta(t_1 - t_2).$$

Thus if we interpret eqn. (5.134) as describing a mechanical system with y being the displacement of a particle from equilibrium, then clearly the random coefficient $g_1(t)$ can be regarded as being equivalent, on average, to a negative friction $\frac{1}{2}\sigma_1^2$ despite the fact that its mean value is zero. Thus there is a destabilizing effect introduced by the random excitation, a fact to which we will refer again in connection with power reactor stability.

Two further points are of interest in connection with this method. The first is that it may be applied quite readily to non-linear systems using the same techniques to calculate A_i and B_{ij}. However, it is worth noting that, in non-linear systems, the resulting moment equations, i.e. for $\langle y^n \rangle$, are not closed and therefore some closure approximations are required.

The second point concerns non-white noise. If the noise causes only a small deviation about the mean then perturbation theory leads to closed form solutions for the moments and correlation functions (Samuels and Eringen, 1959; Bharucha-Reid, 1963). Alternatively, we can retain the Fokker–Planck formalism if the non-white noise in question can be produced by passing white noise through a linear filter. In such a case the non-white noise, whilst non-Markovian itself, may be regarded as the projection of a Markoff process (Bogdanoff and Kozin, 1962). As an example, let us consider the pair of equations

$$\frac{dy_1}{dt} = y_2, \tag{5.135}$$

$$\frac{dy_2}{dt} = -a_1 y_1 - a_2 y_2 - \gamma(t) y_1, \tag{5.136}$$

where $\gamma(t)$ is Gaussian non-white noise. Let us now suppose that $\gamma(t)$ can be produced by passing the white noise, $\alpha(t)$, through a filter, described, for example, by the equation

$$\frac{d\gamma}{dt} = -b\gamma + \alpha(t). \tag{5.137}$$

Now if we set $\gamma = y_3$, the three eqns. (5.135), (5.136) and (5.137) can be written

$$\frac{dy_1}{dt} = y_2, \tag{5.138}$$

$$\frac{dy_2}{dt} = -a_1 y_1 - a_2 y_2 - y_1 y_3, \tag{5.139}$$

$$\frac{dy_3}{dt} = -b y_3 + \alpha(t), \tag{5.140}$$

which is now a Markoff process for the three random variables y_1, y_2, y_3. Unfortunately, the system is now non-linear but it can at least be used to construct a Fokker–Planck equation.

It is interesting to note that, in this particular case, the auto-correlation function of $\gamma(t)$ is

$$\langle\gamma(t_1)\gamma(t_2)\rangle = \frac{\sigma_\alpha^2}{2}e^{-b|t_1-t_2|} \tag{5.141}$$

so that its correlation time is $1/b$. The corresponding p.s.d. is given by the Fourier transform, viz.

$$\Phi_\gamma(\omega) = \frac{\sigma_\alpha^2}{\omega^2+b^2}. \tag{5.142}$$

Thus a variety of non-white noise sources can be included in the Fokker–Planck system. This is a useful feature in reactor noise problems when mechanical excitation is prevalent.

Alternative methods of dealing with non-white Gaussian noise may be found in Williams (1971). Experimental confirmation of the applicability of the Fokker–Planck equation to a non-linear oscillator has been found by Morton and Corrsin (1969) who artificially excite the equation for a cubic spring oscillator.

5.9. *The extinction probability*

A further application of the Fokker–Planck technique is to the calculation of extinction probability. As an example, consider the simple birth and death problem in which the variable is continuous. The probability density function $f(n,t)$ is given by the Fokker–Planck equation as

$$\frac{\partial f}{\partial t} = -\frac{\partial}{\partial n}(af)+\frac{1}{2}\frac{\partial^2}{\partial n^2}(bf), \tag{5.143}$$

where a and b are the mean and variance of the transition probability. For birth and death problems it is reasonable to assume that $a = \alpha n$ and $b = \beta n$, i.e. both are proportional to population size.

Subject to an initial condition $n = n_0$ at $t = 0$, the solution of eqn. (5.143) can be shown to be

$$f(n,t) = \frac{2\alpha}{\beta(e^{\alpha t}-1)}\left(\frac{n_0 e^{\alpha t}}{n}\right)^{\frac{1}{2}} \exp\left\{-\frac{2\alpha(n_0 e^{\alpha t}+n)}{\beta(e^{\alpha t}-1)}\right\} I_1\left(\frac{4\alpha(n_0 n e^{\alpha t})^{\frac{1}{2}}}{\beta(e^{\alpha t}-1)}\right), \tag{5.144}$$

where $I_1(x)$ is a modified Bessel function.

We see that the mean value

$$\bar{n}(t) = \int_0^\infty nf(n,t)\,dn = n_0 e^{\alpha t} \tag{5.145}$$

as expected, and also that the variance is given by

$$\sigma^2(t) = \left(\frac{\beta}{\alpha}\right) n_0 e^{\alpha t}(e^{\alpha t}-1). \tag{5.146}$$

Both of these expressions have the same form as the corresponding values given in Section 2.3 for the discrete variable problem. In that case we can identify $\alpha = (\lambda-\mu)$ and $\beta = (\lambda+\mu)$.

We may also inquire about the probability of extinction according to the Fokker–Planck equation.

This can be obtained by calculating the total probability that there are particles in the system at time t, viz.

$$\int_0^\infty f(n,t)\,dn = 1-\exp\left\{-\frac{2\alpha n_0 e^{\alpha t}}{\beta(e^{\alpha t}-1)}\right\}. \tag{5.147}$$

Since this quantity is less than unity, the probability of no particles in the system at time t is

$$p_0(t) = \exp\left\{-\frac{2\alpha n_0 e^{\alpha t}}{\beta(e^{\alpha t}-1)}\right\}. \tag{5.148}$$

The extinction probability, which is the limit of $p_0(t)$ as $t \to \infty$, is given by

$$\left.\begin{aligned} p_0(\infty) &= 1 && (\alpha \leqslant 0),\\ &= \exp\{-2\alpha n_0/\beta\} && (\alpha > 0),\end{aligned}\right\} \tag{5.149}$$

which should be compared with eqn. (2.23) of Section 2.3.

The continuous approximation implied by the Fokker–Planck technique has been applied to neutronics problems by Dalfes (1963). Dalfes assumes that for a sufficiently large population size the variables N, C_i, Z in eqn. (2.45) can be considered as continuous. He then argues that the variations in the value of $\mathbf{y} = (N, C_i, Z)$ over a generation due to immigration, fission and absorption are small compared with $\mathbf{y}$ and therefore they can be considered as infinitesimal variations and can be treated as Markoff processes. Thus the right-hand side can be expanded in a Taylor series to give the following result:

$$\frac{\partial P(\mathbf{y},t)}{\partial t} = -\sum_{k=0}^{m}\frac{\partial}{\partial y_k}m_k(\mathbf{y},t)P(\mathbf{y},t)+\frac{1}{2}\sum_{k=0}^{m}\sum_{l=0}^{m}\frac{\partial^2}{\partial y_k\,\partial y_l}\sigma_{kl}(\mathbf{y},t)P(\mathbf{y},t), \tag{5.150}$$

where m_k and σ_{kl} are given averages over the nuclear properties of the system.

The full details of the calculation will not be given here but it is worth noting the case for no delayed neutrons, when eqn. (5.150) becomes

$$\frac{\partial P(n,t)}{\partial t} = -\frac{\partial}{\partial n}(m(n,t)P(n,t))+\frac{1}{2}\frac{\partial^2}{\partial n^2}(\sigma(n,t)P(n,t)) \tag{5.151}$$

with

$$m(n,t) = [\bar{\nu}\lambda_f-\lambda_a]n+S, \tag{5.152}$$

and

$$\sigma(n,t) = [\lambda_f\overline{\nu(\nu-1)}-\bar{\nu}\lambda_f+\lambda_a]n+S. \tag{5.153}$$

This is in agreement with our earlier assumption and gives explicit expressions for α and β. Moreover, in the case when a source is present it shows that the mean and variance must be augmented by the immigration rate S.†

We note that in terms of m and σ, the extinction probability becomes

$$p_0(\infty) = \exp\left\{\frac{-2n_0\rho}{\dfrac{\overline{\nu(\nu-1)}}{\bar{\nu}}-\rho}\right\}, \tag{5.154}$$

where

$$\rho = (k_\infty-1)/k_\infty \quad \text{and} \quad k_\infty = \bar{\nu}\lambda_f/\lambda_a.$$

† See Appendix for a more detailed explanation of the structure of eqn. (5.151).

5.10. *Artificial generation of random noise*

In dealing with the theoretical studies of noise in the previous sections, we have implicitly assumed that a white noise source is available: this has enabled us to use the approximation of a Dirac delta function for the input correlation function. However, white noise is a physically unrealistic entity, since it would result only from discrete independent events extending over an infinite period of time. Certainly the simulation of discrete events can be done via thermionic valves or radioactive decay; however, the period of time over which a particular simulation can be done must be finite. As a result the auto-correlation function of this noise input has sidebands which can lead to difficulties. Yet another viewpoint as to the impossibility of producing pure, white noise arises if one considers the total power from such a device. If the frequency range is flat and extends from zero to infinity, then the area under the curve, which is a measure of the power, is infinite. Thus there is always a natural frequency in a system which introduces damping and reduces the high-frequency portion of the noise spectrum.

Because of the uncertainties and difficulties associated with the use of "natural" white noise sources, it has become customary to introduce artificial periodic disturbances which have statistical characteristics very similar to those of white noise. Full details of these techniques are described by Uhrig (1970) and so we shall not go into much explanation here. However, the general ideas may be illustrated through a simple example.

If a series of square wave pulses with random intervals between events is used to generate the input signal to the system, i.e.

$$y(t) = \sum_k F(t-\epsilon_k), \tag{5.155}$$

where $F(t)$ is the pulse shape and ϵ_k a random time interval, then the associated auto-correlation function

$$\phi_{yy}(\tau) = \langle y(t)y(t+\tau)\rangle \tag{5.156}$$

may be shown to take the form of a triangle defined by

$$\phi_{yy}(\tau) = bS^2(1-b)\left(1-\frac{|\tau|}{2W}\right) \quad (0 \leqslant |\tau| \leqslant 2W), \tag{5.157}$$

where the average pulse rate is m, and b, the duty cycle, is equal to mW. W is the pulse width and bS is the distance below the zero of y, S being the total pulse height. The crucial point in this white noise simulator is that the area of the pulse for $y > 0$ is just equal to the area below the $y = 0$ axis between pulses.

If the relaxation constant of the impulse response function of the system is α, then the pseudo-random signal so generated will appear virtually as a delta function provided that $W \ll 1/\alpha$ and also, to avoid the chance of pulse overlapping, we should have $W \ll 1/m$.

A large number of other types of pseudo-random signal generator are in existence (Stern *et al.*, 1962; Stern, 1964) for producing white noise. However, a completely different approach has been adopted by Smith and Williams (1972) who have constructed a stochastic generator which will produce a noise signal which is Gaussian, stationary and has a prescribed power spectral density. It has therefore an advantage over those methods which are restricted to white noise production.

The method is based upon the fact that any stationary noise signal can be decomposed into a Fourier series of the following form:

$$X_m(t) = \sum_{k=1}^{\infty} \{a_{mk} \cos(k\omega_0 t) + b_{mk} \sin(k\omega_0 t)\}, \tag{5.158}$$

where X_m is a member of an ensemble of functions where the coefficients a_{mk}, b_{mk} are distributed according to a normal distribution with the desired spectral content. The sequence (5.158) is periodic, repeating at intervals

$$\theta = \frac{2\pi}{\omega_0} S \quad (S = 1,2,3,\ldots).$$

In addition we assume that X has zero mean, thus, for a particular ensemble member,

$$\bar{a}_k = \bar{b}_k = 0. \tag{5.159}$$

Similarly, the a_k's and b_k's are independent of each other for different values of k, such that

$$\overline{a_k a_l} = \overline{b_k b_l} = \sigma_k^2 \delta_{kl} \tag{5.160}$$

and

$$\overline{a_k b_k} = 0,$$

the bar denoting an average over a Gaussian distribution.

In the frequency domain, the sampling theorem (Hancock, 1961) requires that for the spectral content of a record of length θ to be completely represented, it is necessary to sample the spectral density function at frequency intervals equal to or less than $1/\theta$ Hz.

The variance of the contribution of a given frequency is now made proportional to the amplitude of the spectral density function at that frequency by selecting the coefficients a_k, b_k, $k = 1,2,3,\ldots,n$, from Gaussian distributions with variances defined by

$$\sigma_k^2 = \frac{1}{\theta} G(kf_{\min}), \tag{5.161}$$

where $G(f)$ is the desired p.s.d., normalized such that

$$\int_{f_{\min}}^{f_{\max}} G(f)\, df = 1 \tag{5.162}$$

and $f_{\max}$ and $f_{\min}$ are cut-off frequencies with $f_{\max} = nf_{\min}$. The variance of the resulting sequence is then

$$\sigma^2 = \sum_{k=1}^{n} \sigma_k^2 = 1. \tag{5.163}$$

The values of a_k and b_k are selected from Gaussian distributions by choosing random numbers and using the Monte Carlo philosophy (Schreider, 1966). Two tests have been made of the method using the p.s.d. functions shown in Figs. 5.2 and 5.3. The full line is the exact analytic form of $G(f)$ and the circles denote values obtained by generating sequences $X_m(t)$ and performing the auto-correlations and subsequent Fourier transforms numerically. The close agreement gives confidence that the

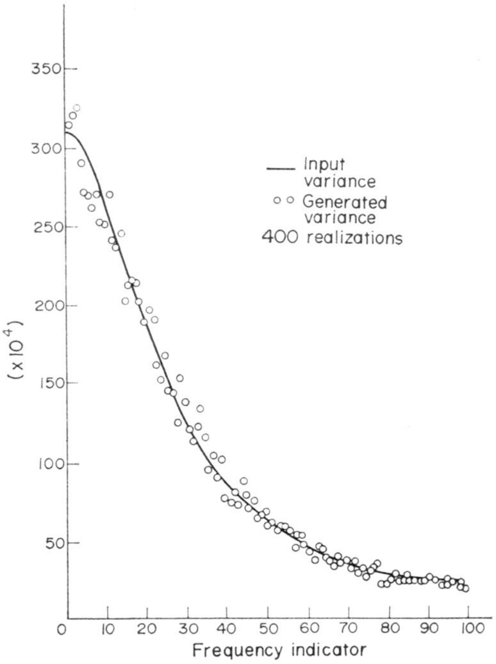

FIG. 5.2. **Artificially generated and exact p.s.d., case 1. (From Smith and Williams (1972) *J. Nucl. Energy*, 26, 525.)**

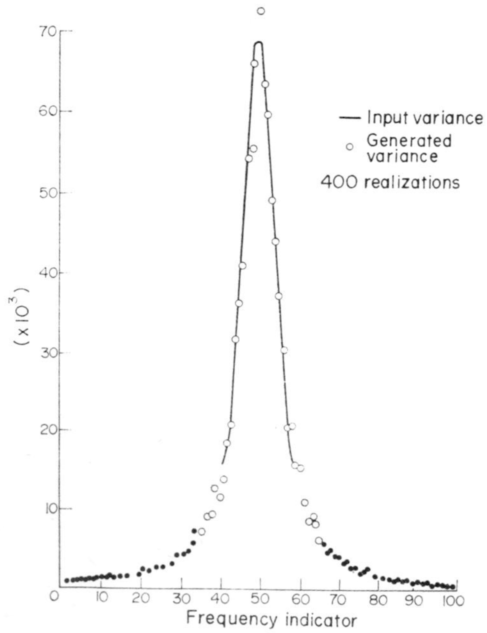

FIG. 5.3. **Artificially generated and exact p.s.d., case 2. (From Smith and Williams (1972) *J. Nucl. Energy*, 26, 525.)**

stochastic generator will simulate any desired Gaussian noise spectrum. Applications of the method to a space-dependent problem involving the correlations between points which are at distances x_1 and x_2 from a randomly varying plane source can be found in Section 7.14.

References

ADOMIAN, G. (1963) Linear stochastic operators, *Rev. Mod. Phys.* **35**, 185.

ADOMIAN, G. (1965) Stochastic Green's functions, *Proc. Symp. Appl. Math.* XVI, 1, Am. Math. Soc., Prov., R.I.

ARIARATNAM, S. T. and GRAEFE, P. W. U. (1965) Linear systems with stochastic coefficients, *Int. J. Control*, **1**, 239; **2**, 161; **2**, 205.

BARTLETT, M. S. (1962) *An Introduction to Stochastic Processes*, Cambridge University Press.

BHARUCHA-REID, A. T. (1963) On the theory of random equations, *Proc. Symp. Appl. Math.* XVI, 40, Am. Math. Soc. Prov., R.I.

BOGDANOFF, J. L. and KOZIN, F. (1962) Moments of the output of linear random systems, *J. Acoust. Soc. Am.* **34**, 1063.

CAUGHEY, T. K. and DIENES, J. K. (1962) The behaviour of linear systems with random parametric excitation, *J. Math. Phys.* (MIT), **41**, 300.

DALFES, A. (1963) The Fokker–Planck and Langevin equations of a nuclear reactor, *Nukleonik*, **5**, 348.

GRAEFE, P. W. U. (1966) *Linear Stochastic Systems*, Nat. Res. Council of Canada, Mech. Eng. Rep. MK-20, NRC No. 9526.

GRAY, A. H. and CAUGHEY, T. K. (1965) A controversy in problems involving random parametric excitation, *J. Math. Phys.* (MIT), **44**, 288.

HANCOCK, J. C. (1961) *An Introduction to the Principles of Communication Theory*, McGraw-Hill, N.Y.

KLAHR, C. N. (1958) Calculations of neutron distributions by the methods of stochastic processes, *Nucl. Sci. Engng.* **3**, 269.

LEIBOWITZ, M. A. (1963) Statistical behaviour of linear systems with randomly varying parameters, *J. Math. Phys.* **4**, 852.

MIDDLETON, D. (1960) *An Introduction to Statistical Communication Theory*, McGraw-Hill, N.Y.

MORTON, J. B. and CORRSIN, S. (1969) Experimental confirmation of the applicability of the Fokker–Planck equation to a non-linear oscillator, *J. Math. Phys.* **10**, 361.

SAMUELS, J. C. and ERINGEN, A. C. (1959) On stochastic linear systems, *J. Math. Phys.* (MIT), **38**, 83.

SCHREIDER, YU. A. (1966) *The Monte Carlo Method*, Pergamon Press, Oxford.

SMITH, P. R. and WILLIAMS, M. M. R. (1972) The application of a stochastic generator to determine the kinetic response of a neutronic system with randomly fluctuating parameters, *J. Nucl. Energy*, **26**, 525.

SRINIVASAN, S. K. and VASUDEVAN, R. (1971) *Introduction to Random Differential Equations and Their Applications*, Elsevier, N.Y.

STERN, T. E. (1964) *Reactor Impulse Response using Random Source Excitation and Cross Correlation, Noise Analysis in Nuclear Systems*, AEC Symp. Ser. No. 4.

STERN, T. E. *et al.* (1962) Reactivity measurement using pseudo-random source excitation, *J. Nucl. Energy*, pts. A/B, **16**, 499.

UHRIG, R. E. (1970) *Random Noise Techniques in Nuclear Reactor Systems*, Ronald Press, N.Y.

WAX, N. (1954) *Selected Papers on Noise and Stochastic Processes*, Dover, N.Y.

WILLIAMS, M. M. R. (1971) The kinetic behaviour of simple neutronic systems with randomly fluctuating parameters, *J. Nucl. Energy*, **25**, 563.

CHAPTER 6

Point Model Power Reactor Noise

6.1. *Introduction*

Whilst studies of zero-power noise are of basic academic interest, it is clear that if reactor noise is to have any real future in nuclear engineering it must lead to results of practical engineering value. Thus it must either be of aid in diagnosing existing faults, or, better still, help predict these faults before they become serious.

In the treatment of zero-power problems we have concentrated on simple models. In practice, however, because the noise sources in zero-power systems are exactly known, it is possible to write down an exact equation for the noise distribution in position and velocity and we may regard the calculation of zero-power noise as a completely solved problem: only computational difficulties remain [see eqn. (4.25)]. Power reactor noise, on the other hand, is very different. There are so many possible noise sources, whose statistical natures are unknown, that it is impossible to write down a probability balance equation with any certainty that it covers all contingencies.

In order to appreciate the magnitude of the task confronting would-be power reactor noise analysts, we list below a few of the possible noise sources that can arise and the range of frequencies over which they are important (these may be considered to affect the reactivity directly):

1. Ageing processes: change of nuclear properties by irradiation ($<10^{-4}$ Hz).
2. Xenon poisoning (10^{-4}–10^{-2} Hz).
3. Control-rod movements either operator induced or in response to automatic control-system signals (0·001–0·1 Hz).
4. Delayed neutron effects (0·01–1·0 Hz).
5. Characteristic reactor cycle times such as coolant circulation period (0·01–1·0 Hz).
6. Mechanical vibration of reactor components, e.g. control rods, fuel elements and core assembly due to excitation by coolant flow (frequency range depends on natural modes of components but may extend upwards from 0·1 Hz).
7. Fuel non-uniformity in molten fuel reactors, also local heterogeneities in enrichment in solid-fuel elements leading to variations in power output (~ 1 Hz).
8. Fluctuations in the inlet coolant temperature and flow rate ($\sim$0·1–1·0 Hz).
9. Large-scale flow instabilities and pressure fluctuations in the coolant, also

turbulence causing fluctuations of the heat-transfer characteristics of fuel element surfaces (0·01 Hz, upwards).

10. In boiling-water reactors there exists the statistical problem of bubble formation, collapse, gas entrainment and cavitation due to pumps (10 Hz upwards).

In addition to these direct effects on reactivity one must also account for the acoustic noise field in a power reactor due to the boiling of liquid coolants. In addition, blower and pump noise and turbulence-induced pressure fluctuations can be very high, leading to component fatigue and failure if a lightly damped structural resonance is strongly excited, hence the need to know probability distributions (see Chapter 9). It should be noted, however, that acoustic noise is generally significant in the range 10^2–10^4 Hz and has little effect on reactivity in the frequencies of interest in power reactors—although it may be a useful diagnostic tool in its own right (Saxe, 1967).

Despite the number of these different noise sources in power reactors, it is encouraging to learn that a number of successful experiments have been carried out on various reactors in which the effects of some noise sources have been isolated or have proved to be dominant. For example, on the high flux isotope reactor at Oak Ridge, it has been demonstrated clearly that there is a strong correlation between control-rod vibrations and neutron flux noise in the range 0·1–20 Hz (Robinson, 1967; Fry, 1971). Correlations have also been obtained in the molten salt reactor experiment between gas pressure fluctuations and neutron flux noise (Robinson, 1970). The cause of the latter is attributed to gas bubbles, entrained in the fuel, vibrating and causing reactivity perturbations (Saxe, 1967). Further evidence of coolant flow fluctuations causing random control-rod movement has been found by Yamada and Kage (1967).

The sodium reactor experiment suffered from bowing of the fuel elements which led to a resonance in the neutron flux noise. However, it was noted that by wrapping wire around the fuel elements, thus making them more rigid, the resonance was significantly reduced in amplitude (Thie, 1963).

Tests on the Nuclear Ship *Savannah* have shown that rough seas cause control-rod movements leading to resonances in the power spectral density (Ball and Batch, 1964).

Randall and Griffin (1964) have shown that low-frequency power fluctuations can arise from coolant flow fluctuations caused by voltage variations in pump supplies.

Boardman (1964) has attributed resonances in the neutron flux noise measured on the Dounreay fast reactor to interactions between inlet coolant temperature fluctuations and the reactor core temperature.

Many other experiments on power reactors have been performed and are reported in the AEC Symposium series No. 4 (1963) and No. 9 (1967) and also in the Japan–U.S. Symposium (1968).

More recently Greef (1971), at the Central Electricity Generating Board's Research Centre at Berkeley (U.K.), has shown how measurements of temperature noise and calculation of the correlation function between coolant and fuel temperatures have enabled the associated transfer function to be obtained, a knowledge of which is valuable for reactor-fault protection studies. This method was employed instead of the usual rod oscillator technique and resulted in a much simpler method of measurement which did not disturb the reactor in any way.

At this point it must be admitted that the list of successful applications mentioned above tends to give a misleading impression of the acceptance of noise analysis as a diagnostic tool in day-to-day reactor operation. Most utilities are still reluctant to accept it as a foolproof method to detect faults such as hot spots (particularly important in fast reactors) or local boiling. This attitude is not surprising in view of the rather limited understanding of the noise sources involved and the associated rather limited mathematical models. However, the advantages to be gained by reliable noise source measurements in place of the present control-rod oscillation techniques and the use of impulsive or ramp disturbances to measure reactor kinetic parameters are very great indeed. In all of these latter techniques, for example, it is necessary to either shut down or reduce reactor power and thereby lead to economic penalties. There is also the problem of the investigation of high-frequency phenomena which are restricted by the slow response of artificial sources of excitation such as the periodic operation of control rods and blowers; noise measurements, on the other hand, can detect high-frequency components of the transfer function with relative ease.

6.2. *The probability balance method for power reactor noise studies*

Undoubtedly the credit for the first serious mathematical study of power reactor noise must go to D. R. Harris who, in 1958, developed a probability balance equation for the noise in a power reactor, excluding boiling. Harris considered the equation for

$$P(N, C_1, \ldots, C_J, \theta, Z, t), \tag{6.1}$$

where N, C_i and Z are defined as in the zero-power case, but

$$\theta = \frac{C_R}{q}\,\theta_R, \tag{6.2}$$

where θ_R is the temperature of the solid portions of the reactor, C_R is the heat capacity of the solid parts and q is the quantity of heat produced in the solid parts when a fission occurs. Thus θ is a discrete variable but, owing to the numerical value of C_R/q, is very nearly continuous. It is assumed in this analysis that the coolant temperature θ_c does not change, i.e. the system is connected to an infinite heat reservoir.

Finally, the reciprocal of the mean neutron lifetime λ_c (including corrections for leakage) is assumed to depend on the temperature in the following manner:

$$\lambda_c = \lambda_{c0} - \alpha(\theta_R - \theta_{R0}), \tag{6.3}$$

where θ_{R0} is a convenient reference temperature and α is the temperature coefficient.

We can also write (6.3) in terms of θ as

$$\lambda_c = \lambda_{c0} - \gamma(\theta - \theta_0), \tag{6.4}$$

where $\gamma = q\alpha/C_R$.

In terms of λ_c, therefore, the reactivity $\rho(t)$ can be written as the random variable

$$\rho(t) = 1 - \frac{\lambda_f + \lambda_c(t)}{\bar{\nu}\lambda_f}, \tag{6.5}$$

where $\lambda_c(t)$ is random because of temperature fluctuations.

Making use of the generating function technique, it is possible to find equations for the various moments $\langle N \rangle$, $\langle C_i \rangle$, $\langle \theta \rangle$, $\langle Z \rangle$, $\langle N^2 \rangle$, $\langle N\theta \rangle$, $\langle \theta^2 \rangle$, etc. However, certain difficulties now arise which may be exemplified by consideration of the equation for the mean value $\langle N \rangle$, which becomes

$$\frac{d\langle N \rangle}{dt} = S - \langle \lambda_c \rangle \langle N \rangle + \lambda_f [\bar{\nu}(1-\beta)-1] \langle N \rangle + \sum_{j=1}^{J} \lambda_j \langle C_j \rangle + \alpha \sigma_{N\theta}^2, \tag{6.6}$$

where

$$\sigma_{N\theta}^2 = \langle N\theta \rangle - \langle N \rangle \langle \theta \rangle. \tag{6.7}$$

The equation for the average density of delayed neutron precursors is the conventional one.

We note the addition of a term $\alpha\sigma_{N\theta}^2$ in eqn. (6.6) which does not appear in the usual deterministic equations. In order to solve, therefore, an additional equation for $\sigma_{N\theta}^2$ is required which in turn introduces new unknowns, e.g. $\langle N^2\theta \rangle$, etc. Thus it is necessary either to neglect certain cross-terms or at least relate them to lower-order ones. We see therefore that we again meet the unpleasant problem of closure. In fact in most practical solutions $\alpha\sigma_{N\theta}^2 \ll \langle \lambda_c \rangle \langle N \rangle (\sim 10^{-3})$ and this term may be neglected from the point of view of average reactor behaviour. However, when the fluctuations themselves are of interest, more careful consideration of the second-order moments is required.

Second- and higher-order moments may be obtained by appropriate differentiation of the equation for the generating function. If the variance $\bar{\mu}_{ZZ}$ is found it is possible to obtain the correlation function $\phi_{NN}(\tau)$ as explained earlier. However, a rather more convenient method has been described by Greef (1971) which enables equations for all of the relevant correlation functions (auto and cross) to be obtained directly without the need to use the Pluta formula. Greef defines the new *joint* probability distribution function

$$P(\mathbf{A}_0, t_0; \mathbf{A}, t), \tag{6.8}$$

where $\mathbf{A}$ is a vector denoting all dependent variables at time t, viz. (N, C_i, θ, etc.), and $\mathbf{A}_0$ is at another time t_0.

Then, by definition, $P(\mathbf{A}_0, t_0; \mathbf{A}, t)$ is the joint probability that if the reactor is in a state $\mathbf{A}_0$ at time t_0 it will be in a state $\mathbf{A}$ at time t. The arguments leading to the appropriate balance equation for $P(\mathbf{A}_0, t_0; \mathbf{A}, t)$ are identical to those for $P(\mathbf{A}, t)$; the advantage is that moments corresponding to different time intervals may be calculated directly. For example, if we require $\langle N(t_0) N(t) \rangle$, we can write it in terms of the probability function as

$$\langle N(t_0) N(t) \rangle = \sum_{A_0} \sum_{A} N(t_0) N(t) P(\mathbf{A}_0, t_0; \mathbf{A}, t). \tag{6.9}$$

Now since the generating function

$$F(\mathbf{A}_0, t_0; \mathbf{x}, t) \equiv \sum_{A} P(\mathbf{A}_0, t_0; \mathbf{A}, t) x_1^N x_2^{C_1} \ldots x_7^{C_J} x_8^{\theta}, \tag{6.10}$$

we can obtain $\langle N(t_0) N(t) \rangle$ directly by differentiating once with respect to x_1, multiplying by $N(t_0)$ and summing over all $\mathbf{A}_0$, i.e.

$$\langle N(t_0) N(t) \rangle \equiv \sum_{A_0} N(t_0) \left. \frac{\partial F}{\partial x_1} \right|_{x=1}. \tag{6.11}$$

In order to close the set of coupled moment equations, Greef uses the result of Harris and Prescop (1969) that the probability function is normally distributed. Thus if x_1, x_2 and x_3 are three random variables normally distributed, it is easy to show that

$$\langle x_1, x_2, x_3 \rangle = \langle x_1 \rangle \langle x_2 \rangle \langle x_3 \rangle + \langle x_1 \rangle \sigma_{23}^2 + \langle x_2 \rangle \sigma_{31}^2 + \langle x_3 \rangle \sigma_{12}^2, \tag{6.12}$$

where

$$\sigma_{ij}^2 = \sigma_{ji}^2 = \langle (x_i - \langle x_i \rangle)(x_j - \langle x_j \rangle) \rangle.$$

In this way equations for $\phi_{A_i A_j}(\tau)$ can be obtained, and solved by Laplace transform.

Even in this simple analysis, however, we can observe the onset of fundamental difficulties, for example in the knowledge of the probability $p(q)$ which governs the amount of heat transferred between coolant and solid material. Similarly if we introduce other noise sources covering boiling and vibration phenomena we shall meet this problem again. Thus we have a method of computing the functional form of the various correlation functions but they contain unknown parameters describing the relative strengths of the noise sources. An additional limitation of this method is the difficulty of extending it to include space and energy dependence: this can be done (Matthes, 1962) but leads to a very complicated formalism which lacks appeal to the practising nuclear engineer.

6.3. *The Langevin technique: application to power reactors*

Because of the difficulties mentioned above, it is useful to see whether an application of the Langevin technique will not be more profitable. Certainly the Langevin technique is more easily handled mathematically than the probability balance method despite the latter's more fundamental approach. As we have seen in Chapter 5, the starting point is simply the set of equations for the mean values of the variables with certain parameters "earmarked" as random. Thus the method can deal readily with space- and energy-dependent problems and, in particular, the important case of control-rod and fuel-element vibration and also, as we shall show, the effects of boiling. Moreover, whilst the statistics of the noise sources are not known, this is also true in the probability balance method, the difference being merely conceptual rather than real since, in both cases, guesses at the noise source characteristics must eventually be made. It is for this reason that we shall concentrate on the Langevin technique as far as practical applications are concerned and illustrate its use in power reactor problems by considering a number of important practical problems and examples.

6.4. *Calculation of the transfer function between coolant temperature and fuel temperature in a gas-cooled, graphite-moderated reactor*

This problem has been dealt with recently by Greef (1971) and demonstrates very markedly the versatility of noise analysis.

In this particular problem only temperature fluctuations are studied since there are no suitable neutron flux detectors in the fuelled region of the Berkeley reactor core which was the system under investigation. The purpose of the experiment was as follows:

(*a*) To survey the temperature noise and ascertain its nature and primary cause.

(*b*) To determine the validity of a theoretical model for the fuel–coolant temperature transfer function.

(*c*) To study the response of the channel gas-outlet thermocouple relative to changes in coolant temperature. A knowledge of this response expressed in terms of the model parameters is valuable for reactor fault protection studies and a method of measuring it without disturbing the reactor is required.

The general practical situation is indicated in Fig. 6.1, which shows the cross-section of part of a particular channel in the reactor core.

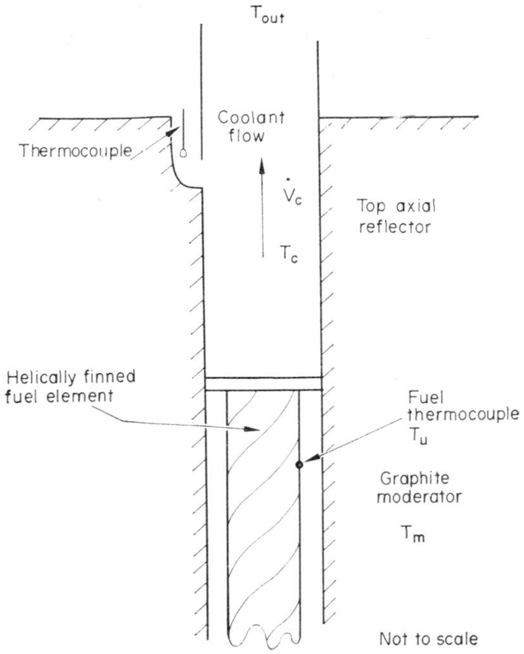

FIG. 6.1. Schematic diagram of the top section of a Berkeley reactor fuel channel.

The equations used to describe this situation are the following point model reactor kinetics equations using one delayed neutron group.

$$\frac{dN}{dt} = (-\lambda_c+\lambda_f[\nu(1-\beta)-1])N+\lambda C+S, \tag{6.13}$$

$$\frac{dC}{dt} = \lambda_f \nu\beta N-\lambda C. \tag{6.14}$$

These are the basic neutronics equations. The associated heat transfer equations are approximated as follows:

$$C_u V_u \frac{dT_u}{dt} = \lambda_f \mu_f N-K_{uc}(T_u-T_c), \tag{6.15}$$

$$C_m V_m \frac{dT_m}{dt} = K_{cm}(T_c-T_m), \tag{6.16}$$

$$C_c V_c \frac{dT_c}{dt} = K_{uc}(T_u-T_c)-K_{cm}(T_c-T_m)-2C_c\dot{V}_c(T_c-T_i). \tag{6.17}$$

C and V are the specific heats and volumes of the components, fuel (u), moderator (m) and coolant (c). K denotes the corresponding heat transfer coefficient which we assume to be temperature independent over the fluctuation range. $\dot{V}_c$ is the volumetric flow rate of coolant through the reactor and T_i is the inlet coolant temperature.

We have defined T_c as

$$T_c = \tfrac{1}{2}(T_{\text{out}}+T_i) \tag{6.18}$$

and the term $2C_c\dot{V}_c(T_c-T_i)$ arises from the point model approximation of the mass transport term

$$V_c C_c U_c \frac{\partial T_c}{\partial z}, \tag{6.19}$$

where U_c is the coolant velocity.

μ_f in eqn. (6.15) is the amount of heat released per fission. The feedback mechanism is by the fuel and moderator temperature changes affecting the neutron lifetime λ_c, viz.

$$\lambda_c = \lambda_{c0}-\alpha(T_u-T_{u0})-\gamma(T_m-T_{m0}). \tag{6.20}$$

It should be noted that in the above equations we have not followed the same notation as Greef in his original work.

We now assume that the following coefficients in eqns. (6.13)–(6.20) are stochastic random variables such that

$$\frac{\mu_f}{C_u V_u} = h_f = \bar{h}_f+\delta h_f, \tag{6.21}$$

$$\frac{K_{uc}}{C_u V_u} = h_u = \bar{h}_u+\delta h_u, \tag{6.22}$$

$$\frac{K_{cm}}{C_m V_m} = h_m = \bar{h}_m+\delta h_m, \tag{6.23}$$

$$\frac{2\dot{V}_c}{V_c} = h_c = \bar{h}_c+\delta h_c, \tag{6.24}$$

$$\nu = \bar{\nu}+\delta\nu, \quad h_c T_i \equiv \Delta = \bar{\Delta}+\delta\Delta. \tag{6.25}$$

Fluctuations δh, $\delta\nu$, etc., in these quantities lead to corresponding fluctuations in N, C, T_u, T_c and T_m. Thus we write

$$N = \bar{N}+\delta N, \quad C = \bar{C}+\delta C, \quad T = \bar{T}+\delta T. \tag{6.26}$$

Inserting (6.21)–(6.26) into the eqns. (6.13)–(6.20), using the steady-state conditions and neglecting products of second-order terms, e.g. $\delta h\,\delta T$, we find that the equations for the fluctuating components become, with

$$\bar{\lambda}_c = \lambda_{c_0}-\alpha(\bar{T}_u-T_{u0})-\gamma(\bar{T}_m-T_{m0}), \tag{6.27}$$

$$\frac{d\delta N}{dt} = \{-\bar{\lambda}_c+\lambda_f[\bar{\nu}(1-\beta)-1]\}\delta N+\lambda\delta C+\alpha\bar{N}\delta T_u+\gamma\bar{N}\delta T_m+\lambda_f\bar{N}(1-\beta)\delta\nu, \tag{6.28}$$

$$\frac{d\delta C}{dt} = \lambda_f\bar{\nu}\beta\delta N-\lambda\delta C+\lambda_f\beta\bar{N}\delta\nu, \tag{6.29}$$

$$\frac{d\delta T_u}{dt} = \bar{h}_f\delta N-\bar{h}_u\delta T_u+\bar{h}_u\delta T_c+\bar{N}\delta h_f+(\bar{T}_u-\bar{T}_c)\delta h_u, \tag{6.30}$$

$$\frac{d\delta T_m}{dt} = \bar{h}_m\delta T_c-\bar{h}_m\delta T_m+(\bar{T}_c-\bar{T}_m)\delta h_m, \tag{6.31}$$

$$\frac{d\delta T_c}{dt} = \bar{h}_u\delta T_u-(\bar{h}_u+\bar{h}_m+\bar{h}_c)\delta T_c+\bar{h}_m\delta T_m+Q(t), \tag{6.32}$$

where

$$Q(t) = (\bar{T}_u-\bar{T}_c)\delta h_u+(\bar{T}_c-\bar{T}_m)\delta h_m-\bar{T}_c\delta h_c-\delta\Delta, \tag{6.33}$$

which can be written in matrix notation in the following manner:

$$\frac{d}{dt}\begin{bmatrix}\delta N\\ \delta C\\ \delta T_u\\ \delta T_m\\ \delta T_c\end{bmatrix} + \begin{bmatrix} \\ \\ \{B_{ij}\} \\ \\ \\ \end{bmatrix}\begin{bmatrix}\delta N\\ \delta C\\ \delta T_u\\ \delta T_m\\ \delta T_c\end{bmatrix} = \begin{bmatrix} \\ \\ \{\delta F_i\} \\ \\ \\ \end{bmatrix}, \tag{6.34}$$

where

$$\{B_{ij}\} = \begin{bmatrix} \bar{\lambda}_c-\lambda_f[\bar{\nu}(1-\beta)-1] & -\lambda & -\bar{N}\alpha & -\bar{N}\gamma & 0\\ -\lambda_f\bar{\nu}\beta & \lambda & 0 & 0 & 0\\ -h_f & 0 & \bar{h}_u & 0 & -\bar{h}_u\\ 0 & 0 & 0 & \bar{h}_m & -\bar{h}_m\\ 0 & 0 & -\bar{h}_u & -\bar{h}_m & (\bar{h}_u+\bar{h}_m+\bar{h}_c) \end{bmatrix} \tag{6.35}$$

and

$$\{\delta F_i\} = \begin{bmatrix} \lambda_f(1-\beta)\bar{N}\delta\nu\\ \lambda_f\beta\bar{N}\delta\nu\\ \bar{N}\delta h_f+(\bar{T}_u-\bar{T}_c)\delta h_u\\ (\bar{T}_c-\bar{T}_m)\delta h_m\\ (\bar{T}_u-\bar{T}_c)\delta h_u+(\bar{T}_c-\bar{T}_m)\delta h_m-\bar{T}_c\delta h_c-\delta\Delta \end{bmatrix}. \tag{6.36}$$

We may write this equation more concisely as follows:

$$\left(\mathbf{I}\frac{d}{dt}+B\right)\delta\mathbf{A} = \delta\mathbf{F}, \tag{6.37}$$

which is the form of the generalized set of coupled Langevin equations for a random vector $\delta\mathbf{A}$ subject to a random source vector $\delta\mathbf{F}$. $\mathbf{I}$ is the unit matrix and $\mathbf{B}$ is defined by eqn. (6.35). Notice that there is no parametric excitation so that the corresponding Fokker–Planck equation could be found and solved very easily via eqns. (5.78) and (5.80) (Chandrasekhar, 1943).

Equation (6.37) may be solved by diagonalizing the matrix $\mathbf{B}$, viz.

$$\mathbf{B} = \mathbf{S\Lambda T}, \tag{6.38}$$

$$\mathbf{\Lambda} = \begin{bmatrix}\lambda_1 & 0 & 0 & 0 & 0\\ 0 & \lambda_2 & 0 & 0 & 0\\ 0 & 0 & \lambda_3 & 0 & 0\\ 0 & 0 & 0 & \lambda_4 & 0\\ 0 & 0 & 0 & 0 & \lambda_5\end{bmatrix} \tag{6.39}$$

and $\mathbf{ST} = \mathbf{I}$. The λ_i are the eigenvalues of $\mathbf{\Lambda}$.

Then it is easy to show that a typical component of $\delta\mathbf{A}$ is given by Goertzel and Tralli (1960):

$$\delta A_i(t) = \sum_{k,j} S_{ik}e^{\lambda_k t}T_{kj}\delta A_j(0)+\sum_{k,j} S_{ik}T_{kj}\int_0^t ds\,\delta F_j(s)e^{\lambda_k(t-s)}, \tag{6.40}$$

which is the multi-component analogue of eqn. (5.3).

Now if we assume that the system was "switched-on" a long time in the past and also recognize that in a stable system $\lambda_k < 0$, we can neglect the initial conditions and rewrite (6.40) as

$$\delta A_i(t) = \sum_{k,j} S_{ik} T_{kj} \int_{-\infty}^{t} ds\, \delta F_j(s) e^{\lambda_k(t-s)} \tag{6.41}$$

$$\equiv \sum_{k,j} S_{ik} T_{kj} \int_0^{\infty} du\, e^{\lambda_k u} \delta F_j(t-u). \tag{6.42}$$

We are now in a position to find the cross-correlation function between any two δA_i, e.g.

$$\langle \delta A_i(t_1) \delta A_l(t_2) \rangle = \sum_{k,j} \sum_{m,n} S_{ik} T_{kj} S_{lm} T_{mn} \int_0^{\infty} du \int_0^{\infty} du'\, e^{\lambda_k u + \lambda_m u'} \times \langle \delta F_j(t_1 - u) \delta F_n(t_2 - u') \rangle. \tag{6.43}$$

But if the noise sources are stationary random processes

$$\langle \delta F_j(t_1) \delta F_n(t_2) \rangle \equiv \phi_{jn}(t_1 - t_2) = \phi_{jn}(\tau). \tag{6.44}$$

Thus with

$$\langle \delta A_i(t_1) \delta A_l(t_2) \rangle \equiv \psi_{il}(\tau) \tag{6.45}$$

we find

$$\psi_{il}(\tau) = \sum_{jkmn} S_{ik} S_{lm} T_{kj} T_{mn} \int_0^{\infty} du \int_0^{\infty} du'\, e^{\lambda_k u + \lambda_m u'} . \phi_{jn}(\tau - u + u'). \tag{6.46}$$

Thus we have put the output noise cross-correlation functions in terms of the input noise cross-correlation functions.

The corresponding cross-p.s.d. can be obtained in the usual way by defining

$$\Psi_{il}(\omega) = \int_{-\infty}^{\infty} d\tau\, e^{-i\omega\tau} \psi_{il}(\tau) \tag{6.47}$$

and

$$\Phi_{jn}(\omega) = \int_{-\infty}^{\infty} d\tau\, e^{-i\omega\tau} \phi_{jn}(\tau), \tag{6.48}$$

whence from (6.46) we obtain

$$\Psi_{il}(\omega) = \sum_{kjmn} S_{ik} S_{lm} T_{kj} T_{mn} \frac{\Phi_{jn}(\omega)}{(\lambda_k + i\omega)(\lambda_m - i\omega)} \tag{6.49}$$

or in matrix notation:

$$\mathbf{\Psi}(\omega) = (\mathbf{B} - i\omega\mathbf{I})^{-1} \mathbf{\Phi}(\omega) (\mathbf{B} + i\omega\mathbf{I})^{-1+}, \tag{6.50}$$

where $(\mathbf{B} - S\mathbf{I})^{-1}$ is the transfer function.

Thus we have relationships between the system parameters and the input and output noise phenomena. Generally, one wishes to find values of the system parameters by comparing input and output. Unfortunately, in our case, the nature of the noise sources due to heat transfer, etc., are not known and are likely to depend on the operating state of the reactor.

Returning to the basic problem studied by Greef on the CEGB's Berkeley reactor we note that he made the following assumptions regarding the noise sources:

1. The dominant noise source was due to heat transfer fluctuations caused by coolant turbulence.
2. The noise sources are "white" over the frequency ranges of interest.
3. The noise sources are uncorrelated.
4. The Schottky formula can be used to estimate the components of $\Phi(\omega)$.

These assumptions typify the state of the art in power noise analysis and indicate a need for more fundamental studies on individual noise sources.

On the basis of the above assumptions the cross-p.s.d.'s depend only on the transfer function of the system and some unknown constants representing the strength of the noise sources, viz. according to the above assumptions 1–3:

$$\Phi_{jn}(\omega) = \sigma_j^2 \delta_{jn}, \tag{6.51}$$

then

$$\Psi_{il}(\omega) = \sum_{km} \frac{S_{ik} S_{lm}}{(\lambda_k + i\omega)(\lambda_m - i\omega)} \sum_j \sigma_j^2 T_{kj} T_{mj} \tag{6.52}$$

In the particular example chosen, measurements were made of correlations between fuel-element can temperature and coolant temperature changes, viz. $\Psi_{T_c T_u}(\omega)$. For this, the transfer function $[(\mathbf{B}+s\mathbf{I})^{-1}]_{T_c T_u}$ is given by the block diagram in Fig. 6.2.

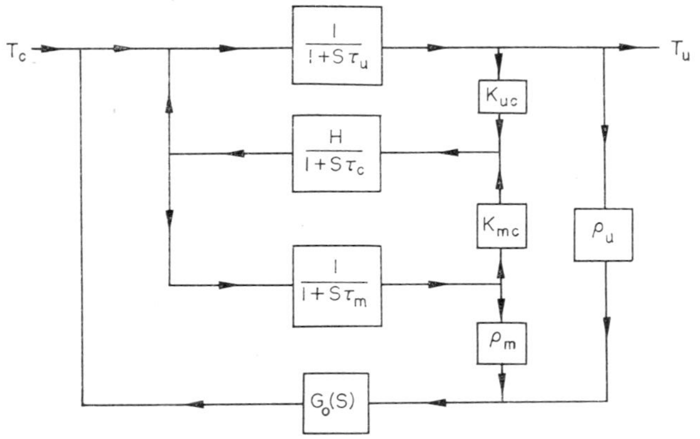

FIG. 6.2. Block diagram of transfer function between coolant and fuel temperatures.

In this figure $G_0(s)$ is the zero-power transfer function, defined by

$$G_0(s) = \left[s - \frac{\rho}{\Lambda} + \frac{1}{\Lambda} \frac{\beta s}{s+\lambda} \right]^{-1}. \tag{6.53}$$

The other symbols are

$$\tau_u = \bar{h}_u^{-1} = C_u V_u / K_{uc}, \tag{6.54}$$

$$\tau_m = \bar{h}_m^{-1} = C_m V_m / K_{mc}, \tag{6.55}$$

$$\tau_c = (\bar{h}_u + \bar{h}_m + \bar{h}_c)^{-1} = (K_u / C_c V_c + K_m / C_c V_c + 2\dot{V}_c / V_c)^{-1}, \tag{6.56}$$

$$H = \tau_c / C_c V_c = (K_{uc} + K_{mc})^{-1}, \tag{6.57}$$

$$\rho_u = \alpha \lambda_f \bar{\mu}_f \bar{N} C_u V_u / K_{uc}, \tag{6.58}$$

$$\rho_m = \gamma \lambda_f \bar{\mu}_f \bar{N} C_m V_m / K_{mc}. \tag{6.59}$$

In principle, therefore, by measurement of the p.s.d., we can by a least squares fitting procedure obtain the various time constants of the system. Figure 6.3 shows the noise spectrum for T_u (i.e. the can thermocouple output) whilst Fig. 6.4 shows the gain and phase spectra of the cross-p.s.d. of the coolant and fuel temperatures, i.e. $T_c T_u$.

Comparison with experiment and theory is given in Table 6.1.

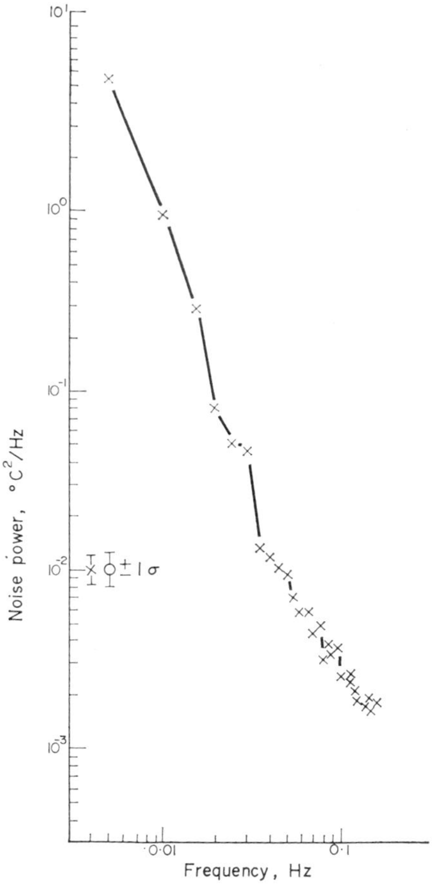

FIG. 6.3. Power spectral density of the can thermocouple temperature fluctuations. (From Greef, 1971.)

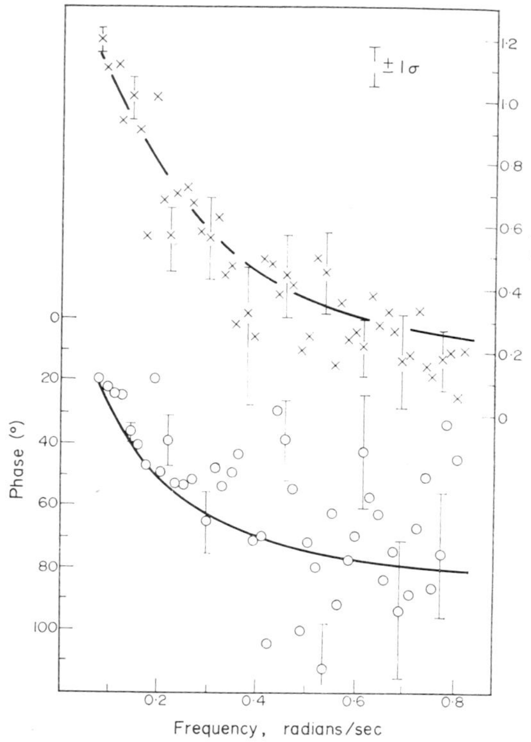

FIG. 6.4. Gain and phase of the power spectral density o the cross-correlation function between coolant and fue temperature fluctuations. (From Greef, 1971.)

TABLE 6.1

		Expected value from other methods	Expt.
Fuel time constant	τ_u (sec)	11·3	5·22
Moderator time constant	τ_m (sec)	370	66·6
Fuel feedback coefficient	ρ_u	−0·38	−0·30
Moderator feedback coefficient	ρ_m	+2·86	+0·15
Heat transfer factor	H	0·13	0·12

The large discrepancies in τ_m and ρ_m are thought to be due to the limitations of the point model, which is unable to account for the spatial variation of temperature in the moderator and fuel. An improved model will be discussed in a later section.

One of the main purposes of the noise investigation described above was to see whether the method would yield information which, hitherto, depended on the use of a pile oscillator. If such was the case then considerable economic advantages were to be derived. To judge whether this objective has been achieved consider a typical experimental result, shown in Fig. 6.5, which compares the phase and gain of the transfer function between coolant temperature and the response of the gas outlet

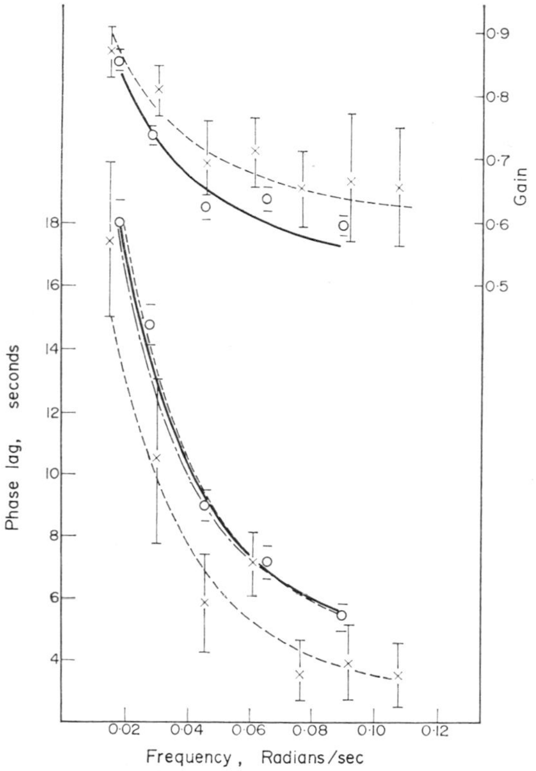

FIG. 6.5. Experimentally obtained gain and phase of the fuel–coolant temperature transfer function. The crosses are obtained from noise data and the circles from the rod oscillator technique. (From Greef, 1971.)

thermocouple as measured by noise methods and by the pile oscillator. The noise data are not as accurate as those from the oscillator but it is clear that for many practical purposes the agreement is sufficiently close to make the noise method preferable when the economic penalties are considered as well. Undoubtedly, further work on this problem will yield more accurate results.

6.5. *Fluctuations due to multiple random inputs*

As a further example of the Langevin technique we mention the rather convenient formalism given by Seifritz (1970) for the point model with multiple random inputs.

The basic neutronic equations are

$$\frac{dN}{dt} = \frac{\rho_{\text{tot}}(t)-\beta}{\Lambda} N+\sum_i +\lambda_i C_i+S, \tag{6.60}$$

$$\frac{dC_i}{dt} = -\lambda_i C_i+\frac{\beta_i}{\Lambda} N, \tag{6.61}$$

$$\rho_{\text{tot}}(t) = \sum_{l=1}^{L} \rho_l(t)+\sum_{m=1}^{M}\int_0^\infty g_m(\tau)[N(t-\tau)-\langle N\rangle]d\tau, \tag{6.62}$$

where $\rho_l(t)$ are driving reactivities due to various independent noise sources and the other terms are due to feedback effects, $g_m(\tau)$ denoting the corresponding feedback kernel. $S(t)$ is the noise equivalent source due to fission branching processes (see Appendix).

The form in which the reactivity appears can always be obtained provided that the parameters causing feedback, e.g. temperature changes or voidage variations, are linearly related to the reactor power. We shall assume this to be the case.

Let us now linearize these equations about the mean value $\langle N\rangle$ such that $N(t) = \langle N\rangle + n(t)$ and $C_i(t) = \langle C_i\rangle + c_i(t)$. Thus we find that $n(t)$ is given by the following expression:

$$\overline{n(s)} = \frac{G_0(s)\left\{\Lambda\bar{S}(s)+\langle N\rangle \sum_{l=1}^{L}\bar{\rho}_l(s)\right\}}{1-\langle N\rangle G_0(s)\sum_{m=1}^{N}\bar{g}_m(s)}, \tag{6.63}$$

where $\overline{n(s)}$, $\overline{S(s)}$, $\overline{\rho_l(s)}$ and $\overline{g_m(s)}$ are the Laplace transforms of $n(t)$, $S(t)$, $\rho_l(t)$ and $g_m(t)$, respectively, and $G_0(s)$ is the zero-power transfer function for a critical reactor, viz.

$$\frac{1}{G_0(s)} = s\left\{\Lambda + \sum_i \frac{\beta_i}{s+\lambda_i}\right\}. \tag{6.64}$$

The corresponding cross-correlation function for two random variables $\alpha(t)$ and $\beta(t)$ is, as we have seen in Chapter 3, given by

$$\phi_{\alpha\beta}(\tau) = \langle \alpha(t)\beta(t+\tau)\rangle \tag{6.65}$$

or alternatively, by writing α and β (Rice, see Wax, 1954) as

$$\alpha(t) = \frac{1}{2\pi}\int_{-\infty}^{\infty} d\omega\, e^{i\omega t}\,\overline{\alpha(i\omega)} \tag{6.66}$$

and

$$\beta(t) = \frac{1}{2\pi}\int_{-\infty}^{\infty} d\omega\, e^{i\omega t}\,\overline{\beta(i\omega)}, \tag{6.67}$$

we can show easily that the power spectral density of $\phi_{\alpha\beta}(\tau)$, i.e. $\Phi_{\alpha\beta}(\omega)$, is given by

$$\Phi_{\alpha\beta}(\omega) = \overline{\alpha(i\omega)}\,\overline{\beta(-i\omega)}. \tag{6.68}$$

Therefore if we choose $\overline{\alpha(s)} = \overline{\beta(s)} = \overline{n(s)}$, we can obtain directly the auto-p.s.d. for neutron density fluctuations, viz.

$$\Phi_{nn}(\omega) = |H(i\omega)|^2\left\{\langle N\rangle^2\left(\sum_{l=1}^{L}\overline{|\rho_l(i\omega)|^2} + \sum_{\substack{i,k\\ i\neq k}}^{L}\overline{\rho_i(i\omega)\rho_k(i\omega)^*}\right)+\Lambda^2\overline{|S(i\omega)|^2}\right\}, \tag{6.69}$$

where

$$H(s) = \frac{G_0(s)}{1-\langle N\rangle G_0(s)\sum_{m=1}^{M}\overline{g_m(s)}}, \tag{6.70}$$

and we have assumed that there is no correlation between $\rho_n(t)$ and $S(t)$. Moreover, we could also write

$$\Phi_\rho^{(l)}(\omega) = \overline{|\rho_l(i\omega)|^2} \tag{6.71}$$

which is the auto-p.s.d. of the input reactivity l, and

$$\Phi_\rho^{(ik)}(\omega) = \overline{\rho_i(i\omega)\rho_l(i\omega)^*}, \tag{6.72}$$

the cross-p.s.d. of the input reactivities i and k.

Similarly,

$$\Phi_s(\omega) = |\overline{S(i\omega)}|^2 \simeq \frac{\overline{\nu(\nu-1)}}{\bar{\nu}} \frac{1}{\Lambda} \langle N \rangle \tag{6.73}$$

is the p.s.d. of the fission fluctuations if they are assumed to be white (Cohn, 1960; Sheff, 1965) (see Appendix).

It may also be of value to have the cross-p.s.d. between the neutron density $n(t)$ and any one of the input reactivities $\rho_l(t)$. In that case we readily find that

$$\Phi_{n\rho}^{(l)}(\omega) = \overline{n(i\omega)\rho_l(-i\omega)} \equiv \Phi_{\rho n}^{(l)}(-\omega), \tag{6.74}$$

$$= H(i\omega)\langle N \rangle \{\Phi_{\rho}^{(l)}(\omega) + \sum_{\substack{i,k \\ i \neq k}} \Phi_{n\rho}^{(ik)}(\omega)\}, \tag{6.75}$$

which has a phase-lag term as well as a gain.

6.6. *Application to control-rod vibration*

As an example of the Seifritz technique we will describe a problem studied by Kosály and Williams (1971). These authors were concerned with the effect on the p.s.d. of the neutron noise of the inlet coolant temperature fluctuations and control-rod vibration in the Oak Ridge Research Reactor (ORR). Experiments on this reactor by Stephenson *et al.* (1966) showed that interesting resonance phenomena were present for some coolant flow rates but not for others. An analytical study of this problem was given by Robinson (1967) and the present analysis makes use of some of his results for the temperature fluctuations; however, a rather different approach is used for dealing with the vibrations. Figure 6.6 shows the experimental power spectra obtained by Stephenson *et al.* (1966) for three different coolant flow rates.

The relationship between the reactivity, $\rho_{\text{tot}}(t)$, as given in eqn. (6.62) and the temperature fluctuations and control-rod vibrations can be written as

$$\rho_{\text{tot}}(t) = \alpha_f \Delta T_f(t) + \alpha_c \Delta T_c(t) + \rho_v(t), \tag{6.76}$$

where ΔT_f and ΔT_c are fuel and coolant temperature fluctuations and $\rho_v(t)$ is the reactivity fluctuation due to the vibration of a control rod.

ΔT_c and ΔT_f are related to the power fluctuations $n(\tau)$ and the inlet coolant temperature fluctuations by appropriately spaced averaged heat transfer equations (Robinson, 1967) and it is found that

$$\Delta T_c(t) = \int_0^\infty d\tau\, h_1(\tau)\Delta T_i(t-\tau) + \int_0^\infty d\tau\, g_1(\tau) n(t-\tau) \tag{6.77}$$

and

$$\Delta T_f(t) = \int_0^\infty d\tau\, h_2(\tau)\Delta T_i(t-\tau) + \int_0^\infty d\tau\, g_2(\tau) n(t-\tau), \tag{6.78}$$

where $\rho_l(t) = \int_0^\infty d\tau\, h_l(\tau)\Delta T_i(t-\tau)$ are driving reactivities.

The vibration is rather more difficult to deal with since it is essentially a local effect and therefore spatially dependent. We will describe a method for dealing accurately with local spatial perturbations in a later chapter, but for the present analysis it may be shown that for small, random amplitude vibrations, $\epsilon(t)$, about the mean position, the value of $\rho_v(t)$ is given for most practical purposes by

$$\rho_v(t) = A_0 \epsilon(t), \tag{6.79}$$

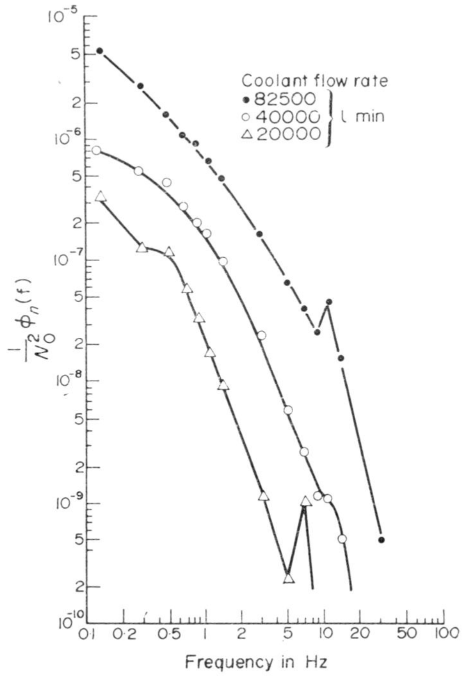

FIG. 6.6. Experimental noise power spectra for the ORR at different coolant flow rates. Reactor operating at 3·75 MW. (From Stephenson *et al.* (1966), ORNL-TM-1401.)

A_0 being a constant which involves the position and absorption cross-section of the rod.

With these facts in mind, and following the general procedure described by Seifritz, we obtain for the neutron density p.s.d. the following expression:

$$\Phi_n(\omega) = \langle N\rangle^2|H(i\omega)|^2|\alpha_f\overline{h_2(i\omega)}+\alpha_c\overline{h_1(i\omega)}|^2\,\Phi_i(\omega)+\langle N\rangle^2|H(i\omega)|^2A_0^2\Phi_v(\omega), \quad (6.80)$$

where

$$H(i\omega) = \frac{G_0(i\omega)}{1-\langle N\rangle G_0(i\omega)[\alpha_f\overline{g_2(i\omega)}+\alpha_c\overline{g_1(i\omega)}]}, \quad (6.81)$$

$$\Phi_i(\omega) = \langle\overline{\Delta T_i(i\omega)}\,\overline{\Delta T_i(i\omega)}^*\rangle, \quad (6.82)$$

$$\Phi_v(\omega) = \langle\overline{\epsilon(i\omega)}\,\overline{\epsilon(i\omega)}^*\rangle, \quad (6.83)$$

$\overline{g_i(s)}$ being the Laplace transform of $g_i(\tau)$ and Φ_i and Φ_v the input p.s.d.'s of the noise sources of the coolant temperature and rod vibrations, respectively. We also assume that zero-power noise due to branching processes is negligible.

Further, assuming white noise sources for the inlet coolant temperature, we can write

$$\Phi_i(\omega) = \langle(\Delta T_i)^2\rangle = \text{mean square temperature fluctuation.} \quad (6.84)$$

The calculation of $\Phi_v(\omega)$ presents more difficulty. It is well known that high velocity coolant, flowing through the reactor core, is a source of energy that can induce and sustain vibration in core components (Wambsganss, 1967). The most common cause of mechanical vibrations is due to the turbulence of the coolant flow resulting in random pressure fluctuations along the surface of a fuel element or control rod. In the case of normal operating conditions the situation is fairly straightforward. If one considers a bar of finite length situated in the axial direction (direction of flow) and attached to fixed supports (see Burgreen *et al.*, 1958 for variable support effects) experiments show (Burgreen *et al.*, 1958) that while the velocity of the flow influences to a large extent the amplitude of the vibrations it does not affect the frequency of motion, which is nearly equal to the natural frequency of the bar as measured in stationary water. Thus in practical cases it may be assumed that the vibration is initiated and sustained by turbulent pressure fluctuations but the rod itself responds to this forcing in its own fundamental bending mode.

Whilst the situation described above is common to all reactors operating normally, from the point of view of diagnostics it is the unusual situation which is of interest—that is, the case of excess vibration due to some failure in the fixing point or bearing of the component. In such cases quite complicated motion of the bar will ensue. We acknowledge this fact in our treatment of $\epsilon(t)$ but argue that whilst the motion of a faulty component may differ from that of its normal counterpart, it can still be described by a dominant single frequency (e.g. the frequency of the bowing rod in the slight clearance caused by the defect) and a fixed damping coefficient. Thus we use for the faulty rod the same functional form of $\Phi_v(\omega)$ and expect the fault to manifest itself in the values of the different parameters in the expression as the coolant velocity increases. Thus, whilst in the normal case, the amplitude of $\Phi_v(\omega)$ would be expected to change with increasing coolant velocity, in the defective case we would expect the fundamental frequency and the damping to change as well. Clearly this is an over-simplification but it is the spirit of the point model philosophy.

In the theory of random vibrations, as we shall see in Chapter 9, the form of $\Phi_v(\omega)$ for a clamped rod is given by ($\omega = 2\pi f$) (Robson, 1963):

$$\Phi_v(\omega) = \frac{A(f)}{(f^2-f_0^2)^2+\eta_0^2 f_0^2}, \tag{6.85}$$

where f_0 is the dominant resonant frequency and η_0 the damping coefficient. Both f_0 and η_0 depend on the properties of the rod and the surrounding medium.

$A(f)$ denotes the influence of turbulent forcing which can be calculated with some degree of precision for a fixed-ended rod (Reavis, 1969); however, for the faulty case no such calculation is available. In view of the uncertainty in $A(f)$ we shall assume that it is white and further assuming that $\eta_0 \ll 1$, we can then rewrite $\Phi_v(\omega)$ approximately as

$$\Phi_v(\omega) = \frac{\Delta^2}{\pi}\left[\frac{W}{(F_0-f)^2+W^2}+\frac{W}{(F_0+f)^2+W^2}\right], \tag{6.86}$$

where Δ is the r.m.s. value of the transverse deflection of the rod averaged over its length, and F_0 and W are given by

$$F_0 = f_0(1+\eta_0^2)^{\frac{1}{4}}\cos(\tfrac{1}{2}\tan^{-1}\eta_0), \tag{6.87}$$

$$W = f_0(1+\eta_0^2)^{\frac{1}{4}}\sin(\tfrac{1}{2}\tan^{-1}\eta_0). \tag{6.88}$$

If we now set $A_v = A_0^2\Delta^2/\pi$ we have from eqns. (6.86) and (6.80) a functional form of the p.s.d. containing parameters $\langle(\Delta T_i)^2\rangle$, A_v, F_0 and W which may be calculated by comparison with experiment.

Rather than compute that part of the p.s.d. due to inlet temperature fluctuations anew, we use the values of $|\alpha_f \overline{h_2(i\omega)} + \alpha_c \overline{h_1(i\omega)}|^2$ and $|H(i\omega)|^2$ as calculated by Robinson (1967). Making use of the phenomenological fact that the transfer function associated with temperature fluctuations vanishes at the peak of $\Phi_v(\omega)$ and that $|H(i\omega)|^2$ is virtually constant in this region we can obtain F_0 and W directly from the experimental curve ($F_0 = 7$ Hz, $W = 0{\cdot}6$ Hz) and then using these values in eqn. (6.80) fit the analytical curves by least squares to the experimental results and obtain $\langle(\Delta T_i)^2\rangle$ and A_v. The results of these calculations for three different coolant flow rates are shown in Table 6.2. In Fig. 6.7, we give the results for the two lower flow rates

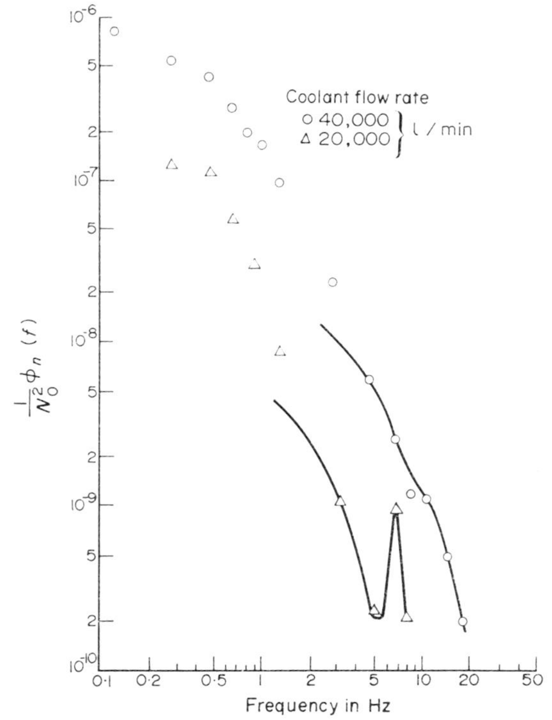

FIG. 6.7. Comparison of the experimental power spectra with the theoretical results. The solid lines show the best fit, according to least squares, of the theoretical expression. (From Kosály and Williams (1971), *Atomkernenergie*, **18**, 203.)

TABLE 6.2

Coolant flow rate, U.S. gal/min	F_0 Hz	$\langle(\Delta T_i)^2\rangle^{\frac{1}{2}}$ °F	A_v sec^{-1}	W Hz
4500	7·0	0·012	$3{\cdot}6 \times 10^{-14}$	0·6
9000	11·8	0·021	$2{\cdot}84 \times 10^{-13}$	3·5
18,500	—	0·043	—	—

only since at the highest rate the method of least squares has not given even a reasonable fit. The authors suspect the cause of this to arise from the two experimental points at high frequencies which may be erroneous.

Inspection of Table 6.2 shows that both the characteristic frequency and the amplitude of the vibration increases with increasing flow velocity. The quantity A_v increases with velocity too, showing that the mean amplitude of the vibrations increases, as was to be expected.

The increase of the r.m.s. inlet temperature fluctuation with flow velocity is roughly linear. It is characteristic of the sensitivity of noise measurements that such extremely small temperature fluctuations can be detected in the noise spectrum.

Finally, we note the "interference" or overlapping of the mechanical vibration and coolant inlet temperature effects as shown by the two graphs in the figure. It should also be pointed out that when the faulty control-rod bearing, which was responsible for the anomalous rod vibration, was replaced, the mechanical resonance disappeared entirely. Thus noise measurements are clearly useful for detecting mechanical faults. However, before they can be of direct practical value it must be shown that they can predict incipient trouble and define its location so that appropriate action may be taken in time. For this to be done effectively, a detailed space-dependent noise theory must be formulated. Nevertheless, even without this sophistication, a number of useful pieces of information have been obtained from point model theories.

It should be added in connection with that part of the calculation concerned with temperature fluctuations that Kosály and Mesko (1972) have studied the spatial behaviour of the coolant flow along the reactor axis and have shown that the associated p.s.d. has sinks at frequencies given approximately by $\omega_n = 2\pi n/\tau$, where τ is the transit time of the coolant through the core. This behaviour is in contrast to the results of Boardman (1964) who mistakenly predicted resonances in this p.s.d. Thus the resonance-like structure actually seen in experimental curves of the p.s.d. on the ORR are actually sink effects and not as previously supposed due to resonances.

6.7. *Application to model of power reactor system with random fluctuations in the coolant flow velocity*

Seifritz (1970) in his original paper on multiple random inputs considered, as an example, the problem of a point model reactor system with forced cooling which is perturbed by irregularities in the coolant flow rate.

The basic equations considered were as follows:

$$\frac{dN}{dt} = \frac{\rho_{\text{tot}}-\beta}{\Lambda}N+\sum_{i=1}^{6}\lambda_i C_i+S, \tag{6.89}$$

$$\frac{dC_i}{dt} = -\lambda_i C_i+\frac{\beta_i}{\Lambda}N, \tag{6.90}$$

$$C_f\frac{dT_f}{dt} = P-K(T_f-T_c) \tag{6.91}$$

(where $P = \mu v\Sigma_f N$ is the power from fission, $\mu = 3{\cdot}2\times 10^{-11}$ Watt sec/fission),

$$C_c\frac{dT_c}{dt} = K(T_f-T_c)-C_c\frac{V}{H}T_c. \tag{6.92}$$

The meaning of many of the parameters in these equations is obvious from physical considerations, with T_f the temperature of the fuel and T_c that of the coolant. H is the core height and V is the coolant velocity, thus H/V is the coolant transit time τ and is an approximation used to represent the space dependence of the flow in a point model scheme.

The zero-energy noise term can generally be neglected for sufficiently high power but for more generality we shall retain it. The feedback reactivity is taken to be of the form

$$\rho_{\text{tot}} = \rho_0 + \alpha_f T_f + \alpha_c T_c, \tag{6.93}$$

where ρ_0 is a static reactivity of the control system which is used to adjust the reactor to criticality at various power levels. If now a small random fluctuation $v(t)$ is allowed in the coolant velocity such that

$$V(t) = V_0 + v(t), \tag{6.94}$$

where V_0 is the mean value, then this will cause a corresponding fluctuation in the heat-transfer coefficient through the dimensionless semi-empirical equations used to represent these factors in various channel configurations (El-Wakil, 1971).

For an ordinary cooling medium such as water in the turbulent-flow region (Re > 2000) the relationship for heat transfer is

$$\text{Nu} = 0{\cdot}023\,(\text{Re})^{0\cdot8}(\text{Pr})^{0\cdot4}, \tag{6.95}$$

where Nu $= K/F$, K being the integral heat-transfer coefficient and F the total cooling surface, Re $= DV\rho/\nu$ (D = effective diameter of coolant channel, ρ = coolant density, ν = viscosity), Pr $= C\nu/\lambda$ (C = specific heat of coolant, λ = thermal conductivity).

For small variations in K and V we find that if $K(t) = K_0 + k(t)$ then the fluctuation, k, in the heat-transfer coefficient will be related to the fluctuation $v(t)$ by

$$\frac{k(t)}{K_0} \simeq 0{\cdot}8\,\frac{v(t)}{V_0}. \tag{6.96}$$

A similar relationship holds for liquid-metal coolants but with a different numerical coefficient. We can therefore write quite generally that

$$\frac{k(t)}{K_0} = A\,\frac{v(t)}{V_0}. \tag{6.97}$$

Inserting these fluctuations into the eqns. (6.89)–(6.92) and writing $N = N_0 + n$, $C_i = C_{i0} + c_i$, $P = P_0 + p$, $T_f = T_{f0} + \gamma_f$, $T_c = T_{c0} + \gamma_c$, we can linearize and use the steady-state equations

$$0 = -\frac{\beta}{\Lambda} N_0 + \sum_{i=1}^{6} \lambda_i C_{i0}, \tag{6.98}$$

$$0 = -\lambda_i C_{i0} + \frac{\beta_i}{\Lambda} N_0, \tag{6.99}$$

$$0 = P_0 - K_0(T_{f0} - T_{c0}), \tag{6.100}$$

$$0 = K_0(T_{f0} - T_{c0}) - C_c T_{c0}/\tau_0 \tag{6.101}$$

with $\tau_0 = H/V_0$. The reactor is then made critical by adjusting ρ_0 such that

$$\rho_0 + \alpha_f T_{f0} + \alpha_c T_{c0} = 0. \tag{6.102}$$

Laplace transformation of the equations for n, γ, C_i, etc., leads to an expression for $\overline{n(i\omega)}$ and hence via eqn. (6.68) to the p.s.d. for the system which we may write as follows:

$$\tilde{\Phi}_{nn}(\omega) = \epsilon \frac{P_0}{\mu} \overline{q^2} + \epsilon^2 \frac{P_0}{\mu} \bar{q}^2 D |H(i\omega)|^2$$
$$+ \epsilon^2 \left(\frac{P_0}{\mu}\right)^2 \bar{q}^2 |H(i\omega)|^2 \left| G(i\omega)[G_1(i\omega) + A(G_2(i\omega) - G_1(i\omega))] \frac{P_0}{K_0} \right|^2 \left| \frac{\overline{v(i\omega)}}{V_0} \right|^2, \tag{6.103}$$

where $D = \overline{\nu(\nu-1)}/(\bar{\nu})^2$ and we have "folded" into this result the detector characteristics, namely its efficiency and the ion chamber output current fluctuations $i(t) = \epsilon \bar{q} n(t)/\bar{\nu}\Lambda$, where $\bar{q}$ is the average charge per detected neutron. Finally, the first term on the right-hand side is due to uncorrelated detections in the ion chamber, $\overline{q^2}$ being the mean square charge per detected neutron.

The various transfer functions G, H, etc., are defined as follows:

$$G(s) = \frac{1}{1 - G_c(s) G_f(s)}; \quad G_1(s) = G_c(s)(\alpha_c + \alpha_f G_f(s)),$$

$$G_c(s) = \frac{1}{1 + \dfrac{\tau_c}{\tau_0} + s\tau}; \quad \tau_c = \frac{C_c}{K_0},$$

$$G_2(s) = G_f(s)(\alpha_f + \alpha_c G_c(s)),$$

$$G_f(s) = \frac{1}{1 + s\tau_f}; \quad \tau_f = \frac{C_f}{K_0},$$

$$H(s) = \frac{G_0(s)}{1 - \dfrac{P_0}{K_0} G(s) G_2(s) G_0(s)},$$

where $G_0(s)$ is the zero-power transfer function of eqn. (6.53) and $|\overline{v(i\omega)}|^2$ is the p.s.d. of the coolant flow fluctuations. This rather complicated feedback circuit can best be understood by a block diagram for which we refer the reader to Seifritz (1970). Some limiting results and general statements may be made about the p.s.d. without actually applying it to a specific system. Firstly, we note that the noise consists of three distinct terms. The first is due to the uncorrelated shot noise in the detector. The second arises from the branching process due to fission as can be seen from the presence of the factor D. The third term is the most important from the point of view of power reactor studies since it arises from the core transit time dynamics, dominated by the transfer function $G_c(s)$, and due to fluctuations in heat transfer associated with the term involving the factor A. By comparison of this term with the general expression of eqn. (7.69), we note that the p.s.d. of the effective driving reactivity takes the form

$$|\rho_{dr}(i\omega)|^2 = \left(\frac{P_0}{K_0}\right)^2 |G[G_1 + A(G_2 - G_1)]|^2 \left| \frac{\overline{v(i\omega)}}{V_0} \right|^2, \tag{6.104}$$

which depends upon the core transit time of the coolant and heat-transfer fluctuations.

Thus to investigate the frequency range in which fluctuations of the coolant flow are likely to become important we can consider, following Seifritz, the cases

$$(1)\ \omega \gg \tau_c^{-1} \text{ and } \tau_f^{-1}, \qquad (2)\ \omega \ll \tau_c^{-1} \text{ and } \tau_f^{-1}.$$

In case (1), the driving reactivity becomes negligible and we are left with the zero power transfer function as represented by the first two terms on the right-hand side of eqn. (6.103) with $H(s)$ replaced by $G_0(s)$. This is not an unexpected result since the assumption implies very rapid attenuation of the driving mechanism at low frequency.

In case (2), the limit of eqn. (6.104) for $\omega\tau_c \ll 1$ and $\omega\tau_f \ll 1$ can be written

$$|\rho_{dr}|^2 \to \left(\frac{P_0}{K_0}\right)^2 \left(\frac{\tau_0}{\tau_c}\right)^2 \left[\alpha_c + \alpha_f\left(1 + A\,\frac{\tau_c}{\tau_0}\right)\right]^2 \Phi_v(\omega), \tag{6.105}$$

where

$$\Phi_v(\omega) = \overline{|v(i\omega)|^2}/V_0^2.$$

Again this can be explained physically in the sense that the response of the temperatures T_f and T_c to fluctuations in the heat-transfer coefficient are so slow that the resulting noise spectrum is white over the frequency range of interest.

From a practical point of view, it is clear that the measured p.s.d. will be sensitive to the values of the time constants of the fuel and coolant and this would have to be noted if the method was to be employed to study any anomalous behaviour in $\Phi_v(\omega)$. On the other hand, it is likely that, in normal operating conditions, $\Phi_v(\omega)$ would be white over a wide frequency range and thereby enable τ_f and τ_c to be extracted. Experiments to test the validity of this and similar models would be invaluable.†

6.8. *Noise in boiling-water reactors*

When boiling-water reactors are operated at high-power levels, the power output record exhibits, superimposed on the normal noise, oscillatory wave packets. These are power oscillations of a distinct frequency whose amplitude exceeds the root mean square value of the normal reactor noise. Figures 6.8 and 6.9 show such a record and other examples may be found in Forbes *et al.* (1956), Maxon *et al.* (1959), Thie (1959, 1963). In the experiments described in the above-quoted references a number of qualitative observations have been made on the behaviour of these oscillatory wave packets; some typical findings are: (*a*) the occurrence of the wave packets is spontaneous and random and occur usually at high power and low pressure; (*b*) for a fixed power level, the amplitude of the wave packets is random; (*c*) the average amplitude of the packets and the oscillation frequency increase with reactor power; (*d*) the waveform of the packets is sinusoidal when their amplitude is small compared with average power fluctuations, but becomes non-sinusoidal for large amplitudes: however, the waveform of the oscillations in the corresponding excess reactivity remains sinusoidal regardless of amplitude, which indicates that the non-sinusoidal effects are caused by non-linearities in the reactor kinetics equations rather than in the mechanism causing the feedback effect. One of the irritations of these spontaneous, large amplitude fluctuations is that they cause the reactor to scram when it is being operated at steady power, simply because the occasional packet exceeds some preset trip level. The

† The above examples, based on the Langevin equation, have recently been shown by Saito (1974) to be special cases of non-equilibrium statistical mechanics. For a further discussion see Appendix.

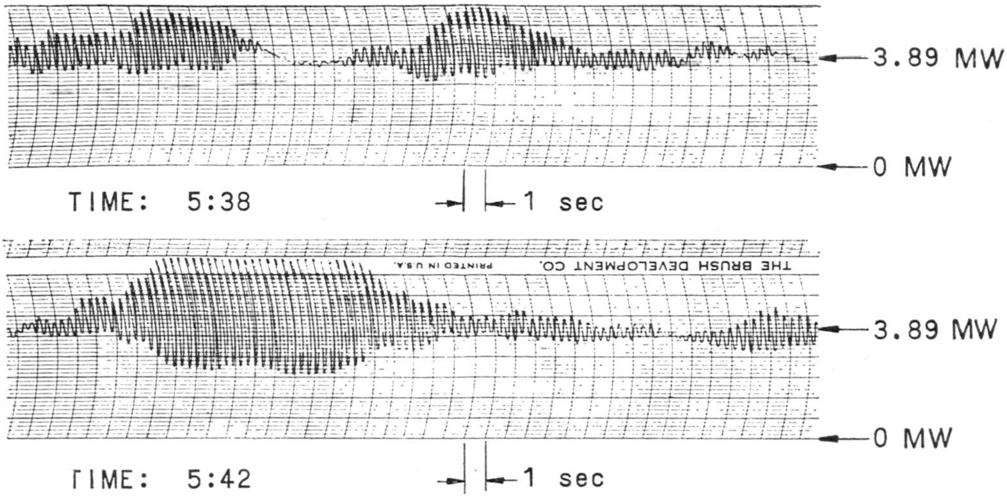

FIG. 6.8. Reactor power fluctuations in BORAX II at 150 p.s.i.g. (From ANL-7533, ANL-5849, ANL-6135.)

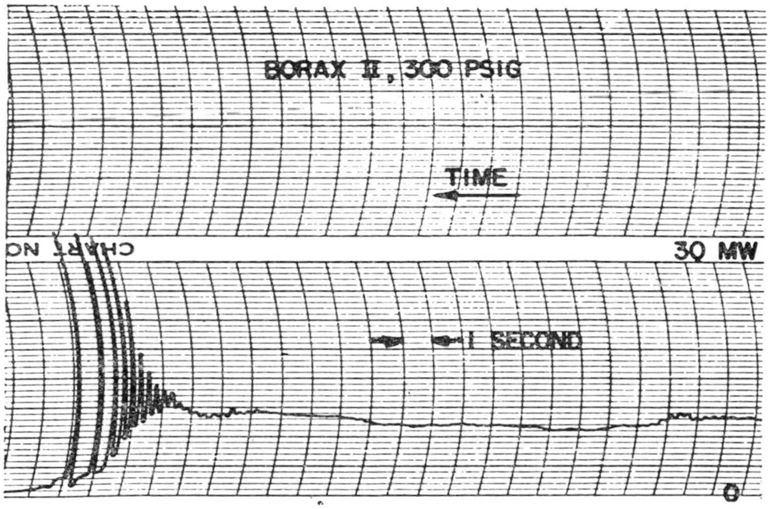

FIG. 6.9. Transition with small oscillatory tendency to violent power oscillations in BORAX II with 72 fuel elements and 4% reactivity in voids. (From ANL-5849.)

general behaviour of the reactor prior to such a trip indicates that it is operating in a stable state so that, providing the random bursts are not of too long a duration, the scram would have been unnecessary.

Some of the above observations can be explained by regarding the reactor as a narrow band pass filter with a constant damping factor which is excited by a random input: this can account for the occurrence of the oscillatory wave trains and the amplitude distribution. The associated theory is given by Rice (Wax, 1954), who shows that the expected number of maxima of the envelopes of these fluctuations per unit time is given by $\mathcal{N} = 3{\cdot}18 f_0\beta$, where f_0 is the midband frequency and β is the damping coefficient. This model fails, however, to account for the spontaneous occurrence of the large oscillatory packets. In the normal course of events the probability, based upon the Rice formalism (see Section 3.11), that the envelope maximum will exceed the r.m.s. value is very small (e.g. 0·2% probability to exceed four times the r.m.s. value).

A method of understanding, if not completely solving, this anomalous behaviour has been given by Akcasu (1961). He notes that in the conventional stability analysis the effect of random variations in the basic system parameters is generally neglected, thus the damping factor will be a deterministic function fixed by average values. However, in practice, owing to mechanical disturbances this may not be a valid assumption. This will be particularly true of boiling-water reactors where we could expect fluctuations in the damping factor to arise from three main sources: (1) random perturbations in the flux shape and importance function caused by irregularities in the void distribution across the core; (2) fluctuations in the steam-production rate due to random variations in the coolant velocity and its enthalpy; (3) random movements in the position of the boiling boundary.

Now the precise way in which these processes affect the reactivity is extremely difficult to assess. Quite clearly a spatially dependent model should be used and, indeed, we shall discuss such a technique in Chapter 8; however, even in the point model approximation it is possible to write down a feasible mechanism of feedback. Thie (ANL 6135), in a general study of boiling-water reactor safety, has proposed that the reactivity $\rho(t)$ be described by the following second-order differential equation:

$$\frac{1}{\omega_0^2}\ddot{\rho}(t)+2\beta(t)\frac{1}{\omega_0}\dot{\rho}(t)+\rho(t) = Z(t), \tag{6.106}$$

where ω_0 is a characteristic frequency, $Z(t)$ is Gaussian white noise with zero mean and $\beta(t)$ is Gaussian noise (not necessarily white) with mean value β.† The narrow band pass filter model discussed above is described by this equation with constant β. In fact, we then have a special case of Seifritz's equation (6.62) with one feedback loop. For a white noise $\beta(t)$ we may also compare eqn. (6.106) directly with eqn. (5.143). Setting $\beta(t) = \beta_0+\Delta\beta$, we note that the equation for the average value of $\rho(t)$ has an effective damping factor equal to $\beta_0-\sigma^2$ where $\langle\Delta\beta(t_1)\Delta\beta(t_2)\rangle = 2\sigma^2\delta(t_1-t_2)$. The parametric excitation therefore has a destabilizing effect which is not predicted by the average value calculations of normal reactor kinetics.

Akcasu solves (6.106), without making the assumption of white noise, by a modified WKB method. His main results can be summarized as follows: for the mean value of $\rho(t)$ to be stable as $t \to \infty$, i.e. $\langle\rho(\infty)\rangle \to 0$, we must have $\beta_0 > R(\infty)$. Similarly, for the system to be stable in the mean square, i.e. $\langle\rho^2(\infty)\rangle \to 0$, it is necessary that $\beta_0 > 2R(\infty)$, where $R(t)$ is defined by

$$R(t) = \omega_0\int_0^t\left(1-\frac{\tau}{t}\right)\phi(\tau)\,d\tau \tag{6.107}$$

$$\equiv 4\omega_0\int_0^\infty W(f)\frac{\sin^2(\pi f t)}{(2\pi f)^2 t}\,df, \tag{6.108}$$

† β should not be confused with delayed neutron fraction. It is used to conform with Akcasu's notation.

where
$$\phi(\tau) = \langle(\beta(t)-\beta_0)(\beta(t+\tau)-\beta_0)\rangle \tag{6.109}$$
and $W(f)$ is its p.s.d.

In general we can show that regardless of the shape of $W(f)$, $R(\infty) = W_0\omega_0/2$ where W_0 is the value of $W(f)$ at $f = 0$.

We see, therefore, that even if the mean value of $\langle\rho\rangle$ is zero, the mean square can be increasing with time and lead to the observed increases in the mean square power level. To test the validity of the model, either the statistical content of the experimentally observed wave packets would have to be analysed or a simulation using an analogue computer could be employed. Owing to a lack of sufficiently accurate experimental data, Akcasu adopted the analogue method. He took the reactor kinetics eqns. (6.60), (6.61) and inserted the solution of eqn. (6.106) for $\rho(t)$ into them. The threshold for mean square instability was calculated to be $\sigma_\beta = (2R(\infty))^{\frac{1}{2}} = 0{\cdot}0225$ for the particular noise source used.

The simulation was performed for $\sigma_\beta = 0$, i.e. no parametric excitation, and the number of expected maxima of the wave packets in the power N was shown to correspond to that for the simple filter. For $\sigma_\beta = 0{\cdot}015$ the wave packets increased in intensity and the number of maxima per unit time was closer to the value $\mathcal{N} = 3{\cdot}18f_0[\beta_0-\sigma_\beta]$; however, there were no violent oscillations observed. For $\sigma_\beta = 0{\cdot}03$, i.e. in excess of the mean square stability threshold, violent oscillations began after 234 sec in one particular run. In other runs, with the same σ_β, violent oscillations occurred as soon as 36 sec. The analogue computer study seems, therefore, to confirm the effect of parametric excitation as the cause of the spontaneous bursts.

A more realistic study using the analogue computer was performed on the equations of a complete reactor system which simulated EBWR at 60 MW and 41 atmospheres using a feedback model due to De Shong (1958). The gain was adjusted so that in the absence of noise the system became unstable at 64 MW. The driving noise $Z(t)$ was then injected at the steam generation point, since variations in this quantity do not affect the damping factor. The parametric noise $\beta(t)$ was superimposed on the power reactivity coefficient. At 60 MW the value of β_0 was 0·014 ($f_0 = 1{\cdot}86$ Hz); the r.m.s. value of the damping noise to cause instability in the mean square was calculated to be $\sigma_\beta = 0{\cdot}024$. The simulation showed that a fluctuation of 12% in the power reactivity coefficient was sufficient to cause instability at 60 MW: this value was confirmed by the associated theory.

More detailed studies of parametric excitation and methods of solving the associated point model equations may be found in Williams (1971).

6.9. *Additional applications*

The examples and techniques discussed in this chapter are far from exhaustive and it is worth noting that the procedure of *deliberately* introducing power fluctuations by perturbing the reactivity in a prescribed manner has been employed in the U.S. nuclear rocket testing programme for the KIWI and NERVA reactor systems. In that case the coolant flow rate is perturbed in a known fashion, i.e. the $\Phi_v(\omega)$ of eqn. (6.105) is known and the response of the system to this random input compared with the theoretical model.

Another problem, which is of the utmost practical importance in fuel-element design and reactor safety, is the prediction of nucleate boiling. Examination of the p.s.d. in the Saxton boiling-water reactor by Rajagopal (1964) shows a well-defined resonance which by varying the operating conditions can, with some confidence, be related to nucleate boiling. The peak resonance frequency in this case occurs at around a frequency of 16 Hz and presumably can be correlated with the frequency of formation and collapse of vapour bubbles on the surface of the fuel elements. A detailed statistical study of this problem would involve a time- and space-dependent study of the nature of steam formation near heated surfaces; this is a difficult problem but clearly worth while in view of the power of the associated experimental method.

We can also mention one potentially important application of zero-power noise theory for estimating parameters of interest in power reactor operation.

It is known that β, the delayed neutron fraction, is different for U^{235} and Pu^{239}. Thus by measuring the break frequency β/Λ of a reactor at various stages in its irradiation history it will be possible to estimate the ratio of U^{235} to Pu^{239} atoms. Thus an estimate of the burn-up can be obtained without the need to unload the core and perform a chemical analysis. The accuracy of this technique is not yet sufficient for an economic analysis (ratio accurate to about $\pm 0{\cdot}005$) but it is a useful indicator of the remaining fuel lifetime.

Finally, we note that the employment of stochastic processes in reactor technology lies in the interpretation of properties not only of the system itself, but also of the power plant. An important example of this has been given by Kiguchi *et al.* (1973) who have studied the stochastic fluctuation in a uranium-enriching cascade using the centrifuge process. The problem is concerned with obtaining engineering tolerances of centrifuge parameters and stage controllers. In order to do this it is necessary to calculate the plant performance via the usual cascade equations. Now in general such a study will yield average plant parameters but makes no allowance for random fluctuations which could well affect plant performance. The processes of interest that may be expected to exhibit some random behaviour are the cut and separation factor of each stage. It was the purpose of the stochastic approach to assess the effect of these random variations on the time behaviour of the flow rate and enrichment.

By assuming that the total uranium flow rate through stage i, $F_i(t)$, and the U^{235} flow rate, $G_i(t)$, are random variables governed by rate equations coupling the various stages, a probability balance equation may be set up with the driving noise sources being the cut $\theta_i(t)$ and the separation factor $\gamma_i(t)$ [$= 1+2\epsilon_i(t)$]. By allowing the discontinuous equations for the coupled stages to go to the continuous limit, it is found that the following stochastic differential equations arise:

$$T\frac{\partial F(\chi,t)}{\partial t}-\frac{1}{2}\frac{\partial^2 F(\chi,t)}{\partial \chi^2}-\frac{\partial F(\chi,t)}{\partial \chi}+2\frac{\partial}{\partial \chi}(\theta(\chi,t)F(\chi,t)) = \delta(\chi-\chi_0)\left\{T\frac{\partial F_f(t)}{\partial t}+F_f(t)\right\}, \tag{6.110}$$

$$\begin{aligned} T\frac{\partial G(\chi,t)}{\partial t}-\frac{1}{2}\frac{\partial^2 G(\chi,t)}{\partial \chi^2}-\frac{\partial G(\chi,t)}{\partial \chi}+2\frac{\partial}{\partial \chi}(\theta(\chi,t)G(\chi,t)) \\ +4\frac{\partial}{\partial \chi}[G(\chi,t)\epsilon(\chi,t)\theta(\chi,t)\{1-\theta(\chi,t)\}] \\ = \delta(\chi-\chi_0)\left[T\frac{\partial}{\partial t}\{F_f(t)N_f(t)\}+F_f(t)N_f(t)\right]. \end{aligned} \tag{6.111}$$

In these equations $F(\chi, t)$ is the continuous approximation to $F_i(t)$ with χ being the coordinate which represents the stage number. Similar definitions apply to the other variables. $F_f(t)$ is the feed rate at the initial stage χ_0 and $N_f(t)$ the enrichment. T is the hold-up time in each stage and the boundary conditions are $F(0,t) = F(l,t) = 0$ and $G(0,t) = G(l,t) = 0$, where $l = M+N+2$, with M the number of stages before the feed stage and N the number stages after.

Assuming that the fluctuation arises from deviations in θ and ϵ at the ith stage the problem is to calculate the fluctuations in the output flow rates at the kth stage. This is accomplished by setting $\theta(\chi,t) = \theta(\chi)+\delta\theta(\chi,t)$, $\epsilon(\chi,t) = \epsilon(\chi)+\delta\epsilon(\chi,t)$, $(\chi,t) = F(\chi)+\delta F(\chi,t)$ and $G(\chi,t) = G(\chi)+\delta G(\chi,t)$ and linearizing as described in the general discussion on Langevin equations given above. We shall not go through the various mathematical procedures, which may be found in the original reference, but simply note that using a calculational model, with the following parameters, a number of interesting conclusions can be drawn. The model parameters are:

1. Constant cut and separation factor over every stage.
2. Feed is natural uranium 0·714% enrichment.
3. The product is uranium enriched to 3·251%.

References

AKCASU, Z. (1961) Mean square instability in boiling reactors, *Nucl. Sci. Engng*, **10**, 337.

BALL, R. M. and BATCH, M. L. (1964) Measurement of noise in three pressurized water reactors, *Noise Analysis in Nuclear Systems*, AEC Symp. Ser. No. 4.

BOARDMAN, F. D. (1964) Noise measurements in the Dounreay fast reactor, *Noise Analysis in Nuclear Systems*, AEC Symp. Ser. No. 4.

BURGREEN, D. *et al.* (1958) Vibration of rods induced by water in parallel flow, *Trans. ASME*, July, p. 991.

CHANDRASEKHAR, S. (1943) Stochastic problems in physics and astronomy, *Rev. Mod. Phys.* **15**, 1.

COHN, C. E. (1960) A simplified theory of pile noise, *Nucl. Sci. Engng*, **7**, 472.

DE SHONG, J. A. (1958) Power transfer functions of the EBWR obtained using a sinusoidal reactivity driving function, USAEC, ANL 5798.

EL-WAKIL, M. M. (1971) *Nuclear Heat Transport*, International Textbook Co., Pennsylvania.

FORBES, S. G. *et al.* (1956) Instability in the SPERT-I reactor, preliminary report, USAEC, IDO-16309.

FRY, D. N. (1971) Experience in reactor malfunction diagnosis using on-line noise analysis, *Nuclear Technology*, **10**, 273.

GOERTZEL, G. and TRALLI, N. (1960) *Some Mathematical Methods of Physics*, McGraw-Hill, N.Y.

GREEF, C. (1971) A Study of Fluctuations in a Nuclear Reactor at Power, Ph.D. Thesis (Council for Nat. Acad. Awards).

HARRIS, D. R. (1958) Stochastic fluctuations in a power reactor, USAEC, WAPD-TM-190.

HARRIS, D. R. and PRESCOP, V. (1969) Stability and stationarity of a reactor as a stochastic process with feedback, *Nucl. Sci. Engng*, **37**, 171.

KIGUCHI, T. *et al.* (1973) Stochastic fluctuation in a uranium enriching cascade using the centrifuge process, *Nuclear Technology*, **17**, 168.

KOSÁLY, G. and MESKO, L. (1972) Remarks on the transfer function relating inlet temperature fluctuations to neutron noise, *Atomkernenergie*, **20**, 33.

KOSÁLY, G. and WILLIAMS, M. M. R. (1971) Point theory of the neutron noise induced by inlet temperature fluctuations and random mechanical vibrations, *Atomkernenergie*, **18**, 203.

MATTHES, W. (1962) Statistical fluctuations and their correlation in reactor neutron distributions, *Nukleonik*, **4**, 213.

MAXON, B. S. *et al.* (1959) Reactivity transients and steady state operation of a thoria–urania fueled direct cycle light water moderated reactor, ANL-5733.

RAJAGOPAL, V. (1964) Reactor noise measurements on Saxton reactor, *Noise Analysis in Nuclear Systems*, AEC Symp. Ser. No. 4.

RANDALL, R. L. and GRIFFIN, C. W. (1964) Application of power spectra to reactor system analysis, *Noise Analysis in Nuclear Systems*, AEC Symp. Ser. No. 4.

REAVIS, J. R. (1969) Vibration correlation for maximum fuel element displacement in parallel turbulent flow, *Nucl. Sci. Engng*, **38**, 63.

RICE, S. O. (1943) See WAX, N. (1954).

ROBINSON, J. C. (1967) Analysis of neutron fluctuation spectra in the Oak Ridge research reactor and the high flux isotope reactor, USAEC, ORNL-4149.

ROBINSON, J. C. (1970) Analytical determination of the neutron flux-to-pressure frequency response: application to M.S.R.E., *Nucl. Sci. Engng*, **42**, 382.

ROBSON, J. D. (1963) *Random Vibration*, Edinburgh University Press.

SAITO, K. (1974) On the theory of power reactor noise, Parts I, II, III, *Ann. Nucl. Sci. Engng* (formerly *J. Nucl. Energy*), **1**, Nos. 1, 2, 3.

SAXE, R. F. (1967) Acoustic characteristics of the Oak Ridge research reactor, *Neutron Noise, Waves and Pulse Propagation*, AEC Symp. Ser. No. 9.

SEIFRITZ, W. (1970) At power reactor noise induced by fluctuation in the coolant flow, *Atomkernenergie*, **16**, 29.

SHEFF, J. R. (1965) The cross-correlation of the neutron density fluctuations at two points in a nuclear reactor, Ph.D. Thesis (University of Washington).

STEPHENSON, S. E. *et al.* (1966) Neutron fluctuation spectra in the Oak Ridge research reactor, ORNL-TM-1401.

THIE, J. (1959) Dynamic behaviour of boiling reactors, USAEC, ANL-5849.

THIE, J. (1963) *Reactor Noise*, Rowman & Littlefield, N.Y.

WAMBSGANSS, M. W. (1967) Vibration of reactor core components, *Reactor and Fuel Processing Technology*, **10**, No. 3, 208.

WAX, N. (1954) *Selected Problems in Noise and Stochastic Processes*, Dover, N.Y.

WILLIAMS, M. M. R. (1971) The kinetic behaviour of simple neutronic systems with randomly fluctuating parameters, *J. Nucl. Energy*, **25**, 563.

YAMADA, S. and KAGE, H. (1967) *Reactor Noise Caused by Coolant Flow Fluctuation*, AEC Symp. Ser. No. 9.

CHAPTER 7

The Spatial Variation of Reactor Noise

7.1. *Introduction*

In our work thus far we have either neglected, or averaged out, the effect of spatial variations in the noise sources and their subsequent influence on the p.s.d. and related statistical measures. We did, however, at one point, viz. eqn. (4.25), admit the possibility of spatial considerations by writing down the basic transport equation for the generating function for zero-power reactor noise. We shall consider this equation in more detail below to investigate the corrections that must be made to account for velocity and spatial effects: these are straightforward if somewhat tedious.

It is in the study of power reactors that the space- and energy-dependent problem causes some difficulty. For example, we have already seen in Chapter 6 that power noise sources are difficult to assign and define even in a point-wise fashion. These problems assume much larger proportions when we also have to describe the spatial variation of the power noise sources. The probability balance method is undoubtedly not suitable for such a study and we are left with the Langevin technique, which at least enables the effect of noise sources to be assessed, even if it leaves open the question of their magnitude. The problem can best be fully appreciated if we consider a power reactor described by the simplest possible one-speed neutronics equations, viz.

$$\frac{1}{v}\frac{\partial\phi(\mathbf{r},t)}{\partial t} = \nabla . D(\mathbf{r},t)\nabla\phi(\mathbf{r},t)-\Sigma_a(\mathbf{r},t)\phi(\mathbf{r},t) +(1-\beta)\bar{\nu}\Sigma_f(\mathbf{r},t)\,\phi(\mathbf{r},t)+\sum_i \lambda_i C_i(\mathbf{r},t), \quad (7.1)$$

$$\frac{\partial C_i(\mathbf{r},t)}{\partial t} = -\lambda_i C_i(\mathbf{r},t)+\beta_i\bar{\nu}\Sigma_f(\mathbf{r},t)\phi(\mathbf{r},t). \quad (7.2)$$

Coupled to these equations are the heat-transfer and fluid flow equations. Now the cross-sections in the above equations have deliberately been made functions of space and time to emphasize their dependence, not only on the heterogeneous structure of the power reactor, but also on the time dependence of the mechanical properties. Two particular situations are examples of such a state, (1) boiling in the moderator of the b.w.r., (2) vibration of reactor components. In boiling, the moderator number density $N_m(\mathbf{r},t)$ is a random function of position and time described by the probability law which governs the statistical problem of bubble formation and collapse. Very little work has been performed on such laws which will depend on the nature of the heated surface, the operational pressure, geometrical considerations and a number of other factors of a thermodynamic nature. Vibrational effects involve mass motion of parts of the core and will generally be excited by the turbulent pressure fluctuations of the rapidly moving coolant. Thus, in principle, a knowledge of the turbulent noise is required followed by a calculation of its effect on the structure in question. In turn, each effect, and others not considered, are linked to reactor power, thereby completing the feedback loop and closing the system.

We can note, therefore, that in this very elementary case we would be faced with the solution of a set of non-linear partial differential equations with random parametric excitation. In turn the statistical properties of the parameters mentioned are unknown or difficult to calculate. In principle, however, it should be possible to obtain statistical information about $\phi(\mathbf{r}, t)$ if we know the driving forces or, alternatively, measurements of the statistics of $\phi(\mathbf{r}, t)$ could help us to understand the statistics of bubble formation and component vibration.

When faced with such a problem, the engineer will attempt to reduce its complexity by a series of rational approximations to a problem that he can solve and it is this practical approach that will be exploited below. In fact we have already studied one method of approach, namely the point model approximation, and have seen that, used properly, it can be a very powerful means of obtaining quantitative information about the system. Nevertheless, there are cases where a spatial description is unavoidable. Two methods have been adopted for this. The first is based upon homogenization of the core; the second is rather moıe sophisticated and takes into account the detailed spatial distribution of the fuel elements and other noise sources; it will be discussed in the following chapter.

7.2. *The generating function equation for zero-power noise*

As we have mentioned in the introduction, there exists a complete formalism for studying the detailed space, time and velocity dependence of the neutron statistics in zero-power systems (Pál, 1958, 1964; Bell, 1965). Other authors have also provided equivalent but less convenient formalisms (Osborne and Natelson, 1965; Matthes, 1966). An approach using the Langevin technique has been developed by Sheff and Albrecht (1966) but their method can only predict mean values and covariance functions. Moreover, whilst the method is sufficiently accurate for the understanding of space-dependent noise it does not deal with energy dependence and also, because of the use of the Schottky formula (Cohn, 1960; Davenport and Root, 1958), it is unable to calculate the noise-equivalent source exactly although the error involved is generally very small. In justification of their use of the Langevin technique for zero-power systems, Sheff and Albrecht suggested that the methods due to Pál and Bell are too complex for analytical study. In fact, as we shall see, this is not necessarily the case and these basic techniques form the cornerstone from which all zero-power noise studies should start.

We shall not go into the full derivation of the probability balance equation derived by Pál and Bell for

$$P_{N_1 N_2} \ldots (R_1, t_1; R_2, t_2; \ldots | \mathbf{r}, \mathbf{v}, t), \tag{7.3}$$

which is the probability that given one neutron at position $\mathbf{r}$ with velocity $\mathbf{v}$ at time t, there will result N_1 neutrons in region R_1 of phase time space at t_1, N_2 neutrons in R_2

at time t_2, etc. The construction of the equation for this quantity follows the same basic ideas as in point model probability balance problems with the exception that additional terms accounting for the time delay for neutrons of a particular velocity to travel from one point to another are included. Having obtained the balance equation for $P_{N_1 N_2} \ldots (\ldots)$, it is then converted to an equation for the generating function

$$G(Z_1, Z_2, \ldots | \mathbf{r}, \mathbf{v}, t) = \sum_{N_1=0}^{\infty} \sum_{N_2=0}^{\infty} \ldots Z_1^{N_1} Z_2^{N_2} \ldots P_{N_1 N_2} \ldots (\ldots) \tag{7.4}$$

from which any desired averages can be obtained by differentiation. For example,

$$\langle N(R, t_f | \mathbf{r}, \mathbf{v}, t) \rangle = \left. \frac{\partial G}{\partial Z_1} \right|_{Z_1 = Z_2 = \ldots = 1} \tag{7.5}$$

is the average number of neutrons in R at time t_f due to one neutron in $(\mathbf{r}, \mathbf{v})$ at time t. Similarly, we have

$$\langle N(R_1, t_1 | \mathbf{r}, \mathbf{v}, t) N(R_2, t_2 | \mathbf{r}, \mathbf{v}, t) \rangle = \left. \frac{\partial^2 G}{\partial Z_1 \partial Z_2} \right|_{Z_1 = Z_2 = \ldots = 1} + \delta_{R_1 R_2} \left. \frac{\partial G}{\partial Z_1} \right|_{Z_1 = Z_2 = \ldots = 1} \tag{7.6}$$

which is the cross-correlation function for neutrons in R_1 at t_1 and R_2 at t_2, given an initial neutron in $(\mathbf{r}, \mathbf{v})$ at t. $\delta_{R_1 R_2}$ is zero when the regions R_1 and R_2 do not overlap and unity when they are identical, in which case $\langle N_1 N_2 \rangle = \langle N^2 \rangle$.

It is important to note in this formalism that $\langle N \rangle$ is not the conventional neutron density as described by the Boltzmann equation. In fact, if we define $n(\mathbf{r}, \mathbf{v}, t \to \mathbf{r}_0, \mathbf{v}_0, t_0)$ as the Green's function of the Boltzmann equation, it is related to $\langle N \rangle$ as follows:

$$\langle N(R, t_0 | \mathbf{r}, \mathbf{v}, t) \rangle = \int_{(\mathbf{r}_0, \mathbf{v}_0) \in R} d\mathbf{r}_0 \int d\mathbf{v}_0 n(\mathbf{r}, \mathbf{v}, t \to \mathbf{r}_0, \mathbf{v}_0, t_0). \tag{7.7}$$

In a similar fashion one defines the doublet density $n_2(\mathbf{r}, \mathbf{v}, t \to \mathbf{r}_1, \mathbf{v}_1, t_1; \mathbf{r}_2, \mathbf{v}_2, t_2)$ as

$$\langle N(R_1, t_1 | \mathbf{r}, \mathbf{v}, t) N(R_2, t_2 | \mathbf{r}, \mathbf{v}, t) \rangle = \int_{(\mathbf{r}_1, \mathbf{v}_1) \in R_1} d\mathbf{r}_1 \int d\mathbf{v}_1 \int_{(\mathbf{r}_2, \mathbf{v}_2) \in R_2} d\mathbf{r}_2 \int d\mathbf{v}_2 n_2(\mathbf{r}, \mathbf{v}, t \to \mathbf{r}_1, \mathbf{v}_1, t_1; \mathbf{r}_2, \mathbf{v}_2, t_2). \tag{7.8}$$

We may calculate higher moments if necessary and they lead to a hierarchy of transport equations which may be solved recursively, the solution of one providing the source for the next. In practice, however, we are interested in the covariance $\langle N_1 N_2 \rangle - \langle N_1 \rangle \langle N_2 \rangle$ which we may obtain readily from the first and second moments defined above.

Before considering any specific problems, let us recast the equation for the generating function into a form which Bell has shown is more convenient for practical application. Indeed, we have already done this for the one-speed approximation, eqn. (4.25*b*), in Section 4.4. A new generating function $\mathscr{G} = 1 - G$ is defined and the equation is written as

$$L^{+}(\mathbf{r}, \mathbf{v}, t) \mathscr{G}(Z_1, Z_2, \ldots | \mathbf{r}, \mathbf{v}, t) = -\sum_{j=2}^{I} \frac{(-)^j}{j!} \overline{\nu(\nu-1) \ldots (\nu-j+1)} v \Sigma_f(\mathbf{r}, \mathbf{v}, t) \times \left[\int d\mathbf{v}' \chi(\mathbf{v}' \mathbf{r}, t) \mathscr{G}(Z_1, Z_2, \ldots | \mathbf{r}, \mathbf{v}, t) \right]^j, \tag{7.9}$$

where $$\overline{\nu(\nu-1)\dots(\nu-j+1)} = \sum_{\nu=j}^{I} \nu(\nu-1)\dots(\nu-j-1)P_\nu(\mathbf{r},\mathbf{v},t), \tag{7.10}$$

$P_\nu(\dots)$ being the probability that a fission at $(\mathbf{r},t)$ induced by a neutron of velocity $\mathbf{v}$ will emit ν neutrons.

L^+ is the Boltzmann equation adjoint operator defined by

$$L^+(\mathbf{r},\mathbf{v},t)g(\mathbf{r},\mathbf{v},t) = \left[-\frac{\partial}{\partial t} - \mathbf{v}.\nabla + \mathbf{v}\Sigma(\mathbf{r},\mathbf{v},t)\right] g(\mathbf{r},\mathbf{v},t)$$
$$-v\int d\mathbf{v}'\{\Sigma(\mathbf{v}\to\mathbf{v}';\mathbf{r},t)+\bar{\nu}\sum_f(\mathbf{r},\mathbf{v},t)\chi(\mathbf{v}',\mathbf{r},t)\}g(\mathbf{r},\mathbf{v}',t), \tag{7.11}$$

where $\Sigma(\mathbf{v}\to\mathbf{v}';\mathbf{r},t)$ is the differential scattering cross-section at $\mathbf{r}$ and t, and $\chi(\mathbf{v},\mathbf{r},t)$ is the fission spectrum (Williams, 1971). The cross-sections are made functions of space and time for generality and not because they themselves are random functions.

Finally, when a non-fission, Poisson source of strength $S(\mathbf{r},\mathbf{v},t)$ is present in the medium, the new value of the generating function, which we denote by $G(Z_1,Z_2,\dots|S)$, is related to $\mathscr{G}$ as follows (Bell, 1965):

$$G(Z_1,Z_2,\dots|S) = \exp\left\{-\int d\mathbf{r}\int d\mathbf{v}\int_{-\infty}^{t_1} dt\,\mathscr{G}(Z_1,Z_2,\dots|\mathbf{r},\mathbf{v},t)\,S(\mathbf{r},\mathbf{v},t)\right\}, \tag{7.12}$$

where $t_1 < t_2 < \dots$.

We have omitted the effects of delayed neutrons in the above equations for notational simplicity. The additional terms are given by Bell and by Pál. With this proviso, however, we can state that subject to the appropriate boundary conditions most zero-power noise problems can be solved via these equations. One of these, the extinction problem, has already been solved (see Chapter 4).

7.3. *The diffusion approximation to the generating function equation*

As in most reactor theory problems, it is useful to reduce the transport equation to diffusion theory. We can show that under the usual limitations (Weinberg and Wigner, 1958), eqn. (7.9) reduces to

$$-\frac{\partial\mathscr{G}_D}{\partial t} - v\nabla.D(\mathbf{r},v)\mathscr{G}_D + v\Sigma(\mathbf{r},v)\mathscr{G}_D - v\int dv'\{\Sigma_s(v\to v';\mathbf{r})+\bar{\nu}\Sigma_f(\mathbf{r},v)\chi(v'\mathbf{r})\}\mathscr{G}_D(\dots v')$$
$$= -\sum_{j=2}^{I}\frac{(-)^j}{j!}\overline{\nu(\nu-1)\dots(\nu-j+1)}v\Sigma_f(\mathbf{r},v)\left[\int dv'\chi(v',\mathbf{r})\mathscr{G}_D(\dots v')\right]^j, \tag{7.13}$$

where $\Sigma_s(\mathbf{v}\to\mathbf{v}';\mathbf{r},t) = \dfrac{1}{4\pi v'^2}\Sigma_s(v\to v',\mathbf{r})$ and $\chi(\mathbf{v},\mathbf{r},t) = \dfrac{1}{4\pi v^2}\chi(v,\mathbf{r})$

and we have assumed that cross-sections are not time-dependent and the obvious dependent variables have been suppressed to ease the notation. In this equation the new generating function $\mathscr{G}_D$ is related to $\mathscr{G}$ as follows:

$$\mathscr{G}_D(Z_1,Z_2,\dots|\mathbf{r},v,t) = \frac{1}{4\pi}\int d\Omega\,\mathscr{G}(Z_1,Z_2,\dots|\mathbf{r},\mathbf{v},t), \tag{7.14}$$

where $\mathbf{v} = v\mathbf{\Omega}$.

Similarly, if the source is isotropic and written as

$$S(\mathbf{r}, \mathbf{v}, t) = \frac{1}{4\pi v^2} S(\mathbf{r}, v, t) \tag{7.15}$$

we have for

$$G_D(Z_1, Z_2, \ldots | S) = \exp\left\{- \int d\mathbf{r} \int d\mathbf{v} \int_{-\infty}^{t_1} dt\, S(\mathbf{r}, v, t) \mathscr{G}_D(Z_1, Z_2, \ldots | \mathbf{r}, v, t)\right\}. \tag{7.16}$$

If account of the velocity must be made then multigroup methods would have to be employed to solve the resulting moment equations for $\langle N \rangle$ and $\langle N_1 N_2 \rangle$. A two-group calculation of this nature has been made by Natelson *et al.* (1966) using their equivalent formalism. Comparison with experiment indicates that in some practical situations space- and energy-dependent effects are very important.

A rather convenient reduction of the energy-dependent diffusion equation given above has been made by Williams (1967), who uses the idea of a slowing-down kernel (Weinberg and Wigner, 1958; Ferziger and Zweifel, 1966). In this way the complications arising from slowing down and diffusion are replaced by an auxiliary problem for $K(\mathbf{r}' \to \mathbf{r}; t' \to t) d\mathbf{r}\, dt$, which is the probability that a fast neutron born at time t' at $\mathbf{r}'$ will become thermal in $d\mathbf{r}$ about $\mathbf{r}$ in the time interval $(t, t+dt)$. The resulting calculation is then concerned only with the thermal neutrons which in practice are those most involved in noise measurements. The drawback of the method is that it prevents any knowledge of energy correlations from being obtained: however, this is generally not an important restriction since experiments involving energy correlations can be dealt with in another way as we shall see later.

Assuming, therefore, that the slowing-down kernel technique is applicable we note that by analogy with conventional methods (Ferziger and Zweifel, 1966) we neglect the scattering terms and replace the fission source term

$$\chi(v) \int_0^{\infty} dv'\, v' \Sigma_f(v') N(\mathbf{r}, v', t) \tag{7.17}$$

by

$$\int_{-\infty}^{t} dt' \int d\mathbf{r}'\, \overline{v\Sigma_f(\mathbf{r}')} K(\mathbf{r}' \to \mathbf{r}; t' \to t) N(\mathbf{r}', t') \tag{7.18}$$

(Krieger and Zweifel, 1959), where N refers to thermal neutrons only and $\overline{v\Sigma_f}$ is the thermally averaged fission reaction rate.

Now in eqn. (7.13) for the generating function $\mathscr{G}_D$ we note that the adjoint of (7.17) divided by $v\Sigma_f(v)$ appears, hence we replace it by the adjoint of (7.18) divided by $\overline{v\Sigma_f}$, viz.

$$\int_t^{\infty} dt' \int d\mathbf{r}'\, K(\mathbf{r} \to \mathbf{r}'; t \to t') N^{+}(\mathbf{r}', t'). \tag{7.19}$$

The new equation for the generating function $\overline{\mathscr{G}}_D$ for thermal neutrons may now be written

$$T^{+}\mathscr{G}_D(Z_1, Z_2, \ldots | \mathbf{r}, t) = - \sum_{j=2}^{I} \frac{(-)j}{j!} \overline{\nu(\nu-1)\ldots(\nu-j+1)}\, \overline{v\Sigma_f(\mathbf{r})}$$
$$\times \left[\int_t^{\infty} dt' \int d\mathbf{r}'\, K(\mathbf{r} \to \mathbf{r}'; t \to t') \mathscr{G}_D(Z_1, Z_2, \ldots | \mathbf{r}', t')\right]^j, \tag{7.20}$$

where

$$T^{+}g(\mathbf{r},t)$$

$$\equiv \left[-\frac{\partial}{\partial t}-\nabla . D\nabla+\overline{v\Sigma_a(\mathbf{r})}\right] g(\mathbf{r},t)-\overline{\bar{\nu} v\Sigma_f(\mathbf{r})}\int_t^{\infty} dt' \int d\mathbf{r}' K(\mathbf{r}\to\mathbf{r}'; t\to t')g(\mathbf{r}',t'). \tag{7.21}$$

We can interpret $\bar{G}_D = 1-\overline{\mathscr{G}_D}$ as the generating function of the distribution $P_{N_1N_2}\dots(R_1,t_1; R_2,t_2; \dots|\mathbf{r},t)$, which is the probability that N_1 *thermal* neutrons are in R_1 at t_1, N_2 in R_2 at t_2, etc., on the condition that there was one thermal neutron at $\mathbf{r}$ at time t.

The corresponding source generating function $\bar{G}_D$ is related to $\overline{\mathscr{G}_D}$ as follows:

$$\bar{G}_D(Z_1,Z_2,\dots|S) = \exp\left\{-\int d\mathbf{r}' \int_{-\infty}^{t_1} dt\, S_T(\mathbf{r}',t)\overline{\mathscr{G}_D}(Z_1,Z_2,\dots|\mathbf{r}',t)\right\}, \tag{7.22}$$

where

$$S_T(\mathbf{r},t) = \int_{-\infty}^{t} dt' \int d\mathbf{r}'\, S_F(\mathbf{r}',t')K_s(\mathbf{r}'\to\mathbf{r}; t'\to t) \tag{7.23}$$

is the thermal source arising from the fast source S_F, $K_s(\dots)$ being the slowing-down kernel for source neutrons.

7.4. *The covariance function*

With our general formalism we can now calculate the space-dependent correlation function or, as we shall call it, the covariance function.

To obtain the mean value we use eqn. (7.5) and find

$$T^{+}\langle N(R_1,t_1|\mathbf{r},t)\rangle = 0 \tag{7.24}$$

subject to the final condition

$$\begin{aligned}\langle N(R_1,t_1|\mathbf{r},t_1)\rangle &= 1 \quad \text{for } \mathbf{r}\in R_1\\ &= 0 \quad \text{for } \mathbf{r}\notin R_2.\end{aligned}$$

Similarly,

$$T^{+}\langle N(R_1,t_1|\mathbf{r},t)N(R_2,t_2|\mathbf{r},t)\rangle$$

$$= \overline{\nu(\nu-1)}\,\overline{v\Sigma_f(\mathbf{r})}\int_t^{\infty} dt' \int d\mathbf{r}' K(\mathbf{r}\to\mathbf{r}'; t\to t')\langle N(R_1,t_1|\mathbf{r}',t')\rangle$$

$$\times \int_t^{\infty} dt' \int d\mathbf{r}' K(\mathbf{r}\to\mathbf{r}'; t\to t')\langle N(R_2,t_2|\mathbf{r}',t')\rangle. \tag{7.25}$$

The final conditions we discuss below.

The solution of eqn. (7.24) may be written in more familiar notation as

$$\langle N(R_1,t_1|\mathbf{r},t)\rangle = \int_{\mathbf{r}_0\in R_1} d\mathbf{r}_0\, G(\mathbf{r},t\to\mathbf{r}_0,t_1),$$

G being the Green's function of the operator T^{+} (and not to be confused with the generating function).

Similarly, we may write the solution of eqn. (7.25) as

$$\begin{aligned}\langle N(R_1,t_1|\mathbf{r},t)N(R_2,t_2|\mathbf{r},t)\rangle &= \int \langle N(R_2,t_2|\mathbf{r}',t_1)\rangle G(\mathbf{r},t\to\mathbf{r}',t_1)\,d\mathbf{r}' \\ &+ \int_t^{t_1} dt_0 \int d\mathbf{r}_0\, G(\mathbf{r},t\to\mathbf{r}_0,t_0)\overline{\nu(\nu-1)}\,\overline{v\Sigma_f(\mathbf{r}_0)} \\ &\times \int_{t_0}^{\infty} dt' \int d\mathbf{r}'\, K(\mathbf{r}_0\to\mathbf{r}';\, t_0\to t')\langle N(R_1,t_1|\mathbf{r}',t')\rangle \\ &\times \int_{t_0}^{\infty} dt'' \int d\mathbf{r}''\, K(\mathbf{r}_0\to\mathbf{r}'';\, t_0\to t'')\langle N(R_2,t_2|\mathbf{r}'',t'')\rangle.\end{aligned} \tag{7.26}$$

The first term on the right-hand side arises from the final condition when $t_1 = t (t_1 < t_2)$, for in that case

$$\begin{aligned}\langle N(R_1,t_1|\mathbf{r},t)N(R_2,t_2|\mathbf{r},t)\rangle &= \int_{\mathbf{r}'\in R_1} d\mathbf{r}'\langle N(R_2,t_2|\mathbf{r}',t)\rangle\delta(\mathbf{r}-\mathbf{r}') \\ &= \langle N(R_2,t_2|\mathbf{r},t)\rangle \quad (\mathbf{r}\in R_1) \\ &= 0 \quad (\mathbf{r}\notin R_1).\end{aligned} \tag{7.27}$$

Since we will be interested from a practical point of view in the more general case when a source is present, we make use of eqn. (7.16) for $\overline{G_D}(\ldots S)$ to obtain

$$\begin{aligned}\langle N(R_1,t_1|S)\rangle &= \int_{-\infty}^{t_1} dt \int d\mathbf{r}\, S_T(\mathbf{r},t)\langle N(R_1,t_1|\mathbf{r},t)\rangle \\ &= \int_{\mathbf{r}_0\in R_1} d\mathbf{r}_0 \int_{-\infty}^{t_1} dt \int d\mathbf{r}\, S_T(\mathbf{r},t)G(\mathbf{r},t\to\mathbf{r}_0,t_1).\end{aligned} \tag{7.28}$$

But the integral of the source over the Green's function is simply the neutron density $f(\mathbf{r}_0,t)$, thus we have

$$\langle N(R_1,t_1|S)\rangle = \int_{\mathbf{r}_0\in R_1} d\mathbf{r}_0 f(\mathbf{r}_0,t_1). \tag{7.29}$$

The covariance, which we define as

$$\operatorname{covar}(N) = \langle N(R_1,t_1|S)N(R_2,t_2|S)\rangle - \langle N(R_1,t_1|S)\rangle\langle N(R_2,t_2|S)\rangle, \tag{7.30}$$

is obtained from $\overline{G_D}$ as

$$\operatorname{covar}(N) = \int_{-\infty}^{t_1} dt \int d\mathbf{r}\, S_T(\mathbf{r},t)\langle N(R_1,t_1|\mathbf{r},t)N(R_2,t_2|\mathbf{r},t)\rangle. \tag{7.31}$$

Inserting eqn. (7.26) and rearranging leads to

$$\begin{aligned}\operatorname{covar}(N) &= \int_{\mathbf{r}_1\in R_1} d\mathbf{r}_1 \int_{\mathbf{r}_2\in R_2} d\mathbf{r}_2 \Big\{ f(\mathbf{r}_1,t_1)G(\mathbf{r}_1,t_1\to\mathbf{r}_2,t_2) \\ &+ \int_{-\infty}^{t_1} dt_0 \int d\mathbf{r}_0\, \overline{\nu(\nu-1)}\,\overline{v\Sigma_f(\mathbf{r}_0)} f(\mathbf{r}_0,t_0) \\ &\times \int_{t_0}^{\infty} dt' \int d\mathbf{r}'\, K(\mathbf{r}_0\to\mathbf{r}';\, t_0\to t')G(\mathbf{r}',t'\to\mathbf{r}_1,t_1) \\ &\times \int_{t_0}^{\infty} dt'' \int d\mathbf{r}''\, K(\mathbf{r}_0\to\mathbf{r}'';\, t_0\to t'')G(\mathbf{r}'',t''\to\mathbf{r}_2,t_2)\Big\};\ t_1 \leqslant t_2.\end{aligned} \tag{7.32}$$

The neutron density correlation function Γ_N is defined in terms of the covariance as follows:

$$\text{covar}(N) = \int_{\mathbf{r}_1 \in R_1} d\mathbf{r}_1 \int_{\mathbf{r}_2 \in R_2} d\mathbf{r}_2\, \Gamma_N(\mathbf{r}_1, \mathbf{r}_2;\, t_1, t_2) \tag{7.33}$$

and is therefore given by the quantity in curly brackets in eqn. (7.32).

7.5. *The detection process*

As we noted when discussing the theory of the point model, the measured quantity must include the effect of the detector in the sense that it has a certain efficiency and response time and also that its presence removes a neutron by the act of detection. It is more correct therefore to consider the function $P_{C_1 C_2}(R_1, t_1;\, R_2, t_2 | \mathbf{r}, t)\, dt_1\, dt_2$ which denotes the probability that a neutron released at $\mathbf{r}$ at time t will lead to C_1 counts in the detector region R_1 between t_1 and $t_1 + dt_1$ and C_2 counts in the detector region R_2 between t_2 and $t_2 + dt_2$. The basic theory does not change very much, except that now we must consider $\langle C \rangle$ in place of $\langle N \rangle$ and $\langle C_1 C_2 \rangle$ in place of $\langle N_1 N_2 \rangle$. If we define $\tilde{\epsilon}(\mathbf{r}_0)$ as the probability per unit time for the detector to capture a neutron and record a count we can write immediately that

$$\langle C(R_1, t_1 | \mathbf{r}, t) \rangle = \int_{\mathbf{r}_0 \in R_1} d\mathbf{r}_0 \tilde{\epsilon}(\mathbf{r}_0)\, G(\mathbf{r}, t \to \mathbf{r}_0, t_1). \tag{7.34}$$

(N.B. In the point model scheme of Chapter 3 we have used λ_d for $\tilde{\epsilon}$.)

The calculation for covar(C) proceeds as for covar(N) and we find that

$$\begin{aligned}\text{covar}(C) = \int_{\mathbf{r}_1 \in R_1} d\mathbf{r}_1 \int_{\mathbf{r}_2 \in R_2} d\mathbf{r}_2\, \{&\tilde{\epsilon}(\mathbf{r}_1) f(\mathbf{r}_1, t_1)\, \delta(\mathbf{r}_1 - \mathbf{r}_2)\, \delta(t_1 - t_2) \\ &+ \tilde{\epsilon}(\mathbf{r}_1) \tilde{\epsilon}(\mathbf{r}_2) \tilde{\Gamma}_N(\mathbf{r}_1, \mathbf{r}_2;\, t_1, t_2)\},\end{aligned} \tag{7.35}$$

where $\tilde{\Gamma}_N$ denotes the second term in the curly brackets in eqn. (7.32) for covar(N). The first term arises from the fact that if a neutron is detected at $t = t_1$ it cannot contribute to the count rate at a later time t_2. $\Gamma_C(\mathbf{r}_1, \mathbf{r}_2;\, t_1, t_2)$ is defined by the quantity in curly brackets in eqn. (7.35).

A further quantity of experimental interest is the variance in the total number of counts recorded in a time t. Thus if $Z(t)$ is the total counts recorded by a detector in a time t, we have

$$Z(t) = \int_0^t C(t')\, dt'. \tag{7.36}$$

Then it may be shown (Pluta, 1962; Osborne, 1965) that if

$$\text{covar}(Z) = \int_{\mathbf{r}_1 \in R_1} d\mathbf{r}_1 \int_{\mathbf{r}_2 \in R_2} d\mathbf{r}_2\, \Gamma_Z(\mathbf{r}_1, \mathbf{r}_2;\, t_1, t_2), \tag{7.37}$$

$$\equiv \langle Z(R_1, t_1 | S) Z(R_2, t_2 | S) \rangle - \langle Z(R_1, t_1 | S) \rangle \langle Z(R_2, t_2 | S) \rangle, \tag{7.38}$$

Γ_Z is related to Γ_C by

$$\Gamma_Z(\mathbf{r}_1, \mathbf{r}_2;\, \tau) = 2 \int_0^\tau dt_2 \int_0^{t_2} dt_1 \Gamma_C(\mathbf{r}_1, \mathbf{r}_2;\, t_1), \tag{7.39}$$

where we have assumed stationary processes such that

$$\Gamma_C(\mathbf{r}_1, \mathbf{r}_2;\, t_1, t_2) = \Gamma_C(\mathbf{r}_1, \mathbf{r}_2;\, t_2 - t_1) \quad \text{and} \quad \Gamma_Z(\mathbf{r}_1, \mathbf{r}_2;\, t_1, t_2) = \Gamma_Z(\mathbf{r}_1, \mathbf{r}_2;\, t_2 - t_1).$$

7.6. *Delayed neutrons*

In most cases it is essential to include delayed neutrons into the theory. The necessary modifications to the above theory are tedious but straightforward and the basic change is to convert terms of the type

$$\int_{t_0}^{\infty} dt' \int d\mathbf{r}'\, K(\mathbf{r}_0 \to \mathbf{r}'; t_0 \to t')\, G(\mathbf{r}', t' \to \mathbf{r}_1, t_1) \tag{7.40}$$

first to the form of time-independent cross-sections, so that

$$G(\mathbf{r}', t' \to \mathbf{r}_1, t_1) \equiv G(\mathbf{r}' \to \mathbf{r}_1;\, t_1 - t') \quad \text{and} \quad K(\mathbf{r}_0' \to \mathbf{r}';\, t_0 \to t') \equiv K(\mathbf{r}_0' \to \mathbf{r}';\, t - t),$$

i.e.

$$\int_0^{\infty} dt_0' \int d\mathbf{r}' K(\mathbf{r}_0 \to \mathbf{r}';\, t_0')\, G(\mathbf{r}' \to \mathbf{r}_1;\, t_1 - t_0 - t_0'), \tag{7.41}$$

and then to

$$(1-\beta) \int_0^{\infty} dt_0' \int d\mathbf{r}'\, K(\mathbf{r}_0 \to \mathbf{r}';\, t_0')\, G(\mathbf{r}' \to \mathbf{r}_1;\, t_1 - t_0 - t_0')$$
$$+ \sum_{i=1}^{l} \beta_i \lambda_i \int_0^{t_1 - t_0} d\xi\, e^{-\lambda_i \xi} \int_0^{\infty} dt_0' \int d\mathbf{r}'\, K_i(\mathbf{r}_0 \to \mathbf{r}';\, t_0')\, G(\mathbf{r}' \to \mathbf{r}_1;\, t_1 - t_0 - \xi - t_0'). \tag{7.42}$$

$K_i(\ldots)$ is the slowing-down kernel for the ith group of delayed neutrons which, in general, is not the same as for prompt neutrons due to the differences in initial energies of the neutrons.

7.7. *The power spectral density*

The p.s.d. can be defined in analogy with the point model theory, viz.

$$\Phi_{12}(\omega) = \int_{-\infty}^{\infty} e^{-i\omega\tau}\, \Gamma_c(\mathbf{r}_1, \mathbf{r}_2;\, \tau)\, d\tau. \tag{7.43}$$

Applying the Fourier transform to Γ_c as defined by eqn. (7.35) leads to

$$\Phi_{12}(\omega) = \tilde{\epsilon}(\mathbf{r}_1)\, \delta(\mathbf{r}_1 - \mathbf{r}_2)\, f(\mathbf{r}_1) + \tilde{\epsilon}(\mathbf{r}_1)\, \tilde{\epsilon}(\mathbf{r}_2) \int d\mathbf{r}_0\, \overline{\nu(\nu-1)}\, \overline{v \Sigma_f(\mathbf{r}_0)}$$
$$\times\, f(\mathbf{r}_0)\, \bar{F}(\mathbf{r}_0 \to \mathbf{r}_1;\, -i\omega)\, \bar{F}(\mathbf{r}_0 \to \mathbf{r}_2;\, i\omega), \tag{7.44}$$

where

$$\bar{F}(\mathbf{r}_0 \to \mathbf{r};\, s) = \int d\mathbf{r}'\, \bar{K}(\mathbf{r}_0 \to \mathbf{r}';\, s)\, \bar{G}(\mathbf{r}' \to \mathbf{r};\, s), \tag{7.45}$$

$\bar{K}$ and $\bar{G}$ denoting Laplace transforms of K and G, respectively.

We note that, except when $\mathbf{r}_1 = \mathbf{r}_2$, the p.s.d. is complex, and therefore a phase angle may be measured as well as the frequency spectrum of the fluctuations.

When delayed neutrons are included, the term $\bar{F}(\ldots; s)$ is replaced by

$$(1-\beta) \int d\mathbf{r}'\, \bar{K}(\mathbf{r}_0 \to \mathbf{r}';\, s)\, G(\mathbf{r}' \to \mathbf{r};\, s) + \sum_{i=1}^{l} \frac{\beta_i \lambda_i}{s + \lambda_i} \int d\mathbf{r}'\, \bar{K}_i(\mathbf{r}_0 \to \mathbf{r}';\, s)\, G(\mathbf{r}' \to \mathbf{r};\, s). \tag{7.46}$$

7.8. *Fluctuations in an infinite medium*

The case of an infinite medium is a useful one to study from the point of view of spatial correlation since it removes the problem of leakage and allows the effect on the p.s.d. of detector shape and spacing to be studied.

A basic assumption to be made in the ensuing work is that the detector itself does not perturb the neutron flux. This is, of course, untrue and we have discussed its limitations in Section 3.12; we shall have more to say about this effect below. For the moment we assume that the neutron density $f(\mathbf{r})$ is a constant and equal to unity. In addition, because the medium is homogeneous, we can write the slowing-down kernel and Green's function as displacement functions, viz. $G(|\mathbf{r}-\mathbf{r}'|; t)$ and $K(|\mathbf{r}-\mathbf{r}'|; t)$.

Defining the Fourier transforms $\bar{P}$ and $\bar{G}$

$$K(|\mathbf{r}|; t) = \left(\frac{1}{2\pi}\right)^3 \int d\mathbf{k}\, \bar{P}(\mathbf{k}; t) e^{i\mathbf{k}\cdot\mathbf{r}} \tag{7.47}$$

and

$$G(|\mathbf{r}|; t) = \left(\frac{1}{2\pi}\right)^3 \int d\mathbf{k}\, \bar{G}(\mathbf{k}; t) e^{i\mathbf{k}\cdot\mathbf{r}} \tag{7.48}$$

and substituting into eqn. (7.35), we find that

$$\Gamma_c(\mathbf{r}_1, \mathbf{r}_2; \tau) = \tilde{\epsilon}(\mathbf{r}_1)\delta(\mathbf{r}_1-\mathbf{r}_2)\delta(\tau) + \tilde{\epsilon}(\mathbf{r}_1)\tilde{\epsilon}(\mathbf{r}_2)\overline{\nu(\nu-1)}\,\overline{v\Sigma_f}\, H(|\mathbf{r}_1-\mathbf{r}_2|; \tau), \tag{7.49}$$

where

$$H(|\mathbf{y}|; \tau) = \left(\frac{1}{2\pi}\right)^3 \int_0^\infty d\tau_0 \int d\mathbf{k} \int_0^\infty dt_0'\, \bar{P}(-\mathbf{k}; t_0')\bar{G}(-\mathbf{k}; \tau_0 - t_0')$$
$$\times \int_0^\infty dt_0''\, \bar{P}(\mathbf{k}; t_0'')\bar{G}(\mathbf{k}; \tau+\tau_0-t_0'') e^{i\mathbf{k}\cdot\mathbf{y}} \tag{7.50}$$

and when delayed neutrons are important the term

$$\int_0^\infty dt_0'\, \bar{P}(-\mathbf{k}; t_0')\bar{G}(-\mathbf{k}; \tau_0 - t_0') \tag{7.51}$$

is replaced by

$$(1-\beta)\int_0^\infty dt_0'\, \bar{P}(-\mathbf{k}; t_0')\bar{G}(-\mathbf{k}; \tau_0-t_0')$$
$$+\sum_{i=1}^{l} \beta_i \lambda_i \int_0^{\tau_0} d\xi\, e^{-\lambda_i \xi} \int_0^\infty dt_0'\, \bar{P}_i(-\mathbf{k}; t_0')\bar{G}(-\mathbf{k}; \tau_0-\xi-t_0'), \tag{7.52}$$

where the Green's function $G(\mathbf{r}_0 \to \mathbf{r}; t)$ is obtained from

$$\left[\frac{\partial}{\partial t} - D_0\nabla^2 - \overline{v\Sigma_a}\right] G(\mathbf{r}_0 \to \mathbf{r}; t)$$
$$= (1-\beta)\bar{\nu}\overline{v\Sigma_f}\int_0^\infty dt_0' \int d\mathbf{r}'\, K(|\mathbf{r}-\mathbf{r}'|; t_0') G(\mathbf{r}_0 \to \mathbf{r}'; t-t_0')$$
$$+\bar{\nu}\overline{v\Sigma_f}\sum_{i=1}^{l}\lambda_i\beta_i \int_0^t d\xi\, e^{-\lambda_i\xi} \int_0^\infty dt_0' \int d\mathbf{r}'\, K_i(|\mathbf{r}-\mathbf{r}'|; t_0')$$
$$\times G(\mathbf{r}_0 \to \mathbf{r}'; t-\xi-t_0') + \delta(\mathbf{r}-\mathbf{r}_0)\delta(t). \tag{7.53}$$

The corresponding value for the p.s.d. is readily found to be

$$\Phi_{12}(\omega) = \tilde{\epsilon}(\mathbf{r}_1)\delta(\mathbf{r}_1-\mathbf{r}_2)+\tilde{\epsilon}(\mathbf{r}_1)\tilde{\epsilon}(\mathbf{r}_2)\overline{\nu(\nu-1)}\,\overline{v\Sigma_f}\bar{H}(|\mathbf{r}_1-\mathbf{r}_2|\,;\,\omega) \tag{7.54}$$

with

$$\bar{H}(|\mathbf{y}|\,;\,\omega) = \left(\frac{1}{2\pi}\right)^3 \int d\mathbf{k}\, e^{i\mathbf{k}\cdot\mathbf{y}}\, |Q(\mathbf{k};\, i\omega)|^2 \tag{7.55}$$

where, including delayed neutrons,

$$Q(\mathbf{k};\, i\omega) = \left[(1-\beta)\bar{\bar{P}}(\mathbf{k};\, i\omega)+\sum_{j=1}^{l} \frac{\beta_j\lambda_j}{i\omega+\lambda_j}\,\bar{\bar{P}}_j(\mathbf{k};\, i\omega)\right] G(\mathbf{k};\, i\omega), \tag{7.56}$$

where $\bar{\bar{P}}$ and $\bar{\bar{G}}$ are Laplace transforms with respect to time of $\bar{P}$ and $\bar{G}$ with $s = i\omega$.

These expressions are clearly rather complicated and bear little relation to the point model approximations which one might expect to simulate, in some respects, an infinite medium. The crux of the matter lies in the detector shape and the assumption of constant flux.

7.9. *Detector shape effects*

We shall concentrate here on p.s.d. functions as opposed to correlation functions simply for convenience. Thus if we consider the final form of p.s.d. that is actually measured by a detector we find that it is given by

$$\Phi_{V_1V_2}(\omega) = \int_{V_1} d\mathbf{r}_1 \int_{V_2} d\mathbf{r}_2 \Phi_{12}(\omega), \tag{7.57}$$

where the integrations imply an average over the detector volumes V_1 and V_2.

Inserting (7.54) into (7.57) we find that

$$\Phi_{V_1V_2}(\omega) = \tilde{\epsilon}V_1\Delta(V_1, V_2)+\overline{\nu(\nu-1)}\,\overline{v\Sigma_f}\left(\frac{1}{2\pi}\right)^3 \int d\mathbf{k}|\,Q(\mathbf{k};\, i\omega)|^2 g(\mathbf{k}, V_1, V_2), \tag{7.58}$$

where

$$g(\mathbf{k}, V_1, V_2) = \int_{V_1} d\mathbf{r}_1 \int_{V_2} d\mathbf{r}_2 \tilde{\epsilon}(\mathbf{r}_1)\tilde{\epsilon}(\mathbf{r}_2)\, e^{i\mathbf{k}\cdot(\mathbf{r}_1-\mathbf{r}_2)}. \tag{7.59}$$

$\Delta(V_1, V_2)$ is unity for a single detector and zero for two non-overlapping detectors. $g(\ldots)$ is a geometrical factor which depends on the shape and sensitivity of the detector.

As an example, consider the case of an infinite detector for which

$$g(\mathbf{k}) = (2\pi)^3\tilde{\epsilon}^2\delta(k_x)\delta(k_y)\delta(k_z) \tag{7.60}$$

and hence

$$\begin{aligned}\Phi_\infty(\omega) &= \lim_{V\to\infty} \frac{1}{V}\Phi_{VV}(\omega)\\ &= \tilde{\epsilon}+\tilde{\epsilon}^2\overline{\nu(\nu-1)}\,\overline{v\Sigma_f}|\,Q(0;\, i\omega)|^2.\end{aligned} \tag{7.61}$$

For one-speed theory, both $\bar{\bar{P}}$ and $\bar{\bar{P}}_i$ are unity; similarly from the equation for the Green's function

$$\bar{\bar{G}}(\mathbf{k};\, s) = \{D_0k^2-s+\overline{v\Sigma_a}-\bar{\nu}\overline{v\Sigma_f}\Theta(s)\}^{-1}, \tag{7.62}$$

where

$$\Theta(s) = 1-\sum_{i=1}^{l} \frac{s\beta_i}{s+\lambda_i}. \tag{7.63}$$

Thus
$$\Phi_\infty(\omega) = \tilde{\epsilon}+\tilde{\epsilon}^2\overline{\nu(\nu-1)}\,\overline{v\Sigma_f}\,\frac{|\Theta(i\omega)|^2}{|i\omega+\overline{v\Sigma_a}-\bar{\nu}\overline{v\Sigma_f}\Theta(i\omega)|^2}, \tag{7.64}$$

which is in complete agreement with the exact result based on point model probability balance. We can note, however, that in practice we are interested in frequencies such that $\omega \gg \lambda_i$, therefore $\Theta(s) \simeq 1-\beta$. Thus eqn. (7.64) reduces to the form

$$\Phi_\infty(\omega) = \tilde{\epsilon}+\frac{\tilde{\epsilon}^2\overline{\nu(\nu-1)}\,\overline{v\Sigma_f}}{\omega^2+\omega_0^2} \tag{7.65}$$

over the interesting range. In this expression

$$\omega_0 = \frac{1-k_\infty(1-\beta)}{l}, \quad \text{with} \quad l = 1/\overline{v\Sigma_a},$$

or $\omega_0 = (\rho-\beta)/\Lambda \equiv \alpha_1$ of eqns. (3.30) and (3.49).

We see therefore that an infinite detector in an infinite medium corresponds to a point model analysis from the onset.

Other detector shapes lead to different values of the auto-p.s.d. as discussed in Section 3.12. Details may be found in Williams (1967) and also in Natelson *et al.* (1966) concerning these calculations, and also reasons why the wavelength and attenuation length of the neutron distribution which arises in the neutron wave experiment are important parameters. Further work on this problem using one-speed transport theory has been performed by Cassell and Williams (1969) who have obtained the transport theory analogue of eqn. (7.58) and have pointed out its limitations. See also Saito and Otsuka (1965*a*, 1965*b*, 1966).

7.10. *Finite medium corrections*

An infinite medium as discussed above can be misleading as regards the importance of detector shape on the noise spectrum. To show this we consider a bare reactor of finite size which may be described by asymptotic reactor theory (Weinberg and Wigner, 1958; Ferziger and Zweifel, 1966). It will be recalled that asymptotic reactor theory enables the displacement kernel in eqn. (7.53) to be retained and enables the Green's function to be written in the form

$$G(\mathbf{r}_0 \to \mathbf{r}; t) = \sum_\mu a_\mu(t)\Psi_\mu(\mathbf{r})\Psi_\mu(\mathbf{r}_0), \tag{7.66}$$

where $\Psi_\mu(\mathbf{r})$ are the eigenfunctions of the Helmholtz equation ($\mu = l,m,n$):

$$\nabla^2\Psi_\mu(\mathbf{r})+B_\mu^2\Psi_\mu(\mathbf{r}) = 0 \tag{7.67}$$

subject to $\Psi_\mu(\mathbf{r})$ being zero on the extrapolated reactor boundary. B_μ^2 are the corresponding eigenvalues with B_0^2 the geometric buckling. The $\Psi_\mu(\mathbf{r})$ also form a complete set and are orthonormal, viz.

$$\int_{V_R}\Psi_\mu(\mathbf{r})\Psi_{\mu'}(\mathbf{r})\,d\mathbf{r} = \delta_{\mu\mu'}, \tag{7.68}$$

V_R being the volume enclosed by the extrapolated surface.

To proceed further we note from asymptotic reactor theory that

$$\int_{V_R} d\mathbf{r}'\,\Psi_\mu(\mathbf{r}')K(|\mathbf{r}-\mathbf{r}'|;t) = \bar{P}(B_\mu^2;t)\Psi_\mu(\mathbf{r}) \tag{7.69}$$

and

$$\int_{V_R} d\mathbf{r}\int_{V_R} d\mathbf{r}'\,\Psi_\mu(\mathbf{r})\Psi_{\mu'}(\mathbf{r}')K(|\mathbf{r}-\mathbf{r}'|;t) = \bar{P}(B_\mu^2;t)\delta_{\mu\mu'}. \tag{7.70}$$

Using these facts in the defining equation of the Green's function leads to the Laplace transform of $a_\mu(t)$:

$$\overline{a_\mu}(s) = \left[s+\frac{1}{l_\mu}-\bar{\nu}v\overline{\Sigma}_f(1-\beta)\bar{\bar{P}}(B_\mu^2;s)-\bar{\nu}v\overline{\Sigma}_f\sum_{j=1}^{l}\frac{\lambda_j\beta_j}{s+\lambda_j}\bar{\bar{P}}_i(B_\mu^2;s)\right]^{-1}, \tag{7.71}$$

where $\quad l_\mu = l_a/(1+L^2B_\mu^2),\quad l_a = (v\overline{\Sigma}_a)^{-1}\quad$ and $\quad L^2 = D_0 l_a.$

Thus from eqn. (7.44) with $\bar{F}$ given by (7.46) we obtain

$$\Phi_{12}(\omega) = \tilde{\epsilon}(\mathbf{r}_1)\delta(\mathbf{r}_1-\mathbf{r}_2)f(\mathbf{r}_1)$$
$$+\tilde{\epsilon}(\mathbf{r}_1)\Phi\tilde{\epsilon}(\mathbf{r}_2)\overline{\nu(\nu-1)}v\overline{\Sigma}_f\sum_{\mu\mu'}F_{\mu\mu'}H_\mu(-i\omega)H_{\mu'}(i\omega)\Psi_\mu(\mathbf{r}_1)\Psi_{\mu'}(\mathbf{r}_2), \tag{7.72}$$

where

$$H_\mu(s) = \frac{\Theta_\mu(s)}{s+\dfrac{1}{l_\mu}-\bar{\nu}v\overline{\Sigma}_f\Theta_\mu(s)}, \tag{7.73}$$

$$\Theta_\mu(s) = (1-\beta)\bar{\bar{P}}(B_\mu^2;s)+\sum_{j=1}^{l}\frac{\beta_j\lambda_j}{s+\lambda_j}\bar{\bar{P}}_j(B_\mu^2;s) \tag{7.74}$$

and

$$F_{\mu\mu'} = \int_{V_R} d\mathbf{r}_0 f(\mathbf{r}_0)\Psi_\mu(\mathbf{r}_0)\Psi_{\mu'}(\mathbf{r}_0).$$

If we now average over detector volumes V_1 and V_2, we find that

$$\Phi_{V_1V_2}(\omega) = \Delta(V_1,V_2)\int_{V_1}\tilde{\epsilon}(\mathbf{r}_1)f(\mathbf{r}_1)\,d\mathbf{r}_1$$
$$+\overline{\nu(\nu-1)}v\Sigma_f\sum_{\mu\mu'}F_{\mu\mu'}g_{\mu\mu'}H_\mu(-i\omega)H_{\mu'}(i\omega), \tag{7.75}$$

where

$$g_{\mu\mu'} = \int_{V_1} d\mathbf{r}_1\int_{V_2} d\mathbf{r}_2\,\tilde{\epsilon}(\mathbf{r}_1)\tilde{\epsilon}(\mathbf{r}_2)\Psi_\mu(\mathbf{r}_1)\Psi_{\mu'}(\mathbf{r}_2) \tag{7.76}$$

is a geometrical factor depending on detector shape. It is the finite medium analogue of $g(\mathbf{k},V_1,V_2)$.

Let us now attempt to simplify eqn. (7.75) by invoking some rational physical approximations. First, we note that in zero-power noise theory the p.s.d. is measured over a limited frequency range, usually well within the limits $l_s^{-1} \gg \omega \gg (\lambda_i)_{\max}$, where l_s is the slowing-down time ($l_s^{-1} \simeq 10^5\ \text{sec}^{-1}$) and $(\lambda_i)_{\max}$ is the largest of the delayed neutron precursor decay constants ($\sim 3\ \text{sec}^{-1}$, but the weighted one-group value is $0{\cdot}08\ \text{sec}^{-1}$). On the basis of instantaneous slowing down we can set $l_s = 0$ and hence approximate $\bar{\bar{P}}(B_\mu^2;s)$ by $\bar{\bar{P}}(B_\mu^2;0)$. Similarly, for $\omega \gg \lambda_i$, we can neglect the delayed neutron sum in $\Theta_\mu(s)$ and set it equal to $(1-\beta)\bar{\bar{P}}(B_\mu^2;0)$.

Thus $H_\mu(s)$ may be approximated as follows:

$$H_\mu(s) \simeq \frac{\bar{\bar{P}}(B_\mu^2;0)}{s+\dfrac{1}{l_\mu}-\dfrac{k_\infty}{l_a}(1-\beta)\bar{\bar{P}}(B_\mu^2;0)}, \tag{7.77}$$

where $k_\infty = \overline{\nu v \Sigma_f}/\overline{v\Sigma_a}$. Note that it is permissible to neglect β with respect to unity in the numerator of (7.77) but not in the denominator. Because of the frequency range of interest it is useful to note that the equation for the Green's function (7.53) may be solved by neglecting the summation term and simply accounting for delayed neutrons by the factor $(1-\beta)$: this accounts for the reduction in the number of prompt neutrons but neglects the time constants of the associated delayed neutrons which are not of relevance to the final result. We shall use this idea occasionally in later work.

Returning to the basic problem and defining break frequencies ω_μ by

$$\omega_\mu = \frac{1}{l_\mu} - \frac{k_\infty}{l_a}(1-\beta)\bar{\bar{P}}(B_\mu^2; 0) \tag{7.78}$$

we may write eqn. (7.75) as follows:

$$\Phi_{V_1 V_2}(\omega) = \Delta(V_1, V_2)\int_{V_1} \tilde{\epsilon}(\mathbf{r}_1) f(\mathbf{r}_1)\, d\mathbf{r}_1 + \overline{\nu(\nu-1)}\,\overline{v\Sigma_f} \sum_{\mu\mu'} F_{\mu\mu'} g_{\mu\mu'} \frac{\bar{\bar{P}}(B_\mu^2; 0)\bar{\bar{P}}(B_{\mu'}^2; 0)}{\omega^2 + \omega_\mu \omega_{\mu'} + i\omega(\omega_\mu - \omega_{\mu'})}. \tag{7.79}$$

The real part of this equation gives the gain, and the imaginary part the phase angle. It should be noted that this expression is much closer in structural form to the point model approximation than is the corresponding infinite medium model with point, plate or line detectors.

We can obtain qualitative information about eqn. (7.79) by consideration of some limiting cases. First, we note that if the detectors have the same volume as the extrapolated reactor $g_{\mu\mu'} = \tilde{\epsilon}^2 \delta_{\mu\mu'}$ and

$$\Phi_{VV}(\omega) = \tilde{\epsilon}\int_{VR} f(\mathbf{r})\, d\mathbf{r} + \overline{\nu(\nu-1)}\,\overline{v\Sigma_f}\,\tilde{\epsilon}^2 \sum_{\mu} \frac{F_{\mu\mu}\bar{\bar{P}}(B_\mu^2; 0)^2}{\omega^2 + \omega_\mu^2}. \tag{7.80}$$

This is not in the point model form, but if the $F_{\mu\mu}$, $(\mu > 0)$, decrease rapidly, as is thought likely, we may write with some confidence

$$\Phi_{\nu\nu}(\omega) = \tilde{\epsilon}\int_{VR} f(\mathbf{r})\, d\mathbf{r} + \frac{\tilde{\epsilon}^2\, \overline{\nu(\nu-1)}\,\overline{v\Sigma_f}\, F_{00}\bar{\bar{P}}(B_0^2; 0)^2}{\omega^2 + \omega_0^2}, \tag{7.81}$$

where ω_0 is now the dominant break frequency. From eqn. (7.78) we may write

$$\omega_0 = \frac{1}{l_0} - \frac{k_\infty}{l_a}(1-\beta)\bar{\bar{P}}(B_0^2; 0). \tag{7.82}$$

Now with $l_0 = l_a P_T(B_0^2)$, with $P_T \equiv (1 + L^2 B_0^2)^{-1}$, this may be rearranged to give

$$\omega_0 = \frac{\beta - \rho}{\Lambda_{\text{eff}}}, \tag{7.83}$$

where $\Lambda_{\text{eff}} = l_0/k_{\text{eff}}$, $k_{\text{eff}} = k_\infty \bar{\bar{P}} P_T$. Thus we have the same functional form as in the point model but with modified parameters.

If the detectors do not fill the reactor, as is usually the case, we may consider them as points such that $\tilde{\epsilon}(\mathbf{r}) = \tilde{\epsilon} V_d \delta(\mathbf{r} - \mathbf{r}_0)$, V_d being the detector volume. Thus for a single detector at $\mathbf{r}_0$,

$$g_{\mu\mu'} = \tilde{\epsilon}^2 V_d^2 \Psi_\mu(\mathbf{r}_0)\Psi_{\mu'}(\mathbf{r}_0) \tag{7.84}$$

and $\Phi_{V_1 V_2}(\omega)$ becomes

$$\Phi_{V_1 V_2}(\omega) = \tilde{\epsilon} V_d f(\mathbf{r}_0) + \overline{\nu(\nu-1)}\, \overline{v\Sigma_f}\, \tilde{\epsilon}^2 V_d^2 \sum_{\mu\mu'} \frac{F_{\mu\mu'}(B_\mu^2; 0)\bar{\bar{P}}(B_{\mu'}^2; 0)\Psi_\mu(\mathbf{r}_0)\Psi_{\mu'}(\mathbf{r}_0)}{\omega^2 + \omega_\mu \omega_{\mu'} + i\omega(\omega_\mu - \omega_{\mu'})}. \quad (7.85)$$

For some systems, water moderated in particular, numerical calculation indicates that the values of ω_μ increase rapidly with μ and that ω_0 is well separated from the higher ω_μ. In such a case we may neglect ω with respect to ω_μ, except for ω_0, and write eqn. (7.79) as

$$\Phi_{V_1 V_2}(\omega) = \Delta(V_1, V_2) \int_{V_1} \tilde{\epsilon}(\mathbf{r}_1) f(\mathbf{r}_1)\, d\mathbf{r}_1 + \frac{\overline{\nu(\nu-1)}\, \overline{v\Sigma_f}\, g_{00} F_{00} \bar{\bar{P}}^2(B_0^2; 0)}{\omega^2 + \omega_0^2}$$

$$+ \text{higher-order terms.} \quad (7.86)$$

For a single detector of arbitrary shape with a spatially uniform value of $\tilde{\epsilon}$ within it we can further write (7.86) as

$$\Phi_V(\omega) = \tilde{\epsilon} \int_{V_d} f(\mathbf{r})\, d\mathbf{r} + \frac{\overline{\nu(\nu-1)}\, \overline{v\Sigma_f}\, \tilde{\epsilon}^2 \left[\int_{V_d} \Psi_0(\mathbf{r})\, d\mathbf{r}\right]^2 \int_{V_R} f(\mathbf{r}) \Psi_0^{*2}(\mathbf{r})\, d\mathbf{r}\, \bar{\bar{P}}^2(B_0^2; 0)}{\omega^2 + \omega_0^2}$$

$$+ \text{higher-order terms.} \quad (7.87)$$

This approximation, however, is not always acceptable and as Natelson *et al.* (1966) have shown, systems that are large compared with a migration length will require several terms in the sum to represent accurately the experimental data.

Finally, we note that the method of expanding in the eigenfunctions of the Green's function may be used for inhomogeneous systems with reflectors and various zones. However, convergence of the sum will be less rapid and the results more difficult to interpret in terms of single break frequencies since these may depend on position. A series of papers by Sheff, Johnson and Macdonald and Natelson *et al.* in the 1966 Florida Conference Proceedings (Uhrig, 1967) discusses in considerable numerical detail the effects of space and energy dependence on the p.s.d. In the case of the paper by Natelson *et al.* it is shown that the experimental measurements of Ricker *et al.* (1965) can only be explained on the basis of a space and energy-dependent model. Figure 7.1 illustrates this effect by showing how neglect of the non-leakage probability affects the calculated p.s.d. Also it should be noted that the reactor under consideration was small enough for only a single term of the sum in eqn. (7.79) to be needed; however, neglecting $\bar{\bar{P}}$ requires sixty-four terms for adequate convergence of the series (albeit to an incorrect value).

7.11. *Variance to mean method*

As we showed in Section 7.5, the covariance of Z, the correlated number of counts in a given time in two detectors at different space points, can be obtained by the use of eqn. (7.39). We can make use of that result from the value of Γ_c given by eqn. (7.35). Let us consider simply the finite medium case of the previous section. To obtain the exact space- and energy-dependent analogue of eqn. (3.6) we should invert the transform $\Phi_{12}(\omega)$ back to the time domain. Whilst this can be done, it is algebraically tedious and

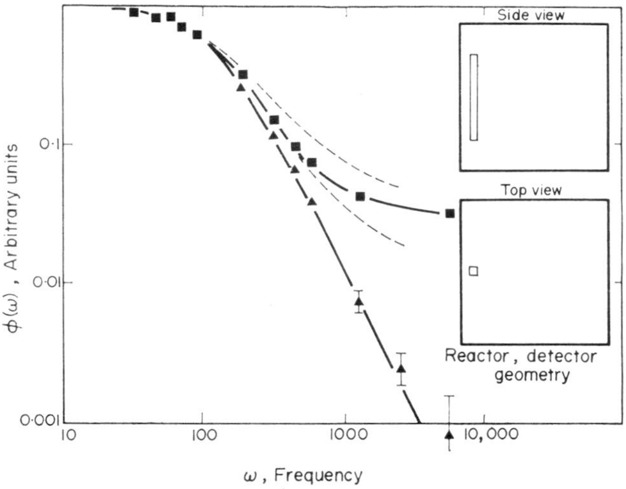

FIG. 7.1. Measured and calculated p.s.d. for a small water-moderated reactor near critical ($\beta/\Lambda = 117 \text{ sec}^{-1}$, $M = 7{\cdot}06$ cm, $L = 2{\cdot}88$ cm). The solid squares are the values measured by Ricker *et al.* (1965). The full line upper curve is calculated from eqn. (7.87) with the integration over detector geometry chosen to represent the experimental situation, viz. a $2 \times 2 \times 30$ cm rectangular parallelepiped detector placed as shown in the sketch. The slowing-down model used assumed that $\bar{\bar{P}}(B^2; 0) = (1+M^2B^2)^{-1}$. The dashed line upper curve is the result obtained from setting $\bar{\bar{P}} = 1$, i.e. neglecting the slowing down. The lower full line in the figure, passing through the solid triangles, is obtained from eqn. (7.87) by neglecting higher terms in (7.87) and subtracting the constant first term. The corresponding dashed line above it results from setting $\bar{\bar{P}}(B^2; 0) = 1$. (After Natelson *et al.* (1966) *J. Nucl. Energy*, **20**, 557.)

a better understanding of the problem can be obtained by consideration of the case described by eqn. (3.18) which corresponds to the approximation involved in eqn. (7.77) for $H_\mu(s)$. Thus we find that the correlation function $\Gamma_c(\mathbf{r}_1, \mathbf{r}_2; \tau)$ is given by

$$\Gamma_c(\mathbf{r}_1, \mathbf{r}_2; \tau) = \frac{1}{2\pi}\int_{-\infty}^{\infty} d\omega\, e^{i\omega\tau}\,\Phi_{12}(\omega) \tag{7.88}$$

which leads to

$$\begin{aligned}\Gamma_c(\mathbf{r}_1, \mathbf{r}_2; \tau) &= \tilde{\epsilon}(\mathbf{r}_1)\delta(\mathbf{r}_1-\mathbf{r}_2)f(\mathbf{r}_1)\delta(\tau)\\ &\quad + \tilde{\epsilon}(\mathbf{r}_1)\tilde{\epsilon}(\mathbf{r}_2)\overline{\nu(\nu-1)}\,\overline{v\Sigma_f}\sum_{\mu\mu'}\frac{F_{\mu\mu'}\Psi_\mu(\mathbf{r}_1)\Psi_{\mu'}(\mathbf{r}_2)}{(\omega_\mu-\omega_{\mu'})}\bar{\bar{P}}(B_\mu^2; 0)\bar{\bar{P}}(B_{\mu'}^2; 0)\\ &\quad \times \begin{pmatrix} e^{-\omega_\mu\tau}; & \tau > 0 \\ e^{\omega_\mu\tau}; & \tau < 0, \end{pmatrix}\end{aligned} \tag{7.89}$$

showing that the correlation function is antisymmetric about $\tau = 0$: a direct result o spatial dependence.

To find the value of Γ_Z we use eqn. (7.39) and obtain

$$\begin{aligned}\Gamma_Z(\mathbf{r}_1, \mathbf{r}_2; \tau) &= \tilde{\epsilon}(\mathbf{r}_1)\delta(\mathbf{r}_1-\mathbf{r}_2)f(\mathbf{r}_1)\tau\\ &\quad + 2\tilde{\epsilon}(\mathbf{r}_1)\tilde{\epsilon}(\mathbf{r}_2)\overline{\nu(\nu-1)}\,\overline{v\Sigma_f}\sum_{\mu\mu'}\frac{F_{\mu\mu'}\Psi_\mu(\mathbf{r}_1)\Psi_{\mu'}(\mathbf{r}_2)}{\omega_{\mu'}(\omega_\mu+\omega_{\mu'})}\\ &\quad \times \bar{\bar{P}}(B_\mu^2; 0)\bar{\bar{P}}(B_{\mu'}^2; 0)\tau\left[1-\frac{1-e^{-\omega_{\mu'}\tau}}{\omega_{\mu'}\tau}\right].\end{aligned} \tag{7.90}$$

Assuming a single detector of volume V_d and integrating Γ_Z accordingly, we find that with

$$\Gamma_Z(V_d, \tau) = \frac{1}{V_d}\int_{V_d} d\mathbf{r}_1 \int_{V_d} d\mathbf{r}_2 \Gamma_Z(\mathbf{r}_1, \mathbf{r}_2; \tau) \tag{7.91}$$

and the corresponding mean value of Z given by

$$\langle Z(V_d, \tau)\rangle = \tau \int_{V_d} d\mathbf{r}\,\tilde{\epsilon}(\mathbf{r}) f(r) \tag{7.92}$$

the variance to mean takes the following form:

$$\text{var}(Z) = \frac{\Gamma_Z(V_d, \tau)}{\langle Z(V_d, \tau)\rangle} = 1 + \frac{2\overline{\nu(\nu-1)}\,\overline{v\Sigma_f}}{V_d \int_{V_d} d\mathbf{r}\,\tilde{\epsilon}(\mathbf{r}) f(\mathbf{r})} \times \sum_{\mu\mu'} \frac{F_{\mu\mu'} g_{\mu\mu'} \bar{\bar{P}}(B_{\mu'}^2; 0)\bar{\bar{P}}(B_{\mu'}^2; 0)}{\omega_{\mu'}(\omega_\mu + \omega_{\mu'})}\left[1 - \frac{1-e^{-\omega_{\mu'}\tau}}{\omega_{\mu'}\tau}\right]. \tag{7.93}$$

If the detector may be regarded as a point of volume V_d located at $\mathbf{r}_0$, eqn. (7.84) applies and if, further, the first term only in the expansion is dominant, then var(Z) becomes, approximately,

$$\text{var}(Z, \mathbf{r}_0) \simeq 1 + \frac{\tilde{\epsilon}\,\overline{v\Sigma_f}\,\overline{\nu(\nu-1)}\,F_{00}\Psi_0^{\dagger 2}(\mathbf{r}_0)\bar{\bar{P}}(B_0^2; 0)\Lambda_{\text{eff}}^2}{f(\mathbf{r}_0)(\beta-\rho)^2}\left[1 - \frac{1-e^{-\omega_0\tau}}{\omega_0\tau}\right], \tag{7.94}$$

where $\omega_0 = (\beta-\rho)/\Lambda_{\text{eff}}$ ($\equiv \alpha_1$ of Chapter 4) and $\tilde{\epsilon} = \epsilon\overline{v\Sigma_f} = \epsilon\lambda_f$.

Thus we see that, compared with the point model approximation of Chapter 3, the amplitude factor Y_1 is substantially modified by spatial corrections and also by slowing-down terms. However, the functional form of var(Z) with τ remains the same as in the point model, with the parameter ω_0 containing effective parameters corrected for leakage. Thus under certain conditions we can expect the point model to yield accurate results for reactivity, $\beta/\Lambda_{\text{eff}}$, and for Y_1. In addition it must always be ascertained that, in a particular case, higher modal effects in the expansion can be neglected.

7.12. *The Rossi-α method*

In principle, the theory of the Rossi-α method which accounts for energy and space dependence, including transport theoretical effects, can be obtained via the generating function equation of Pál (1958, 1964) or Bell (1965). However, we shall adopt a simpler method in this section based upon direct physical argument. We first note that a space-dependent treatment of the Rossi-α method is certainly required since the point model approximation is only valid for systems near criticality where the fundamental mode is dominant. For far sub-critical systems the normal flux shape may deviate considerably from the critical shape and a large number of harmonics may be needed to represent it accurately. The energy dependence is not so important unless the method is to be applied to thermal neutrons; however, it is important to use the correctly energy-spectrum-weighted cross-sections in the analysis. In addition, if the experiments are performed on small, fast systems, transport theory corrections may be needed due to the failure of diffusion theory.

In developing the general theory we will use full energy-dependent transport theory and only as an example introduce the diffusion approximation. We consider a neutron detector D in a medium with a steady-state neutron-density distribution $N(\mathbf{r},\mathbf{v})$, then the counting rate will be

$$C(D) = \int d\mathbf{v} \int_D d\mathbf{r}\,\tilde{\epsilon}(\mathbf{r},\mathbf{v})\,N(\mathbf{r},\mathbf{v}), \tag{7.95}$$

where we neglect detector perturbations. $\tilde{\epsilon}(\mathbf{r},\mathbf{v})$ is the detector reaction rate and integration is over the volume of the detector. Evidently $C(D)\Delta t$ is the total number of counts in the time interval Δt. However, if Δt is chosen to be sufficiently small, this expression gives the probability of obtaining one count in D in the interval Δt.

We can now calculate $P(D_1 D_2;\, t_1,t_2)\Delta t_1 \Delta t_2$ which is the joint probability that we obtain a count with a detector D_1 in the time interval Δt_1 around t_1 and a count with another detector D_2 in Δt_2 at a later time t_2. $P(D_1, D_2;\, t_1,t_2)$ is the generalization of the point model expression $p_c(t_1,t_2)$ of Chapter 4. In calculating $P(D_1,D_2;\, t_1,t_2)$ we will continue to assume that the detector does not perturb the flux and also that it does not emit secondary neutrons (i.e. we do not use a fission counter). Both of these effects can be corrected if necessary. As we have seen in Chapter 3, $P(\ldots)$ can be expressed as the sum of two terms each arising from different aspects of the statistical nature of the process. One arises from random coincidence counts and the other from correlated events as depicted in Fig. 3.6.

The probability of random coincidence is clearly

$$C(D_1)\,C(D_2)\,\Delta t_1 \Delta t_2, \tag{7.96}$$

D_1 and D_2 being the two detectors operating over the ranges Δt_1 and Δt_2.

Correlations arise from nuclear processes that generate several neutrons simultaneously, e.g. neutron-induced fission and spontaneous fission. The latter effect was not mentioned in our point model analysis but for complete generality we shall introduce it here since there are situations where it may be significant. Let us assume that $p_f(\nu)$ is the probability that ν neutrons are emitted from a neutron-induced fission reaction and $p_s(\nu)$ for a spontaneous fission reaction. Let us also assume that $\chi_f(\mathbf{v})$ and $\chi_s(\mathbf{v})$ are the neutron-energy emission spectra of the two processes which we take to be isotropic. Now each of the emitted neutrons will generate descendants through the fission chain that follows. The *average* number resulting from an event taking place at $\mathbf{r}_0$ at time t_0 is given by

$$\tilde{G}_{s,f}(\mathbf{r}_0, t_0 \to \mathbf{r},\mathbf{v},t) = \int d\mathbf{v}_0\,\chi_{s,f}(\mathbf{v}_0)\,G(\mathbf{r}_0,\mathbf{v}_0,t_0 \to \mathbf{r},\mathbf{v},t), \tag{7.97}$$

where $G(\mathbf{r}_0,\mathbf{v}_0,t_0 \to \mathbf{r},\mathbf{v},t)$ is the Green's function defined by the conventional Boltzmann transport equation. The subscript s or f depends on whether the cause of emission is spontaneous or neutron-induced fission.

The probability of obtaining one count with D_1 in Δt_1 by the reaction at $(\mathbf{r}_0,t_0)$ is

$$\nu\tilde{G}_{s,f}(\mathbf{r}_0,t_0;\, D_1,t_1)\Delta t_1, \tag{7.98}$$

where

$$\tilde{G}_{s,f}(\mathbf{r}_0,t_0;\, D_1,t_1) = \int_{D_1} d\mathbf{r}_1 \int d\mathbf{v}_1\,\tilde{\epsilon}(\mathbf{r}_1,\mathbf{v}_1)\,\tilde{G}_{s,f}(\mathbf{r}_0,t_0;\mathbf{r}_1,\mathbf{v}_1,t_1) \tag{7.99}$$

and ν is the number of neutrons emitted by the event. Some of the descendants will also be detected by D_2 and this number is clearly given by

$$(\nu-1)\tilde{G}_{s,f}(\mathbf{r}_0, t_0;\ D_2, t_2)\Delta t_2, \qquad (7.100)$$

where we have accounted for the fact that one neutron was lost on detection in D_1.

The joint probability of detection in D_1 and D_2 during Δt_1 and Δt_2 is now given by

$$\int_{-\infty}^{t_1} dt_0 \int d\mathbf{r}_0\, \{\tilde{G}_f(\mathbf{r}_0, t_0;\ D_1, t_1)\tilde{G}_f(\mathbf{r}_0, t_0;\ D_2, t_2)F_f(\mathbf{r}_0) + \tilde{G}_s(\mathbf{r}_0, t_0;\ D_1, t_1)\tilde{G}_s(\mathbf{r}_0, t_0;\ D_2, t_2)F_s(\mathbf{r}_0)\}\Delta t_1 \Delta t_2, \qquad (7.101)$$

where $F_{s,f}(\mathbf{r}_0)$ are related to the total number of fissions at $\mathbf{r}_0$ and the average of $\overline{\nu(\nu-1)}$ with respect to $p_{s,f}(\nu)$.

For neutron-induced fissions we have that

$$F_f(\mathbf{r}_0) = \int d\mathbf{v}_0\, \{\overline{\nu(\nu-1)}\}_f v_0 \Sigma_f(\mathbf{r}_0, \mathbf{v}_0) N(\mathbf{r}_0, \mathbf{v}_0), \qquad (7.102)$$

where $\{\ldots\}_f$ is the average of $\overline{\nu(\nu-1)}$ over $p_f(\nu)$ and $\Sigma_f(\mathbf{r}_0, v_0)$ is the fission cross-section at $(\mathbf{r}_0, v_0)$. In general $\{\ldots\}_f$ will be a function of v_0 since it depends on the speed of the neutron causing fission.

If the neutrons arise from spontaneous fission then

$$F_s(\mathbf{r}_0) = \{\overline{\nu(\nu-1)}\}_s R_s(\mathbf{r}_0), \qquad (7.103)$$

where $R_s(\mathbf{r}_0)$ is the spontaneous fission rate and is known from basic measurements. It should be noted that if spontaneous fissions are important, a term $\bar{\nu}_s \chi_s(v) R_s(\mathbf{r})/4\pi$ must be added to the Boltzmann equation source term to calculate $N(\mathbf{r}, \mathbf{v})$.

Finally, if we assume that $\chi_f(\mathbf{v}) \equiv \chi_s(\mathbf{v})$, we may write for the total joint probability:

$$P(D_1, D_2;\ t_1, t_2)\Delta t_1 \Delta t_2 = \{C(D_1)C(D_2) + \phi(D_1, D_2;\ t_1, t_2)\}\Delta t_1 \Delta t_2, \qquad (7.104)$$

where

$$\phi(D_1, D_2;\ t_1, t_2) = \int_{-\infty}^{t_1} dt_0 \int d\mathbf{r}_0 \tilde{G}(\mathbf{r}_0, t_0;\ D_1, t_1)\tilde{G}(\mathbf{r}_0, t_0;\ D_2, t_2).\{F_j(\mathbf{r}_0) + F_s(\mathbf{r}_0)\}. \qquad (7.105)$$

Converting to diffusion theory with a slowing-down kernel we find that, over the time scale of prompt neutrons, the probability of correlation for thermal neutrons is given as follows:

$$P(D_1, D_2;\ t_1, t_2) = \int_{D_1} d\mathbf{r}_1 \tilde{\epsilon}(\mathbf{r}_1) f(\mathbf{r}_1) \int_{D_2} d\mathbf{r}_2 \tilde{\epsilon}(\mathbf{r}_2) f(\mathbf{r}_2) + \sum_{\mu\mu'} \frac{\tilde{F}_{\mu\mu'} g_{\mu\mu'}}{\omega_\mu - \omega_{\mu'}} \bar{\bar{P}}(B_\mu^2;\ 0)\bar{\bar{P}}(B_{\mu'}^2;\ 0) e^{-\omega_{\mu'}(t_2 - t_1)} \quad (t_2 > t_1), \qquad (7.106)$$

where
$$\tilde{F}_{\mu,\mu'} = \{\overline{\nu(\nu-1)}\}_f v \Sigma_f \int_{V_R} f(\mathbf{r}_0)\Psi_\mu(\mathbf{r}_0)\Psi_{\mu'}(\mathbf{r}_0)\, d\mathbf{r}_0 + \{\overline{\nu(\nu-1)}\}_s \int_{V_R} R_s(\mathbf{r}_0)\Psi_\mu(\mathbf{r}_0)\Psi_{\mu'}(\mathbf{r}_0)\, d\mathbf{r}_0 \qquad (7.107)$$

and we have used the same approximations regarding slowing down as in previous sections.

We note that whilst the formula given above applies for thermal neutrons, the general expression of eqn. (7.104) can be used for fast neutrons if the appropriate Green's function is used.

Experiments to test the need for a spatial analysis of the Rossi-α method have been made by Ukai *et al.* (1965) who measured the value of α in a light water-moderated natural uranium sub-critical assembly. The geometry was a cylinder with effective diameter 81·7 cm and height 100 cm. The core itself consisted of 294 fuel rods of uranium dioxide in a hexagonal lattice of pitch 45 mm and $H_2O:UO_2$ volume ratio

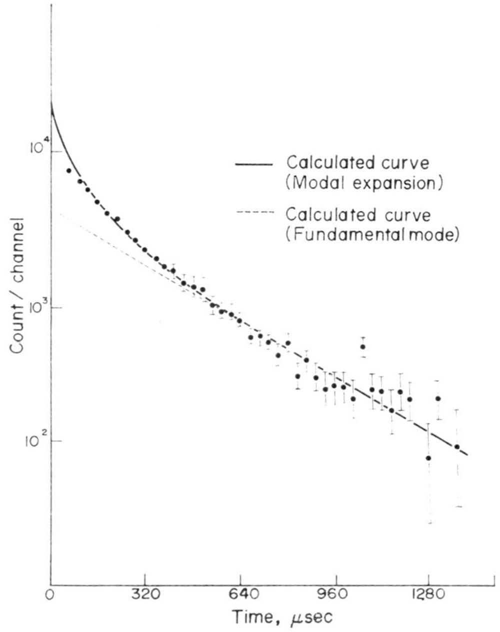

FIG. 7.2. Counts per channel recorded by a single detector placed at position $r = 9{\cdot}0$ cm, $z = 57{\cdot}75$ cm. The experimental results are compared with a full modal analysis and a fundamental mode approximation. (From Ukai *et al.* (1965) *J. Nucl. Sci. Technol.* **2**, 355.)

of 1·196:1. Both single detector correlations were made and two detector ones. The effective multiplication factor of the system was about 0·83 so that the ratio of correlated to random events was very small even under the minimum fission rate obtained from spontaneous fission only. In this experiment the background had to be determined with great accuracy, and extremely long periods of measurements, of around 30 hours, were required for each curve from which α is to be obtained. A detailed modal analysis was made, based upon the above theory, and it was demonstrated for a number of detector positions that a point model analysis based on the fundamental mode in eqn. (7.106) was unable to account for the experimental results. Figures 7.2 and 7.3 show some typical results. They also demonstrate the time scale which Rossi-α measurements on thermal systems must cover. It should be noted that this long time scale is necessary in order to isolate the fundamental mode and hence the value of ω_0 from which the sub-criticality is obtained.

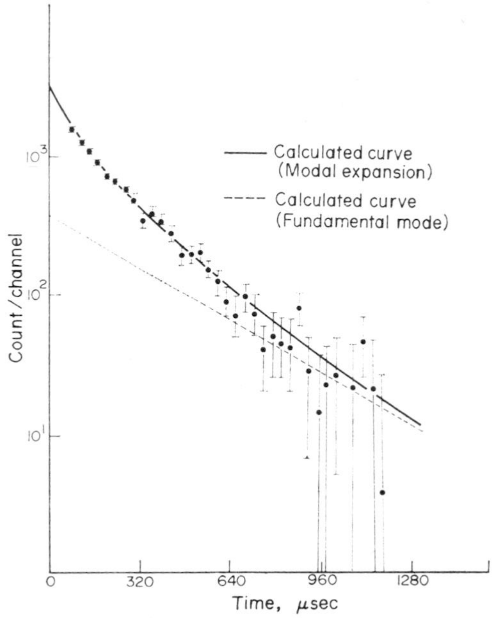

FIG. 7.3. Counts per channel recorded by a single detector placed at position $r = 38{\cdot}97$ cm, $z = 44{\cdot}0$ cm. The curves are as explained in Fig. 7.2. (From Ukai *et al.* (1965) *J. Nucl. Sci. Technol.* **2**, 355.)

7.13. *Energy-dependent fluctuations*

Not only can cross-correlations be made between different space points but, for a given space point, it is possible to derive covariance functions for different energies. Thus, for example, by correlating at different energies we may be able to extract information regarding slowing-down times. Such a result might be accomplished by the use of threshold detectors.

The complete analysis of this problem is clearly difficult although it is relatively easy to reduce eqn. (7.9) to a multi-energy group approximation and calculate correlations between the neutrons in the various groups (Pál, 1964). However, to illustrate the problem in terms of a continuous energy variable we shall neglect spatial dependence completely and consider the energy correlations in an infinite homogeneous medium. In that case the equation for the generating function can be written from eqn. (7.13) as

$$\left[-\frac{\partial}{\partial t}+v\Sigma(v)\right] \mathscr{G}(\mathbf{Z}|v,t)-v\int dv' \{\Sigma_s(v\to v')+\bar{\nu}\Sigma_f(v)\chi(v')\}\, \mathscr{G}(\mathbf{Z}|v',t)$$
$$= -\sum_{j=2}^{I} \frac{(-)^j}{j!}\, \overline{\nu(\nu-1)\ldots(\nu-j+1)}\, v\Sigma_f(v)\left[\int dv'\,\chi(v')\,\mathscr{G}(\mathbf{Z}|v',t)\right]^j \quad (7.108)$$

with $\mathbf{Z}$ denoting $(Z_1, Z_2, \ldots)$. Also we have

$$G(\mathbf{Z}|S) = \exp\left\{-\int dv \int_{-\infty}^{t} dt\, S(v,t)\, \mathscr{G}(\mathbf{Z}|v,t)\right\}. \quad (7.109)$$

Following the procedures outlined above it is readily shown that the covariance of the count rate is given by

$$\begin{aligned}\Gamma_c(v_1, v_2; \tau) &= \tilde{\epsilon}(v_1) f(v_1)\delta(v_1 - v_2)\delta(\tau) \\ &\quad + \tilde{\epsilon}(v_1)\tilde{\epsilon}(v_2)\overline{\nu(\nu-1)}F_0 \int_0^\infty d\tau_0 \int dv' \chi(v') G(v' \to v_1; \tau_0) \\ &\quad \times \int dv'' \chi(v'') G(v'' \to v_2; \tau+\tau_0),\end{aligned} \tag{7.110}$$

where F_0 is the fission rate.

Thus once the appropriate Green's function is known we may obtain the covariance directly. Two interesting examples can be given which indicate how an appropriately designed experiment would extract slowing-down parameters from cross-correlation at different energies. In the first we assume that slowing down is by age theory and that the velocities of correlation v_1 and v_2 are well below fission energies. In such a case the Green's function can be written (for constant Σ_s)

$$\int dv' \chi(v') G(v' \to v; \tau) = \frac{2}{\xi \Sigma_s v^2} \delta\left(\tau - \frac{2}{v \xi \Sigma_s}\right) \tag{7.111}$$

and the cross-covariance using eqn. (7.111) becomes ($v_1 \neq v_2$)

$$\Gamma_c(v_1, v_2; \tau) = \tilde{\epsilon}(v_1)\tilde{\epsilon}(v_2)\overline{\nu(\nu-1)}F_0 \frac{4}{(\xi\Sigma_s)^2 v_1^2 v_2^2} \delta\left(\tau - \frac{2}{\xi\Sigma_s}\left[\frac{1}{v_2} - \frac{1}{v_1}\right]\right). \tag{7.112}$$

Since

$$t_s = \frac{2}{\xi\Sigma_s}\left[\frac{1}{v_2} - \frac{1}{v_1}\right]$$

is the slowing-down time from v_1 to v_2, we should obtain a covariance with a peak at t_s.

The other extreme situation, namely slowing down in a hydrogenous substance, leads for v_1 and v_2 well below fission energies, and constant Σ_s, to (Williams, 1966):

$$\int dv' \chi(v') G(v' \to v; \tau) = \Sigma_s v \tau^2 e^{-v\tau\Sigma_s}. \tag{7.113}$$

Inserting this into eqn. (7.110) we find ($v_1 \neq v_2$)

$$\Gamma_c(v_1, v_2; \tau) = \frac{\tilde{\epsilon}(v_1)\tilde{\epsilon}(v_2)\overline{\nu(\nu-1)}F_0}{\Sigma_s(v_1+v_2)^3} 2v_1 v_2 e^{-\Sigma_s v_2 \tau}\left[\tau^2 + \frac{6\tau}{\Sigma_s(v_1+v_2)} + \frac{12}{\Sigma_s^2(v_1+v_2)^2}\right],$$

which could be measured using detectors of the same type as employed in the slowing-down time spectrometer (Williams, 1966).

This technique might well be useful for measurements of the inelastic scattering properties of various nuclei and in the interpretation of $(n, 2n)$ reactions.

A further example of the energy-dependent equation is that of extinction probability which we have already discussed from the space-dependent viewpoint. Using the alternative form of the generating function and remembering that the extinction probability is given by $q(v) = G(0, v, -\infty)$, we can write

$$\Sigma(v)q(v) = \Sigma_c(v) + \int_0^\infty \Sigma(v \to v')q(v')\,dv' + \Sigma_f(v)\sum_{\nu=1}^{\infty} p(\nu)\zeta^\nu, \tag{7.114}$$

where

$$\zeta = \int_0^\infty \chi(v)q(v)\,dv. \tag{7.115}$$

This non-linear integral equation may be numerically integrated to assess the effect of energy-dependent cross-sections and to compute the error in one-speed theory.

Further problems involving the continuous energy variable are discussed by Matthes (1962) and also a reduction to multigroup form. See also Pál (1964).

7.14. *The Langevin method: a random source problem*

Our considerations thus far have concerned zero-power noise only. As we intimated in earlier chapters, the extension of the generating function technique to power reactor problems is not very convenient. Thus we consider in this section the Langevin method and attempt to use it to solve some of the problems outlined in the introduction to this chapter.

Let us first consider a very simple problem in which we have an infinite plane source situated in an infinite, non-multiplying medium. The source is assumed to be time-dependent and to fluctuate about its mean value in a random manner. The problem is to calculate the corresponding covariance of the neutron density at different space points.

For convenience, we will assume one-speed theory and infinite plate detectors that do not perturb the neutron distribution. The basic equation for the neutron density $N(x,t)$ at position x, time t, can therefore be written

$$\frac{\partial N(x,t)}{\partial t} = D_0 \frac{\partial^2 N(x,t)}{\partial x^2} - \frac{1}{l} N(x,t) + S(t)\delta(x), \tag{7.116}$$

where D_0 is the diffusion coefficient and l the mean lifetime. We write the fluctuating density and source as

$$N(x,t) = N_0(x) + n(x,t), \tag{7.117}$$

$$S(t) = S_0 + \mathscr{S}(t), \tag{7.118}$$

where N_0 and S_0 are the steady mean values that would be established if the fluctuations were absent.

Inserting eqns. (7.117) and (7.118) into (7.116) and subtracting the corresponding steady-state equation, we find that the stochastic parameters $n(x,t)$ and $\mathscr{S}(t)$ are related by

$$\frac{\partial n(x,t)}{\partial t} = D_0 \frac{\partial^2 n(x,t)}{\partial x^2} - \frac{1}{l} n(x,t) + \mathscr{S}(t)\delta(x). \tag{7.119}$$

If the source has been operating for a long time compared with the characteristic response time of the system, we can write

$$n(x,t) = \int_{-\infty}^{t} \mathscr{S}(t')G(x,t-t')\,dt', \tag{7.120}$$

$$\equiv \int_0^{\infty} \mathscr{S}(t-t_0)G(x,t_0)\,dt_0, \tag{7.121}$$

where the Green's function $G(x,t)$ is given by

$$G(x,t) = \frac{e^{-t/l}}{2(\pi D_0 t)^{\frac{1}{2}}} \exp\left(-\frac{x^2}{4D_0 t}\right). \tag{7.122}$$

Now the cross-correlation function as calculated by two plane detectors in the planes $x = x_1$ and $x = x_2$ is

$$R_{x_1 x_2}(t_1 - t_2) = \langle n(x_1, t_1) n(x_2, t_2) \rangle \tag{7.123}$$

$$= \int_0^\infty dt' \int_0^\infty dt'' \, G(x_1, t') G(x_2, t'') \langle \mathscr{S}(t_1 - t') \, \mathscr{S}(t_2 - t'') \rangle. \tag{7.124}$$

If the source fluctuation is a stationary random variable, we may write

$$\langle \mathscr{S}(t_1 - t') \, \mathscr{S}(t_2 - t'') \rangle = R_s(t_1 - t_2 - t' + t''), \tag{7.125}$$

$R_s(\tau)$ being the source auto-covariance function.

We may calculate also the cross-p.s.d. by taking the Fourier transform of eqn. (7.124), then we find

$$\Phi_{x_1 x_2}(\omega) = \frac{\Phi_s(\omega)}{4\pi D_0} \theta(x_1, \omega) \theta^*(x_2, \omega), \tag{7.126}$$

where $\Phi_s(\omega)$ is the p.s.d. of $R_s(\tau)$ and

$$\theta(x, \omega) = \left\{ \frac{\pi}{\kappa^2 D_0 - i\omega} \right\}^{\frac{1}{2}} \exp \left\{ - \left[\kappa^2 - \frac{i\omega}{D_0} \right]^{\frac{1}{2}} x \right\} \tag{7.127}$$

with $\kappa^2 = 1/(D_0 l)$.

Thus we can write

$$\Phi_{x_1 x_2}(\omega) = \frac{\Phi_s(\omega)}{4 D_0^2} \frac{\exp\{ -\alpha(x_1 + x_2) + i\beta(x_1 - x_2) \}}{\left\{ \kappa^2 + \left(\frac{\omega}{D_0} \right)^2 \right\}^{\frac{1}{2}}}, \tag{7.128}$$

where

$$\alpha = \left\{ \frac{1}{2} \left(\left[\kappa^4 + \left(\frac{\omega}{D_0} \right)^2 \right]^{\frac{1}{2}} + \kappa^2 \right) \right\}^{\frac{1}{2}}, \tag{7.129}$$

$$\beta = \left\{ \frac{1}{2} \left(\left[\kappa^4 + \left(\frac{D_0}{\omega} \right)^2 \right]^{\frac{1}{2}} - \kappa^2 \right) \right\}^{\frac{1}{2}}. \tag{7.130}$$

It should be noted here that $1/\alpha$ is the attenuation length of a neutron-wave disturbance and $2\pi/\beta$ is its wavelength. This suggests that a randomly modulated source could provide much the same information as does the conventional neutron-wave experiment (Perez and Uhrig, 1967). Moreover, a cross-correlation between source and neutron density fluctuations would lead, if the noise spectrum of the source is white, to the Green's function directly.

An interesting analytical solution can be obtained from eqn. (7.124) if the input noise source is white, i.e.

$$R_s(\tau) = \frac{\sigma_s^2}{\gamma} \delta(\tau). \tag{7.131}$$

Then we obtain

$$R_{x_1 x_2}(0) = \frac{\sigma_s^2}{2\pi D_0 \gamma} K_0 \left\{ \frac{\sqrt{\{2(x_1^2 + x_2^2)\}}}{L} \right\}, \tag{7.132}$$

where $L = (D_0 l)^{\frac{1}{2}}$ is the thermal diffusion length and $K_0(y)$ is the modified Bessel function.

In the paper by Smith and Williams (1972) the Gaussian stochastic generator of Section 6.7 has been used to excite eqn. (7.119). The resulting auto-correlation function for the case $x_1 = x_2 = x$ is shown in Fig. 7.4 along with the analytical solution.

The excellent agreement lends confidence to the use of the stochastic generator in realistic noise problems.

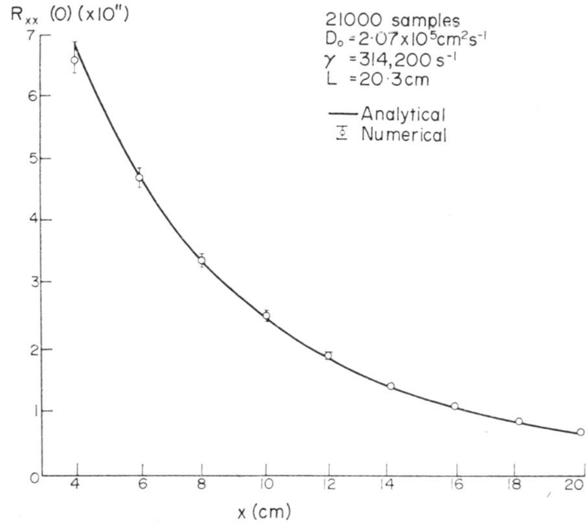

FIG. 7.4. Zero lag auto-correlation function; comparison of analytical and numerical simulation solutions. (From Smith and Williams (1972), *J. Nucl. Energy*, **26**, 525.)

7.15. *Power reactor noise: the homogeneous approximation*

Whilst a reactor is in practice a complicated heterogeneous assembly, often a reasonable first step in its analysis is to assume that the core can be homogenized. We shall follow this procedure in this section and examine the effect of finite core size on the noise spectrum.

We can incorporate slowing down and spatial behaviour into the problem by assuming that the modified one-group diffusion equation is valid in the reactor. Thus we may write for the neutronics equations the following expressions:

$$l\frac{\partial\phi(\mathbf{r},t)}{\partial t} = M^2\nabla^2\phi(\mathbf{r},t)+(k_\infty-1-\beta)\phi(\mathbf{r},t)+\sum_i\frac{\lambda_i}{\Sigma_a}C_i(\mathbf{r},t), \tag{7.133}$$

$$\frac{\partial C_i(\mathbf{r},t)}{\partial t} = \beta_i\Sigma_a k_\infty\phi(\mathbf{r},t)-\lambda_i C_i(\mathbf{r},t), \tag{7.134}$$

where l is the thermal lifetime and M^2 is the migration area.

Associated with these equations are those for heat transfer and fluid flow. On the assumption that we have a cell of the type shown in Fig. 7.5 we can write the spatially averaged heat balance equations as follows (per unit volume of cell):

For the fuel:

$$C_u\frac{\partial T_u}{\partial t} = \alpha g\phi-h_{ua}(T_u-T_a), \tag{7.135}$$

α = fraction of fission heat deposited in fuel,
g = heat produced/fission/unit flux/unit volume of cell,
h_{ua} = heat-transfer coefficient between fuel and cladding.

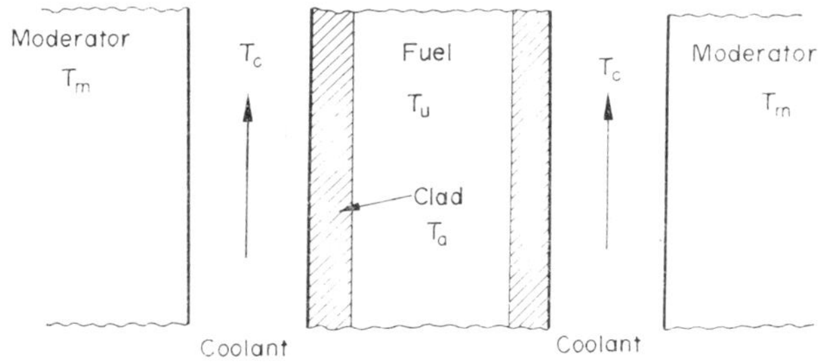

FIG. 7.5. Schematic diagram of a reactor lattice cell.

For the cladding: $$C_a \frac{\partial T_a}{\partial t} = h_{ua}(T_u - T_a) - h_{ac}(T_u - T_c), \tag{7.136}$$

h_{ac} = heat-transfer coefficient between cladding and coolant.

For the coolant:

$$C_c \left\{ \frac{\partial T_c}{\partial t} + u \frac{\partial T_c}{\partial z} \right\} = h_{ac}(T_a - T_c) + h_{mc}(T_m - T_c), \tag{7.137}$$

where we note the term $u \partial T_c / \partial z$ due to mass motion along the axis.

h_{mc} = heat-transfer coefficient between moderator and coolant.

For the moderator:

$$C_m \frac{\partial T_m}{\partial t} = (1-\alpha) g \phi - h_{mc}(T_m - T_c). \tag{7.138}$$

These heat-transfer equations are coupled to the neutronics equations by the reactivity feedback, viz.

$$k_\infty = k_{\infty 0} + a_u(T_u - T_{u0}) + a_m(T_m - T_{m0}) + a_v(f - f_0) + \ldots, \tag{7.139}$$

where a_u and a_m are temperature coefficients, f is the void fraction (in boiling-water reactors) and a_v the void coefficient.

7.16. *Noise sources*

Having established the basic equations of the system, we now have the difficult task of identifying the important noise sources. In the present state of power reactor analysis this depends markedly on the particular reactor type. For example, in a boiling-water reactor, we can expect the random motion, formation and collapse of steam bubbles to lead to randomly fluctuating diffusion and absorption properties of the moderator. Thus, in the basic neutronics equations, l, M and Σ_a could all be noise sources. However, it is generally found that the dominant effects of such changes can be incorporated into the feedback term involving k_∞. Thus by making f, the void fraction, a random variable we can study its effect on the reactor dynamic behaviour. At the same time, of course, the heat-transfer coefficients will also be random variables, as also will the coolant velocity.

In a gas-cooled reactor, the problem of steam voids is absent but the other noise sources remain.

It should also be noted that our neutronic equations refer to a homogenized system and cannot therefore give any detailed information regarding the noise arising from specific components in the system—for example, the vibration of control rods or fuel elements or similar mechanical components. It is with this limitation in mind that we develop in a later chapter a more refined method of spatial noise analysis which is based upon the so-called "Feinberg–Galanin" method of heterogeneous reactor theory. We shall be discussing this method and its application later and will therefore defer detailed comments until then. Basically, however, the method allows the effect of each individual fuel element and control rod to be included in the noise analysis. Thus the mechanical vibrations and any other random process associated with a particular rod or localized component may be assessed.

7.17. *Effects of noise on spatial distribution of power*

The point model description, by definition, prevents any investigation into the spatial effects of noise. However, the homogenized eqns. (7.133) and (7.134) enable us to relax this restriction somewhat and at least study the spatial effect in a macroscopic sense—with the noise sources themselves homogenized.

The procedure is basically the same as that for the point model. We decide on the noise sources and set, for example, $h = \bar{h}+\delta h$, etc., and then write the dependent variables, T, ϕ, C, in a similar fashion, viz. $T = \bar{T}+\delta T$, $\phi = \bar{\phi}+\delta\phi$, $C = \bar{C}+\delta C$.

Neglecting second-order terms leads to a set of linear equations for $\delta\phi(\mathbf{r},t)$, $\delta T(\mathbf{r},t)$, etc., which may be solved by the standard methods of space-dependent reactor kinetics.

As an example of this technique we can consider a gas-cooled, graphite-moderated reactor with a fuel and moderator temperature coefficient feedback, and noise sources arising from fluctuations in the heat-transfer coefficients—that is, eqns. (7.133)–(7.139), but with $a_v = 0$. Linearized, these equations become

$$l\frac{\partial\delta\phi}{\partial t} = M^2\nabla^2\delta\phi+[k_{\infty 0}(1-\beta)-1]\delta\phi+(a_u\delta T_u+a_m\delta T_m)g\bar{\phi}+l\lambda_i\delta C_i, \quad (7.140)$$

$$\frac{\partial\delta C_i}{\partial t} = \beta_i\frac{k_{\infty 0}}{l}\delta\phi-\lambda_i\delta C_i, \quad (7.141)$$

$$C_m\frac{\partial\delta T_m}{\partial t} = (1-\alpha)g\delta\phi-\bar{h}_{mc}(\delta T_m+\delta T_c)-\underline{\delta h_{mc}(\bar{T}_m-\bar{T}_c)}, \quad (7.142)$$

$$C_c\frac{\partial\delta T_c}{\partial t} = \bar{h}_{ua}\delta T_u+\bar{h}_{mc}\delta T_m-\bar{h}_{\text{eff}}\delta T_c+\underline{\delta h_{ua}\bar{T}_u+\delta h_{mc}\bar{T}_m-\delta h_{\text{eff}}\bar{T}_c}. \quad (7.143)$$

[N.B. $h_{\text{eff}}T_c$ accounts approximately for the $u\partial T_c/\partial z$ term in the coolant equation.]

$$C_u\frac{\partial\delta T_u}{\partial t} = \alpha g\delta\phi-\bar{h}_{ua}(\delta T_u-\delta T_c)+\underline{\delta h_{ua}(\bar{T}_u-\bar{T}_c)}. \quad (7.144)$$

[Noise sources have been underlined.]

To solve these equations we consider the critical equation for no fluctuations, viz.

$$0 = M^2\nabla^2\phi_0+[k_{\infty 0}-1]\phi_0, \quad (7.145)$$

which, if we write it in the form

$$\nabla^2\phi_0+B^2\phi_0 = 0, \tag{7.146}$$

subject to $\phi_0 = 0$ on the reactor boundary, constitutes an equation for a set of spatial eigenfunctions $\phi_\nu(\mathbf{r})$ corresponding to eigenvalues B_ν^2.

In the steady state, $\phi_0(\mathbf{r})$ is the fundamental flux shape and B_0^2 is the geometric buckling. However, in a situation where the reactor is perturbed from equilibrium, by time-dependent noise sources for example (or by a deterministic effect), the flux will become a function of time and $\delta\phi(\mathbf{r},t)$ and δT, etc., will depend on all of the $\phi_\nu(\mathbf{r})$ and B_ν^2.

Thus to solve the general, noise-excited equations for $\delta\phi$, δC and δT we assume that the following expansions are valid:

$$\delta\phi(\mathbf{r},t) = \sum_\nu P_\nu(t)\phi_\nu(\mathbf{r}), \tag{7.147}$$

$$\delta T_l(\mathbf{r},t) = \sum_\nu Q_\nu^{(l)}(t)\phi_\nu(\mathbf{r}), \tag{7.148}$$

$$\delta C_i(\mathbf{r},t) = \sum_\nu D_\nu^{(i)}(t)\phi_\nu(\mathbf{r}). \tag{7.149}$$

It is well known that the eigenfunctions of eqn. (7.146) $\phi_\nu(\mathbf{r})$ form an orthonormal set, viz.

$$\int_V \phi_\nu(\mathbf{r})\phi_{\nu'}(\mathbf{r})\,d\mathbf{r} = \delta_{\nu\nu'}. \tag{7.150}$$

Thus if we insert (7.147)–(7.149) into (7.140)–(7.144), multiply by $\phi_{\nu'}(\mathbf{r})$ and integrate over the volume V of the reactor, the equations reduce to:

$$l\frac{\partial P_\nu(t)}{\partial t} = -M^2B_\nu^2P_\nu(t) + [k_{\infty 0}(1-\beta)-1]P_\nu(t)+(a_u Q_\nu^{(u)}(t)+a_m Q_\nu^{(m)}(t))g\delta_{\nu,0}+l\lambda_i D_\nu^{(i)}(t) \tag{7.151}$$

[assume $\bar{\phi} \equiv \phi_0$],

$$\frac{\partial D_\nu^{(i)}(t)}{\partial t} = \beta_i\frac{k_{\infty 0}}{l}P_\nu(t)-\lambda_i D_\nu^{(i)}(t), \tag{7.152}$$

$$C_m\frac{\partial}{\partial t}Q_\nu^{(m)}(t) = (1-\alpha)gP_\nu(t)-\bar{h}_{mc}(Q_\nu^{(m)}(t)+Q_\nu^{(c)}(t))-\delta h_{mc}\Delta_{mc}^\nu \tag{7.153}$$

with

$$\Delta_{mc}^\nu = \int_V d\mathbf{r}\,\phi_\nu(\mathbf{r})(\bar{T}_m(\mathbf{r})-\bar{T}_c(\mathbf{r})), \tag{7.154}$$

$$C_c\frac{\partial}{\partial t}Q_\nu^{(c)}(t) = \bar{h}_{ua}Q_\nu^{(u)}(t)+\bar{h}_{mc}Q_\nu^{(m)}(t)-\bar{h}_{\text{eff}}Q_\nu^{(c)}(t)-\Delta_{umc}^\nu \tag{7.155}$$

with

$$\Delta_{umc}^\nu = \int_V d\mathbf{r}\,\phi_\nu(\mathbf{r})\,\{\delta h_{ua}\bar{T}_u+\delta h_{mc}\bar{T}_m-\delta h_{\text{eff}}\bar{T}_c\}, \tag{7.156}$$

$$C_u\frac{\partial Q_\nu^{(u)}(t)}{\partial t} = \alpha gP_\nu(t)+\bar{h}_{ua}(Q_\nu^{(u)}(t)-Q_\nu^{(c)}(t))+\delta h_{ua}\Delta_{uc}^\nu, \tag{7.157}$$

where we have assumed for simplicity that the heat-transfer coefficients are space independent but which is not necessarily a good assumption (Tyror and Vaughn, 1970).

By this procedure we have reduced the problem to a set of coupled time-dependent equations with random sources, i.e. Langevin's equations. When these are solved, the values of P, Q, D can be replaced in eqns. (7.147)–(7.149) to obtain the general solution for the space-dependent fluctuations. In this way we can calculate cross-correlation in time and space, e.g.

$$\langle \delta\phi(\mathbf{r}_1, t_1)\,\delta\phi(\mathbf{r}_2, t_2)\rangle, \tag{7.158}$$

$$\langle \delta T_i(\mathbf{r}_1, t_1)\,\delta T_j(\mathbf{r}_2, t_2)\rangle \tag{7.159}$$

and obtain information about the effect on one region of the reactor arising from noise sources in another.

Another interesting consequence of a space-dependent analysis is the way in which the noise sources can give rise to changes in the steady-state shape of the reactor.

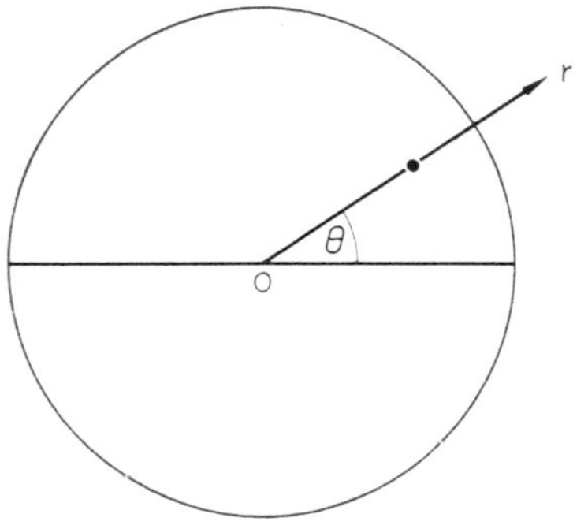

FIG. 7.6. Coordinate system in the plane of a reactor core.

For example, if we consider a cylindrical reactor and study the flux in the radial and angular variables, i.e. $\phi(r,\theta)$ [coordinates defined in Fig. 7.6], then superimposed upon the fundamental mode, which we denote by $\phi_{00}(r,\theta) = \phi(r)$, there will be the effects of the higher harmonics in the expansion of $\delta\phi(\mathbf{r}, t)$, e.g.

$$\delta\phi(\mathbf{r}, t) = \sum_{\nu} P_\nu(t)\,\phi_\nu(\mathbf{r}), \tag{7.160}$$

where in our case $\phi_\nu(\mathbf{r}) \equiv \phi_{ij}(r,\theta)$ with (i,j) corresponding to radial and azimuthal modes, respectively. Thus the noise sources will excite the types of spatial modes shown in Fig. 7.7.

It is because of the necessity to control these spatial modes that a number of independently acting sector control rods are built into large modern power stations. In fact, the larger the reactor core compared with a migration area the larger the number of modes that are excited.

Of course, not only noise sources excite these modes. There are xenon tilt effects and fuel-changing schemes which also affect the spatial behaviour. Nevertheless, the random effects of mechanical noise is an important and, at present, a rather poorly understood factor in reactor stability.

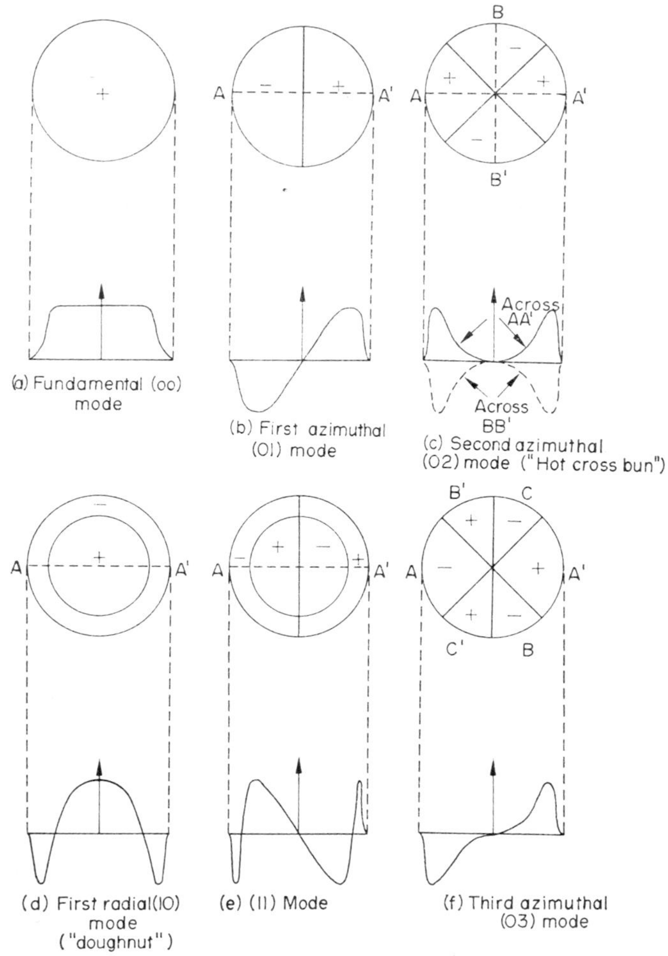

FIG. 7.7. Plan and side view of some radial and azimuthal spatial modes excited in a cylindrical reactor. (From Hitchcock (1960) *Nuclear Reactor Stability*, Temple Press, London.)

References

BELL, G. I. (1965) On the stochastic theory of neutron transport, *Nucl. Sci. Engng*, **21**, 390.

CASSELL, J. S. and WILLIAMS, M. M. R. (1969) Space-dependent noise spectra via the transport equation, *J. Nucl. Energy*, **23**, 575.

COHN, C. E. (1960) A simplified theory of pile noise, *Nucl. Sci. Engng*, **7**, 472.

DAVENPORT, W. B. and ROOT, W. L. (1958) *An Introduction to the Theory of Random Signals and Noise*, McGraw-Hill, N.Y.

FERZIGER, J. and ZWEIFEL, P. F. (1966) *The Theory of Neutron Slowing Down in Nuclear Reactors*, Pergamon, Oxford.

HITCHCOCK, A. (1960) *Nuclear Reactor Stability*, Temple Press, London.

KRIEGER, T. J. and ZWEIFEL, P. F. (1959) Theory of pulsed neutron experiments in multiplying media, *Nucl. Sci. Engng*, **5**, 21.

MATTHES, W. (1962) Statistical fluctuations and their correlation in reactor neutron distributions, *Nukleonik*, **4**, 213.

MATTHES, W. (1966) Theory of fluctuations in neutron fields, *Nukleonik*, **8**, 87.

NATELSON, M. *et al.* (1966) Space and energy effects in reactor fluctuation experiments, *J. Nucl. Energy*, **20**, 557.

OSBORNE, R. K. (1965) Speculations on the interpretation of neutron noise experiments, USAEC, ORNL 3757.

OSBORNE, R. K. and NATELSON, M. (1965) Kinetic equations for neutron distributions, *J. Nucl. Energy*, **19**, 619.

PÁL, L. (1958) On the theory of stochastic processes in nuclear reactors, *Il Nuovo Cimento Supplemento*, **7**, 25.

PÁL, L. (1964) Statistical theory of neutron chain reactors, *Proc. Int. Conf. Geneva*, Vol. 2, p. 218.

PEREZ, R. B. and UHRIG, R. E. (1967) Development of techniques of neutron wave and neutron pulse propagation, in *Neutron Noise, Waves and Pulse Propagation*, AEC Symposium Series No. 9.

PLUTA, P. R. (1962) *An Analysis of Nuclear Reactor Fluctuations by Methods of Stochastic Processes*, Ph.D. Thesis, University of Michigan, Ann Arbor.

RICKER, C. W. *et al.* (1965) Measurement of reactor fluctuation spectra and subcritical reactivity, USAEC, ORNL-TM-1066.

SAITO, K. and OTSUKA, M. (1965*a*) Space–time correlations in neutron distributions in a multiplying medium, *J. Nucl. Sci. Technol.* **2**, 191.

SAITO, K. and OTSUKA, M. (1965*b*) Theory of statistical fluctuations in neutron distributions, *J. Nucl. Sci. Technol.* **2**, 304.

SAITO, K. and OTSUKA, M. (1966) Transfer function and power spectral density in a zero power reactor, *J. Nucl. Sci. Technol.* **3**, 45.

SHEFF, J. R. and ALBRECHT, R. W. (1966) The space dependence of reactor noise. I. Theory. II. Calculations, *Nucl. Sci. Engng*, **24**, 246; **26**, 207.

SMITH, P. R. and WILLIAMS, M. M. R. (1972) The application of a stochastic generator to determine the kinetic response of a neutronic system with randomly fluctuating parameters, *J. Nucl. Energy*, **26**, 525.

TYROR, J. G. and VAUGHN, R. I. (1970) *An Introduction to the Neutron Kinetics of Nuclear Power Reactors*, Pergamon Press, Oxford.

UHRIG, R. E. (1967) *Neutron Noise, Waves and Pulse Propagation*, USAEC, AEC Symposium Series No. 9.

UKAI, S. *et al.* (1965) A generalized analysis of Rossi-α experiment, *J. Nucl. Sci. Technol.* **2**, 355.

WEINBERG, A. M. and WIGNER, E. P. (1958) *The Physical Theory of Neutron Chain Reactors*, Chicago University Press.

WILLIAMS, M. M. R. (1966) *The Slowing Down and Thermalization of Neutrons*, North-Holland.

WILLIAMS, M. M. R. (1967) An application of slowing down kernels to thermal neutron density fluctuations in nuclear reactors, *J. Nucl. Energy*, **21**, 321.

WILLIAMS, M. M. R. (1971) *Mathematical Methods in rticle Transport Theory*, Butterworth, London.

CHAPTER 8

Random Phenomena in Heterogeneous Reactor Systems

8.1. *Introduction*

This chapter may be regarded as of a pioneering nature in so far as we shall discuss techniques that, in principle, are truly representative of the system and yet in many cases are difficult to apply or, at the very least, have not been compared with experiment. We shall attempt to discuss neutron noise in a detailed manner as opposed to the previous chapters where either point model, or homogenized, systems have been studied. Here we argue that, since a reactor is heterogeneous, then so are the noise sources that are produced within it. This being the case, a formalism must be found for a detailed space–time representation of two-phase flow, as also must an accurate representation of the random vibration of the mechanical parts of the core structure, e.g. fuel elements and control rods. We should also study how random fluctuations in the heat-transfer coefficient at a surface affect the propagation of the associated temperature fluctuations within the adjacent medium. Other problems of an important nature are the effects of non-uniform fuel enrichment both within a fuel element and in different fuel elements since these may give rise to local hot spots leading to statistical problems in burn-out. The statistical distribution in the packing and concentration of fuel particles in a pebble-bed reactor is clearly another problem that will lead to local power fluctuations. In the sections that follow we shall describe techniques for dealing with some of these problems and at the same time point out where development will be necessary.

8.2. *Diffusion in media with random physical properties*

In Section 7.1 we discussed the ways in which equations involving random parametric coefficients can arise, due, for example, to boiling or non-uniform fuel concentration. We did not, however, deal with the associated theory and we shall attempt to remedy that omission in this section.

First, it is important to note that the problem of random equations is not confined to the field of neutronics, indeed reactor physics is a relative newcomer to the field. We can find problems involving a wide variety of processes in classical physics in which an equation of motion or a field equation contains a random source or coefficient. Typical of such processes are wave propagation in a duct or guide of randomly varying cross-section, the motion of a charged particle in a randomly inhomogeneous magnetic or electric field, wave propagation in a turbulent medium or one with a random refractive index, electromechanical systems with a random parameter, convective transport of material in a turbulent fluid, the Schrödinger wave equation with a random potential, buffeting problems in aeronautics, the flow of fluid through channels with random surface topography, and many other physical and engineering problems. A full bibliography may be found in Bharucha-Reid (1968) and additional references are given in this book.

Very general methods for obtaining approximate solutions of random differential and integral equations may be found in the treatises of Adomian (1963, 1964, 1970, 1971*a*, *b*) and of Bharucha-Reid (1968). However, we shall not discuss this basic theory but rather illustrate the technique by considering an application to a very simple problem in reactor physics.

8.3. *A simple critical problem*

Let us assume that we have a bare reactor that can be described by one-speed diffusion theory. The diffusion coefficient D is taken to be uniform across the system whilst the number density $N_F(\mathbf{r})$ of fuel nuclei per unit volume is randomly dispersed such that $N_F(\mathbf{r})$ is a stationary random function of position. We write the corresponding diffusion equation therefore as follows

$$D\nabla^2\phi(\mathbf{r})+\{\bar{\nu}\Sigma_f(\mathbf{r})-\Sigma_a(\mathbf{r})-\Sigma_m\}\phi(\mathbf{r}) = 0, \tag{8.1}$$

where $\Sigma_f(\mathbf{r}) = N_F(\mathbf{r})\sigma_f$, $\Sigma_a(\mathbf{r}) = N_F(\mathbf{r})\sigma_a$ and $\Sigma_m = N_m\sigma_m$.

N_m is the number density of moderator nuclei and σ_f, σ_a and σ_m are microscopic cross-sections for fission and absorption in the fuel and absorption in the moderator, respectively. We now abbreviate as follows, viz.

$$\frac{1}{D}\{\bar{\nu}\Sigma_f(\mathbf{r})-\Sigma_a(\mathbf{r})-\Sigma_m\} \equiv B_0^2\{1+\epsilon(\mathbf{r})\}, \tag{8.2}$$

where B_0^2 is the average value of the left-hand side of eqn. (8.2) and $B_0^2\epsilon(\mathbf{r})$ is the fluctuating part. Clearly, $\langle\epsilon(\mathbf{r})\rangle \equiv 0$ by definition. Equation (8.1) for $\phi(\mathbf{r})$ is now rewritten as follows:

$$\nabla^2\phi(\mathbf{r})+B_0^2\phi(\mathbf{r}) = -B_0^2\epsilon(\mathbf{r})\phi(\mathbf{r}). \tag{8.3}$$

If we now define the Green's function $G(\mathbf{r}_0 \to \mathbf{r})$ by the equation

$$\nabla^2 G(\mathbf{r}_0 \to \mathbf{r})+B_0^2 G(\mathbf{r}_0 \to \mathbf{r})+\delta(\mathbf{r}_0-\mathbf{r}) = 0 \tag{8.4}$$

we can write the differential equation for $\phi(\mathbf{r})$ in the following form,

$$\phi(\mathbf{r}) = \phi_0(\mathbf{r})-B_0^2\int_V d\mathbf{r}_0\, G(\mathbf{r}_0 \to \mathbf{r})\epsilon(\mathbf{r}_0)\phi(\mathbf{r}_0), \tag{8.5}$$

where $\phi_0(\mathbf{r})$ is a solution of the unperturbed equation.

The only way to make progress is to use an iterative technique. Thus first we statistically average eqn. (8.5) to get

$$\langle\phi(\mathbf{r})\rangle = \phi_0(\mathbf{r})-B_0^2\int_V d\mathbf{r}_0\, G(\mathbf{r}_0 \to \mathbf{r})\langle\epsilon(\mathbf{r}_0)\phi(\mathbf{r}_0)\rangle. \tag{8.6}$$

Now the approximation $\langle \epsilon\phi \rangle \simeq \langle \epsilon \rangle \langle \phi \rangle$ leads to a trivial solution $\langle \phi(\mathbf{r}) \rangle = \phi_0(\mathbf{r})$, thus we multiply eqn. (8.5) by $\epsilon(\mathbf{r})$ and average again, thereby obtaining

$$\langle \epsilon(\mathbf{r})\phi(\mathbf{r}) \rangle = -B_0^2 \int_V d\mathbf{r}_0 \, G(\mathbf{r}_0 \to \mathbf{r}) \langle \epsilon(\mathbf{r})\epsilon(\mathbf{r}_0)\phi(\mathbf{r}_0) \rangle. \tag{8.7}$$

Substitution back into eqn. (8.6) still leaves the unknown $\langle \epsilon\epsilon\phi \rangle$. However, at this point, we can either continue the iteration or assume that $\epsilon(\mathbf{r})\epsilon(\mathbf{r}_0)$ and $\phi(\mathbf{r}_0)$ are locally independent, i.e.

$$\langle \epsilon(\mathbf{r})\epsilon(\mathbf{r}_0)\phi(\mathbf{r}_0) \rangle \simeq \langle \epsilon(\mathbf{r})\epsilon(\mathbf{r}_0) \rangle \langle \phi(\mathbf{r}_0) \rangle. \tag{8.8}$$

This is an often-used but relatively untested hypothesis: nevertheless, we shall use it here and write eqn. (8.6) in the form

$$\langle \phi(\mathbf{r}) \rangle = \phi_0(\mathbf{r}) + B_0^4 \int_V d\mathbf{r}_0 \, G(\mathbf{r}_0 \to \mathbf{r}) \int_V d\mathbf{r}_0' \, G(\mathbf{r}_0' \to \mathbf{r}_0') \, R(\mathbf{r}_0 - \mathbf{r}_0') \langle \phi(\mathbf{r}_0') \rangle, \tag{8.9}$$

where $R(\mathbf{r}_0 - \mathbf{r}_0') \equiv \langle \epsilon(\mathbf{r}_0)\epsilon(\mathbf{r}_0') \rangle$ is the auto-correlation function of $\epsilon(\mathbf{r})$.

Finally, operating on eqn. (8.9) with $(\nabla^2 + B_0^2)$, we get

$$(\nabla^2 + B_0^2)\langle \phi(\mathbf{r}) \rangle = -B_0^4 \int_V d\mathbf{r}_0 \, G(\mathbf{r}_0 \to \mathbf{r}) \, R(\mathbf{r} - \mathbf{r}_0) \langle \phi(\mathbf{r}_0) \rangle, \tag{8.10}$$

which is an integro-differential equation for the average flux $\langle \phi \rangle$. Assuming that the auto-correlation function has the Markoff form

$$R(\mathbf{r}) = \langle \epsilon^2 \rangle e^{-|\mathbf{r}|/l}, \tag{8.11}$$

where $\langle \epsilon^2 \rangle$ is the mean square fractional fluctuation and l is a correlation length, we can apply asymptotic reactor theory to the integral equation and note that solutions of the type $\langle \phi(\mathbf{r}) \rangle = \exp(iB \cdot \mathbf{r})$ satisfy it, provided the following critical equation is obeyed:

$$B_0^2 - B^2 = \frac{\langle \epsilon^2 \rangle B_0^4 l^2}{B^2 l^2 + (1 - iB_0 l)^2}, \tag{8.12}$$

where we have used

$$G(\mathbf{r}_0 \to \mathbf{r}) = \frac{1}{4\pi|\mathbf{r}_0 - \mathbf{r}|} \exp\{iB_0|\mathbf{r} - \mathbf{r}_0|\}. \tag{8.13}$$

As we shall see below, $\mathrm{Re}(B)^2$ is the critical buckling in the presence of the random homogeneity and B_0^2 is the conventional value given by the smallest eigenvalue of the Helmholtz equation $(\nabla^2 + B_0^2)\phi_0 = 0$.

We note that in general the buckling is complex. Indeed for small $\langle \epsilon^2 \rangle \ll 1$, we can write $B = B_r + iB_i$, where $B_r = B_0(1 - \frac{1}{8}\langle \epsilon^2 \rangle)$ and $B_i = \frac{1}{4}\langle \epsilon^2 \rangle B_0^2 l$. This shows that the critical flux shape has the form, for a slab reactor (width $2a$),

$$\langle \phi(x) \rangle \propto \cos(B_r x) \cosh(B_i x), \tag{8.14}$$

thus random dispersal tends to destroy the conventional normal mode shape. Such effects might well be detectable in certain reactor types, e.g. pebble bed or even boiling water. A more detailed analysis should be carried out, however, since in general the diffusion coefficient is also a random variable. Finally, applying the boundary condition to eqn. (8.14) leads to the critical condition noted earlier, viz.: $B_r = \pi/2a$.

It is interesting to note from a general point of view that random parametric disturbances affect the mean value; this is a consequence of randomness that is rarely considered in the context of reactor theory and deserves further investigation.

We shall not go into the calculation of cross-covariances here since the techniques follow those described above and can be found in the references cited. However, the reader is particularly advised to consult the works by Bourret (1962, 1965) which gives applications to a number of physical problems. The work on point model kinetics with randomly fluctuating parameters by Williams (1971) is also of interest in this respect, as also is that of Stepanov (1965, 1968) who applies many of Bourret's techniques in the calculation of cross-covariances in space. Finally, we mention a deterministic solution of eqn. (8.1) by Thie (1958), who used the WKB method.

8.4. *Application to self-shielding in random media*

It is often desirable to fabricate reactor fuel elements from heterogeneous mixtures of ceramics and metals. Also of interest are systems in which the fuel is in the form of loose pebbles mixed randomly in a liquid moderator-coolant. We also encounter in thermal neutron shields the substance known as boral, which is a heterogeneous mixture of commercial-grade boron carbide (B_4C) and aluminium, sandwiched between aluminium plates. The total sandwich is generally rolled to a thickness of about 3 mm. It is apparent from photomicrographs that the mixture is not uniform and aluminium regions are seen to be present between the lumps of B_4C in the B_4C–Al conglomerate.

Each of the situations discussed above (and others could arise) contains a degree of randomness that makes the usual methods of homogenization to find effective cross-sections of little use, unless the volume fractions of all but one of the phases is very small. In general the presence of two or more phases of comparable volume fraction introduces microscopic effects which are difficult to analyse. Nevertheless, it is essential that a method be developed that will allow the average nuclear properties of the proposed reactor components to be predicted. The basic problem arises from the spatial randomness and the poorly defined geometry of the lumps themselves; this makes it very difficult to predict with certainty the properties of this random heterogeneous mixture. If, however, we can obtain some knowledge of the statistical laws obeyed by the system then it may be possible to use these to obtain an average equation for the system. The same ideas will apply whether we are interested in the equation for transmission of neutrons through the medium or in the equation describing its average thermal, electrical, magnetic or any other physical property.

The ultimate goal of such an analysis is to replace the random heterogeneous microstructure by an effective homogeneous medium in the sense that the actual properties of the microstructure are reproduced using the effective parameters of the homogenized system. Moreover, it is to be hoped that these effective parameters will be independent of any particular problem and simply characterize the medium. In practice, this latter requirement may not be possible to achieve.

In a very detailed treatment of this problem, Randall (1960) has studied the effect of random microstructure on the integral transport equation and has derived a modified form of this equation containing a transport kernel involving the probability distributions which describe the randomness of the microstructure. Let us consider this technique, for it involves some of the ideas discussed earlier and has important applications.

We may write the integral form of the neutron transport equation for a medium of arbitrary composition in the following manner:

$$\phi(\mathbf{r},E,\mathbf{\Omega}) = \int_0^\infty dl \exp\left\{-\int_0^l \Sigma(E,\mathbf{r}-l'\mathbf{\Omega})\,dl'\right\} q(\mathbf{r}-l\mathbf{\Omega},E,\mathbf{\Omega}), \tag{8.15}$$

where $\phi(\mathbf{r},E,\mathbf{\Omega})$ is the angular flux and

$$\begin{aligned} q(\mathbf{r},E,\mathbf{\Omega}) &= S(\mathbf{r},E,\mathbf{\Omega})+\int d\mathbf{\Omega}'\int dE'\Sigma(E'\to E;\mathbf{\Omega}'\to\mathbf{\Omega};\mathbf{r})\phi(\mathbf{r}',E',\mathbf{\Omega}') \\ &\quad +\frac{1}{4\pi}\chi(E,\mathbf{r})\int d\mathbf{\Omega}'\int dE'\bar{\nu}\Sigma_f(E',\mathbf{r})\phi(\mathbf{r},E',\mathbf{\Omega}') \\ &\equiv S(\mathbf{r},E,\mathbf{\Omega})+\int d\mathbf{\Omega}'\int dE'\phi(\mathbf{r},E',\mathbf{\Omega})\,Q(E'\to E;\mathbf{\Omega}'\to\mathbf{\Omega};\mathbf{r}). \end{aligned} \tag{8.16}$$

In the equation for q, S is an independent source, $\Sigma(\ldots)$ is the differential scattering cross-section and $\chi(\ldots)$ is the fission spectrum.

Let us now consider a heterogeneous medium composed of a random two-phase microstructure as illustrated in Fig. 8.1.

We define $F_0(\mathbf{r})$ to be a function such that it is unity in phase 1 and zero in phase 2. Since the phases are randomly mixed, $F_0(\mathbf{r})$ will be a random function and hence when its statistics are given the statistics of the resulting random solution are also, in principle, calculable. In Fig. 8.1, Δ_i is the phase 1 intercept of the ith lump and we define the phase 1 intercept fraction $F(\mathbf{r},\mathbf{l})$ as

$$\begin{aligned} F(\mathbf{r},\mathbf{l}) &= \frac{1}{l}\sum_i \Delta_i \\ &\equiv \frac{1}{l}\int_0^l ds\,F_0(\mathbf{r}-s\mathbf{\Omega}), \end{aligned} \tag{8.17}$$

where $l = |\mathbf{l}|$.

The calculation of the actual probability distributions for $F(\mathbf{r},\mathbf{l})$ involves knowledge of the physical construction of the material and will not be easy to obtain. However, certain assumptions will be made and we shall find that the distribution function $p(F|\mathbf{r},\mathbf{l})\,dF$, i.e. the probability that $F(\mathbf{r},\mathbf{l})$ lies between F and $F+dF$, will be of interest as also will the joint probability density function $P(F,F_0(\mathbf{r}),F_0(\mathbf{r}-\mathbf{l})|\mathbf{r},\mathbf{l})\,dF_0(\mathbf{r})\,dF_0(\mathbf{r}-\mathbf{l})\,dF$. As an example of the meaning of $p(F,\mathbf{r},\mathbf{l})$ let us assume that $F_0(\mathbf{r}) = 1$ and $F_0(\mathbf{r}-\mathbf{l}) = 0$, then $p\,dF$ is the probability, given $\mathbf{r}$ and $\mathbf{l}$, that the vector $\mathbf{l}$ intercepts between lF and $l(F+dF)$ of phase 1, originates in phase 2 at $(\mathbf{r}-\mathbf{l})$ and terminates in phase 1 at $\mathbf{r}$.

In terms of our two phases and the functions F and F_0, we can rewrite the integral eqn. (8.15) as follows:

$$\phi(\mathbf{r},E,\mathbf{\Omega}) = \int_0^\infty dl \exp\{-l\Sigma_2(E)+F(\mathbf{r},\mathbf{l})\Delta\Sigma(E)\}q(\mathbf{r}-l\mathbf{\Omega},E,\mathbf{\Omega}), \tag{8.18}$$

where

$$\begin{aligned} q(\mathbf{r}-l\mathbf{\Omega},E,\mathbf{\Omega}) &= S(\mathbf{r}-l\mathbf{\Omega},E,\mathbf{\Omega})+\int d\mathbf{\Omega}'\int dE'\phi(\mathbf{r}-l\mathbf{\Omega},E',\mathbf{\Omega}') \\ &\quad \times\{Q_2(E'\to E;\mathbf{\Omega}'\to\mathbf{\Omega})+F_0(\mathbf{r}-l\mathbf{\Omega})\Delta Q(E'\to E;\mathbf{\Omega}'\to\mathbf{\Omega})\}. \end{aligned} \tag{8.19}$$

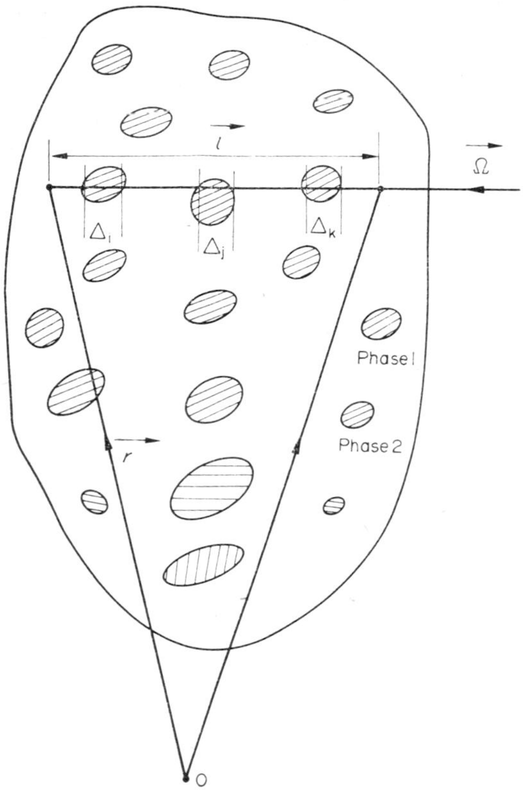

FIG. 8.1. A randomly dispersed two-phase mixture. The vector $\mathbf{\Omega}$ denotes direction of neutron motion and Δ_i is the intercept through the ith component of phase 1 in the direction $\mathbf{\Omega}$.

$\Sigma_2(E)$ is the total cross-section in phase 2 and the difference in cross-sections or sources between the two phases is written $\Delta\Sigma$ or ΔS as the case may be, e.g. $\Delta\Sigma = \Sigma_1 - \Sigma_2$.

As we stated above, given $p(F)$ and $p(F_0)$ we can calculate $p(\phi)$ and hence obtain the mean value as

$$\langle\phi(\mathbf{r},E,\mathbf{\Omega})\rangle = \int d\phi\, p(\phi)\,\phi \tag{8.20}$$

and the covariance functions. In this section, however, we shall concern ourselves only with mean values.

Let us now statistically average eqn. (8.18) as it stands to obtain

$$\begin{aligned}\langle\phi(\mathbf{r},E,\mathbf{\Omega})\rangle &= \int_0^\infty dl\, S(\mathbf{r}-l\mathbf{\Omega},E,\mathbf{\Omega})\,e^{-l\Sigma_2(E)}\langle e^{-lF\Delta\Sigma(E)}\rangle \\ &+ \int_0^\infty dl \int d\mathbf{\Omega}' \int dE'\, e^{-l\Sigma_2(E)}\{Q_2(E'\to E;\mathbf{\Omega}'\to\mathbf{\Omega}) \\ &\times \langle e^{-lF\Delta\Sigma(E)}\phi(\mathbf{r}-l\mathbf{\Omega},E',\mathbf{\Omega}')\rangle + \Delta Q(E'\to E;\mathbf{\Omega}'\to\mathbf{\Omega}) \\ &\times \langle F_0(\mathbf{r}-\mathbf{l})\,e^{-lF\Delta\Sigma(E)}\phi(\mathbf{r}-l\mathbf{\Omega},E',\mathbf{\Omega}')\rangle\}.\end{aligned} \tag{8.21}$$

We observe immediately the problem of closure which bedevils all attempts to solve statistical problems by direct averaging. In this case it is necessary to make some assumptions regarding $\langle \exp\{-lF\Delta\Sigma\}\phi\rangle$. The only case in which an explicit result for $\langle\phi\rangle$ is available from eqn. (8.21) is for a non-scattering, non-fissile material, i.e. a purely absorbing material with an inhomogeneous source term; we then must evaluate $\langle \exp\{-lF\Delta\Sigma\}\rangle$ which, given the statistics of F, is straightforward and unambiguous. In any other situation F and ϕ are not statistically independent variables and it is therefore not necessarily permissible to write

$$\langle F_0 e^{-lF\Delta\Sigma}\phi\rangle = \langle F_0 e^{-lF\Delta\Sigma}\rangle\langle\phi\rangle. \tag{8.22}$$

In view of this difficulty certain physical properties of the system are introduced which enable the following assumptions to be made:

1. Above resonance energies, $\Delta\Sigma$ is so small that $(\phi-\langle\phi\rangle)/\langle\phi\rangle$ is negligible compared with unity (this implies that the microscopic self-shielding factors are nearly unity).
2. The major contribution to the scattering-in integral is due to the scattering of epithermal neutrons. For hydrogenous substances this is certainly valid, although the assumption could not be expected to hold in the thermal neutron region less than 1 eV. On this assumption, however, we can write

$$\langle e^{-l(\Sigma_2(E)+F\Delta\Sigma(E))}\{\Sigma_2(E'\to E;\mathbf{\Omega}'\to\mathbf{\Omega})+F_0(\mathbf{r}-l\mathbf{\Omega})\Delta\Sigma(E'\to E;\mathbf{\Omega}'\to\mathbf{\Omega})\}$$
$$\times\phi(\mathbf{r}-l\mathbf{\Omega},E',\mathbf{\Omega}')\rangle \simeq \langle e^{-l(\Sigma_2+F\Delta\Sigma)}\{\ldots\}\rangle\langle\phi(\mathbf{r}-l\mathbf{\Omega},E',\mathbf{\Omega}')\rangle, \tag{8.23}$$

since, by assumption, at the energies which contribute most to the scattering integral, $\phi\simeq\langle\phi\rangle$. In the absence of a fission source, therefore, we may write a closed equation for $\langle\phi\rangle$ as follows:

$$\langle\phi(\mathbf{r},E,\mathbf{\Omega})\rangle = \int_0^\infty dl\, S(\mathbf{r}-l\mathbf{\Omega},E,\mathbf{\Omega})\,e^{-l[\Sigma_2(E)+f_{mi}(E)V\Delta\Sigma(E)]}$$
$$+\int_0^\infty dl\int d\mathbf{\Omega}'\int dE'\,e^{-l[\Sigma_2(E)+f_{mi}(E)V\Delta\Sigma(E)]}\langle\phi(\mathbf{r}-l\mathbf{\Omega},E',\mathbf{\Omega}')\rangle$$
$$\times\{\Sigma_2(E'\to E;\mathbf{\Omega}'\to\mathbf{\Omega})+Vg_{mi}(E)\Delta\Sigma(E'\to E;\mathbf{\Omega}'\to\mathbf{\Omega})\}, \tag{8.24}$$

where V is the volume fraction of phase 1 and f_{mi} and g_{mi} are *microscopic self-shielding factors* defined as follows:

$$\langle e^{-lF\Delta\Sigma(E)}\rangle = \int_0^1 dF p(F|l)\,e^{-lF\Delta\Sigma(E)}$$
$$\equiv e^{-Vf_{mi}(E)l\Delta\Sigma(E)}, \tag{8.25}$$

$$\langle F_0 e^{-lF\Delta\Sigma(E)}\rangle = \int_0^1 dF\int_0^1 dF_0\, p(F,F_0|l)\,F_0 e^{-lF\Delta\Sigma(E)}$$
$$\equiv Vg_{mi}(E)\,e^{-Vf_{mi}(E)l\Delta\Sigma(E)}. \tag{8.26}$$

In general, f_{mi} and g_{mi} are dependent on l so that the normal transport kernel for a homogeneous medium does not arise in the equation for $\langle\phi\rangle$.

When the assumptions made in 1 and 2 do not hold, then the problem becomes much more difficult. In particular ϕ and Σ_f are closely correlated at thermal energies where their product contributes most towards the fission source $\chi(E)R_f$. To overcome these difficulties, Randall makes the assumption that ϕ can be approximated in the following stepwise manner from phase to phase, viz.

$$\phi \simeq \phi_A \equiv \frac{\langle\phi\rangle}{1-V}[1-\alpha V+F_0(\mathbf{r})(\alpha-1)], \tag{8.27}$$

where α is a non-random variable which may, however, depend upon $\mathbf{r}$, $\mathbf{\Omega}$ and E.

From eqn. (8.27) it follows that

$$\langle\phi-\phi_A\rangle = 0, \tag{8.28}$$

$$\langle\phi_1\rangle = \text{ensemble average of } \phi \text{ in phase 1} \equiv \frac{1}{V}\langle\phi F_0\rangle \simeq \alpha\langle\phi\rangle, \tag{8.29}$$

$$\langle\phi_2\rangle = \text{ensemble average of } \phi \text{ in phase 2} \equiv \frac{1}{1-V}\langle\phi(1-F_0)\rangle \tag{8.30}$$

$$= \frac{(1-\alpha V)}{(1-V)}\langle\phi\rangle \equiv \alpha_2\langle\phi\rangle, \tag{8.31}$$

hence we have for α the definition

$$\alpha = \langle\phi F_0\rangle/V\langle\phi\rangle \tag{8.32}$$

the evaluation of which will be discussed below.

Using the approximation for ϕ given above, we may now write the stochastically averaged equation for $\langle\phi\rangle$ as follows:

$$\begin{aligned}\langle\phi(\mathbf{r},E,\mathbf{\Omega})\rangle = \int_0^\infty dl \exp\{-l[\Sigma_2(E)+f_{mi}(E)V\Delta\Sigma(E)]\}\{S(\mathbf{r}-l\mathbf{\Omega},E,\mathbf{\Omega}) \\ +\int d\mathbf{\Omega}'\int dE'\langle\phi(\mathbf{r}-l\mathbf{\Omega},E',\mathbf{\Omega}')\rangle[\alpha_2(E')Q_2(E'\to E;\mathbf{\Omega}'\to\mathbf{\Omega}) \\ +g_{mi}(E)V\Delta(\alpha(E')Q(E'\to E;\mathbf{\Omega}'\to\mathbf{\Omega}))]\},\end{aligned} \tag{8.33}$$

where $\Delta(\alpha Q) \equiv \alpha Q_1-\alpha_2 Q_2$.

The unknown factor α, in effect, can only be evaluated once $\langle\phi\rangle$ is known: clearly an undesirable situation. However, a reliable estimate of α can be obtained by writing eqn. (8.18) in the integro-differential form and statistically averaging at that stage. We then obtain

$$\begin{aligned}\mathbf{\Omega}.\nabla\langle\phi(\mathbf{r},E,\mathbf{\Omega})\rangle+\langle\Sigma(E,\mathbf{r})\phi(\mathbf{r},R,\mathbf{\Omega})\rangle \\ = S(\mathbf{r},E,\mathbf{\Omega})+\int d\mathbf{\Omega}'\int dE'\langle\phi(\mathbf{r},E',\mathbf{\Omega}')Q(E'\to E;\mathbf{\Omega}'\to\mathbf{\Omega};\mathbf{r})\rangle.\end{aligned} \tag{8.34}$$

Using the approximate form for ϕ in this equation and reconverting it to integral form leads to

$$\begin{aligned}\langle\phi(\mathbf{r},E,\mathbf{\Omega})\rangle = \int_0^\infty dl\exp\{-l[\Sigma_2(E)+\alpha(E)V\Delta\Sigma(E)]\}\{S(\mathbf{r}-l\mathbf{\Omega},E,\mathbf{\Omega}) \\ +\int d\mathbf{\Omega}'\int dE'\langle\phi(\mathbf{r}-l\mathbf{\Omega},E',\mathbf{\Omega}')\rangle[Q_2(E'\to E;\mathbf{\Omega}'\to\mathbf{\Omega}) \\ +\alpha(E')V\Delta Q(E'\to E;\mathbf{\Omega}'\to\mathbf{\Omega})]\}.\end{aligned} \tag{8.35}$$

Comparing the powers of the exponential term in eqn. (8.35) with that in eqn. (8.33) allows us to set $\alpha \simeq f_{mi}$. We therefore have a closed equation for $\langle\phi\rangle$.

Whilst the equation for $\langle\phi\rangle$ would provide an adequate description of neutron transport in the microstructure, it would be more convenient to solve a conventional transport equation for a homogeneous medium with effective cross-sections which themselves take into account the heterogeneity of the medium in a suitable weighted fashion. Thus if we write the homogeneous transport equation as

$$\langle\phi(\mathbf{r},E,\mathbf{\Omega})\rangle = \int_0^\infty dl \exp[-l\Sigma_{\text{eff}}(E)]\{S(\mathbf{r}-l\mathbf{\Omega},E,\mathbf{\Omega}) + \int d\mathbf{\Omega}' \int dE' \langle\phi(\mathbf{r}-l\mathbf{\Omega}',E',\mathbf{\Omega}')\rangle\, Q_{\text{eff}}(E' \to E;\mathbf{\Omega}' \to \mathbf{\Omega})\} \quad (8.36)$$

and compare it with eqn. (8.33), we can identify the effective cross-sections as follows provided that g_{mi} and f_{mi} are independent of l, viz.

$$\Sigma_{\text{eff}}(E) = \Sigma_2(E)[1-Vf_{mi}(E)]+\Sigma_1(E)f_{mi}(E)V, \quad (8.37)$$

$$Q_{\text{eff}}(E' \to E;\mathbf{\Omega}' \to \mathbf{\Omega}) = \frac{[1-Vf_{mi}(E')][1-Vg_{mi}(E)]}{1-V} Q_2(E' \to E;\mathbf{\Omega}' \to \mathbf{\Omega}) + Vf_{mi}(E')g_{mi}(E)Q_1(E' \to E;\mathbf{\Omega}' \to \mathbf{\Omega}). \quad (8.38)$$

In general these effective cross-sections will not be independent of l or indeed of the source S. If the dependence on l is too strong then it may be necessary to solve the "exact" equation for $\langle\phi\rangle$ or to reduce our demands on the effective parameters. For example, by integrating over the angular dependence of $\langle\phi\rangle$ we would have an equation for the total flux, and an attempt to find effective cross-sections that preserved this scalar flux might produce results less sensitive to geometry or source. The effective cross-sections may also be allowed to depend on position in some deterministic way. It is important to ensure, however, that irrespective of the method of obtaining effective cross-sections, the product of cross-section and flux produce the true reaction rate. Thus whilst in principle $\langle\Sigma\phi\rangle$ is the true reaction rate, Σ_{eff} should be arranged so that $\Sigma_{\text{eff}}\langle\phi\rangle$ is as close as possible to this exact value.

8.5. *The microstructure distribution function*

Any further progress in this problem requires the specification of $p(F_0,F|l)$ and $p(F|l)$. These are difficult to obtain and most calculations to date either oversimplify the problem or are intractable. However, some convenient analytical forms, based partly on phenomenological arguments, have been obtained for $p(F_0,F|l)$ and $p(F|l)$ by Randall in terms of the following experimentally obtainable parameters:

V = volume fraction of the lumps,

μ = mean particle intercept of a lump,

σ_Δ^2 = variance of the particle intercepts.

Methods for measuring these parameters can be found in Allio and Randall (1962).

The actual distributions themselves are as follows:

$$p(F|l) = P_0\delta(F)+P_1\delta(F-1)+[1-P_0-P_1]\beta\left[F;\frac{V'l}{\Gamma},(1-V')l/\Gamma\right], \quad (8.39)$$

$$p(F,F_0|l) = \{P_0\delta(F)+(1-P_0-P_1)\beta[F;V'l/\Gamma,(1-V')l/\Gamma](1-F)\}\delta(F_0) + \{P_1\delta(F-1)+(1-P_0-P_1)\beta[F;V'l/\Gamma,(1-V')l/\Gamma]F\}\delta(F_0-1), \quad (8.40)$$

where $P_0 = (1-V)\exp\{-Vl/\mu(1-V)\}$ is the probability of zero coverage,

$$P_1 = V\exp\{-l/\mu\}$$

is an approximation to the probability of full coverage, Γ is a measure of the average particle size defined by

$$\Gamma = \mu(1-V)\left\{1+\frac{\sigma_\Delta^2}{\mu^2}-\frac{Vl}{\mu}\frac{P_0}{(1-P_0)(1-V)^2}\right\}, \tag{8.41}$$

$$V' = \frac{(V-P_0)}{(1-P_0-P_1)} \tag{8.42}$$

and $\beta[F;\, x,y]$ is the β-distribution given by

$$\beta[F;\, x,y] \equiv \frac{F^{x-1}(1-F)^{y-1}}{B[x,y]} \tag{8.43}$$

with $B[x,y]$ the β-function, defined as

$$B[x,y] = \frac{\Gamma(x)\Gamma(y)}{\Gamma(x+y)}. \tag{8.44}$$

These distributions whilst rather complicated have been deduced by the use of exact limiting theorems proved by Halperin (1959) and interpolation using a number of probability arguments based on the "Polya Urn Scheme" (Feller, 1957). It may be verified that $p(F, F_0|l)$ when integrated over F_0 leads to $p(F|l)$. Moreover, the mean and variance of F from eqn. (8.39) are

$$\bar{F} = V \tag{8.45}$$

and

$$\overline{F^2}-\bar{F}^2 = \left\{P_1(1-V')+V(V-V')+\frac{V\Gamma}{l}(1-V)\right\}\Big/(1+\Gamma/l). \tag{8.46}$$

In addition it may be shown that a distribution which preserves the same mean and variance as eqn. (8.39) is

$$p(F|l) \simeq \left(1-\frac{Vl}{\mu}\right)\delta(F)+\left(\frac{Vl}{\mu}\right)\beta\left[F;\left(\frac{\mu^2}{\sigma_\Delta^2}\right),\left(\frac{l}{\mu}-1\right)\left(\frac{\mu^2}{\sigma_\Delta^2}\right)\right], \tag{8.47}$$

which may be easier to handle.

8.6. *Microscopic self-shielding factors*

Using the distributions in the previous sections we can evaluate f_{mi} and g_{mi}. We find after some algebra that

$$\begin{aligned}\langle e^{-lF\Delta\Sigma}\rangle &= P_0+P_1e^{-l\Delta\Sigma}+(1-P_0-P_1)\Phi\left[\frac{V'l}{\Gamma},\frac{l}{\Gamma};\,-l\Delta\Sigma\right]\\ &\equiv e^{-lVf_{mi}\Delta\Sigma}\end{aligned} \tag{8.48}$$

and

$$\begin{aligned}\langle F_0e^{-lF\Delta\Sigma}\rangle &= P_1e^{-l\Delta\Sigma}+(1-P_0-P_1)V'\Phi\left[\frac{V'l}{\Gamma}+1,\frac{l}{\Gamma}+1;\,-l\Delta\Sigma\right]\\ &\equiv g_{mi}Ve^{-lVf_{mi}\Delta\Sigma},\end{aligned} \tag{8.49}$$

where $\Phi(a,b;\,x)$ is the confluent hypergeometric function.

Equation (8.48) has an interesting physical interpretation if we set $\Sigma_2 = 0$ and $\Delta\Sigma = \Sigma_1$, i.e. have a voided system. In that case, P_0 is the probability of neutrons intercepting no phase 1 component, with unity as the transmission factor, $(1-P_0-P_1)$ is the probability of neutrons intercepting some phase 1 with $\Phi[\ldots]$ the transmission factor and, finally, P_1 is the probability of neutrons intercepting phase 1 only and $\exp(-l\Sigma_1)$ their transmission factor.

From the definitions of f_{mi} and g_{mi} we readily find that

$$f_{mi} = -\frac{1}{Vl\Delta\Sigma} \ln \left\{P_0 + P_1 e^{-l\Delta\Sigma} + (1-P_0-P_1)\Phi\left[\frac{V'l}{\Gamma}, \frac{l}{\Gamma}; -l\Delta\Sigma\right]\right\} \quad (8.50)$$

and

$$g_{mi} = \frac{\left\{P_1 e^{-l\Delta\Sigma} + (1-P_0-P_1)V'\Phi\left[\frac{V'l}{\Gamma}+1, \frac{l}{\Gamma}+1; -l\Delta\Sigma\right]\right\}}{V\left\{P_0 + P_1 e^{-l\Delta\Sigma} + (1-P_0-P_1)\Phi\left[\frac{V'l}{\Gamma}, \frac{l}{\Gamma}; -l\Delta\Sigma\right]\right\}}. \quad (8.51)$$

It will be observed that, in general, the self-shielding factors do depend on l.

Before we discuss these factors in any detail, it is worth noting that a much simpler approach to the calculation of transmission factors through boral slabs has been given by Burrus (1960). His arguments are based on a purely absorbing medium and result in a total, orientation-averaged value. He argues that the transmission probability T for a slab of thickness t containing lumps of mean chord length $\bar{l}$ is given by

$$T = \sum_{n=0}^{N} P_n \tau(n),$$

where $N = t/\bar{l}$, P_n is thc probability that a ray will encounter exactly n lumps in traversing the slab and $\tau(n)$ is the probability that a neutron will penetrate n slabs.

Assuming a random distribution, the probability P_n is given by the Bernoulli distribution, viz.

$$P_n = V^n(1-V)^n \binom{N}{n}. \quad (8.52)$$

Now if F is the probability that a neutron will not penetrate a lump, then $(1-F)$ is the probability that it will penetrate. The probability $\tau(n)$ of penetration of n lumps is therefore $(1-F)^n$. The transmission factor T is now given by

$$T = \sum_{n=0}^{N} V^n(1-V)^{N-n} \binom{N}{n} (1-F)^n$$
$$= [1-VF]^N. \quad (8.53)$$

According to transport theory (Case, de Hoffmann and Placzek, 1953) $F = \bar{l}\Sigma(1-P_c)$, where P_c is the self-collision probability of the lump and $\bar{l} = 4 \times$ volume of lump/surface area. P_c is extensively tabulated for various shaped lumps but a reasonable approximation for any shape is given by Wigner's rational approximation $P_c \simeq \Sigma\bar{l}/(1+\Sigma\bar{l})$. If we are to equate T to an expression of the form $T = \exp\{-V\Sigma_{\text{eff}}T\}$, then Σ_{eff} is defined as

$$\Sigma_{\text{eff}} = -\frac{1}{\bar{l}V} \ln(1-VF). \quad (8.54)$$

It should be noted that the material between lumps is taken as void.

Calculations based on the above model have been made by Burrus for a boral sandwich rolled to a thickness of 0·125 in. with the B_4C–Al mixture 0·085 in. in thickness. The overall volume fraction of the absorbing lumps, which were assumed to be spherical in shape, was 25%. For different size groups with average particle diameter varying from 0·009 in. to 0.038 in, the transmission factor calculated for 2200 m/sec neutrons was 0·076. This is to be compared with a value of 0·0015 obtained from the homogeneous approximation of volume weighted cross-sections. Results such as these clearly demonstrate the importance of a detailed treatment of the microstructure.

Let us return now to f_{mi} and g_{mi}. We would, of course, like to find that these quantities are insensitive to l in practical cases and thus only dependent on the microscopic properties of the two phases. A series of numerical calculations of f_{mi} and g_{mi} has been made by Randall for the ranges $0 \leqslant l/\mu \leqslant \infty$, $0 \leqslant \mu\Delta\Sigma \leqslant 3$,

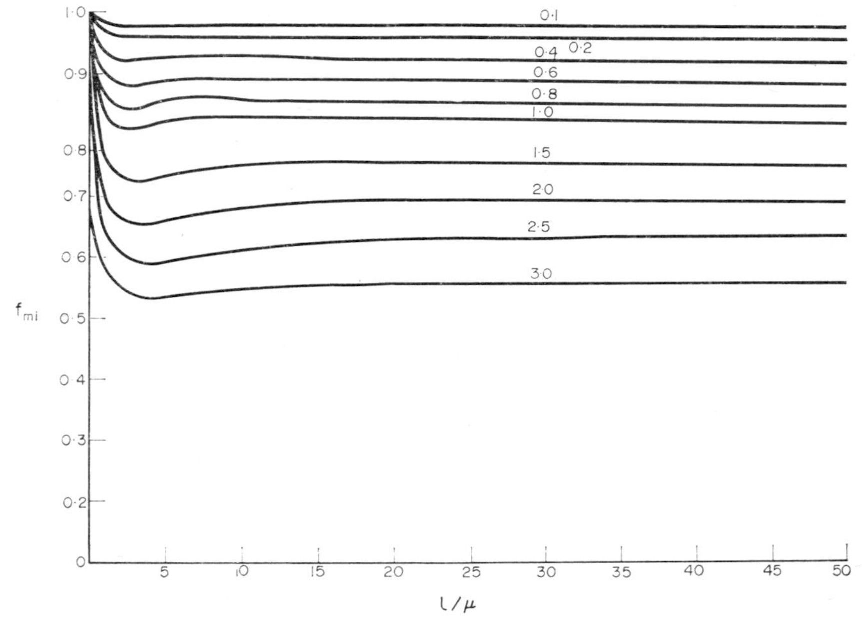

FIG. 8.2. The self-shielding factor f_{mi} as a function of l/μ for various values of $\mu\Delta\Sigma$. The value of $V = 0{\cdot}4$ and $(1+\sigma_\Delta^2/\mu^2) = 1{\cdot}125$. (From Randall (1960) USAEC, KAPL-M-CHR-2.)

$0 \leqslant V \leqslant 1{\cdot}0$. Figures 8.2 and 8.3 show some typical results obtained. Clearly, f_{mi} and g_{mi} are not independent of l; however, the figures show that over an appreciable part of the domain they are relatively insensitive to it, generally for $l/\mu \gtrsim 5$. If therefore we are to apply our effective cross-sections usefully (i.e. independent of l and hence effectively homogeneous) it is necessary to ensure that the effective mean free path, λ_{eff}, of neutrons is very much greater than 5μ; we can then expect most neutrons to have paths greater than 5μ. It is worth while considering the consequences of requiring $l \gg 5\mu$. Certainly if L is some characteristic dimension of the body we should have $L \gg 5\mu$, therefore the body must be large compared with the microstructure. In addition, we know that $\lambda_{\text{eff}} \sim 1/\Sigma_{\text{eff}}$, thus our limitation becomes equivalent to $1 \gg 5\mu\Sigma_{\text{eff}} \simeq 5\mu(\Sigma_2 + Vf_{mi}\Delta\Sigma)$. Considering B_4C and UO_2 dispersion in zircaloy (phase 2, $\Sigma_2 = 0$) and assuming $\mu = 100$ microns, we find that for natural B_4C, $\mu\Delta\Sigma \simeq 0{\cdot}5$ in the thermal group and for highly enriched UO_2 it is about 0·10. For

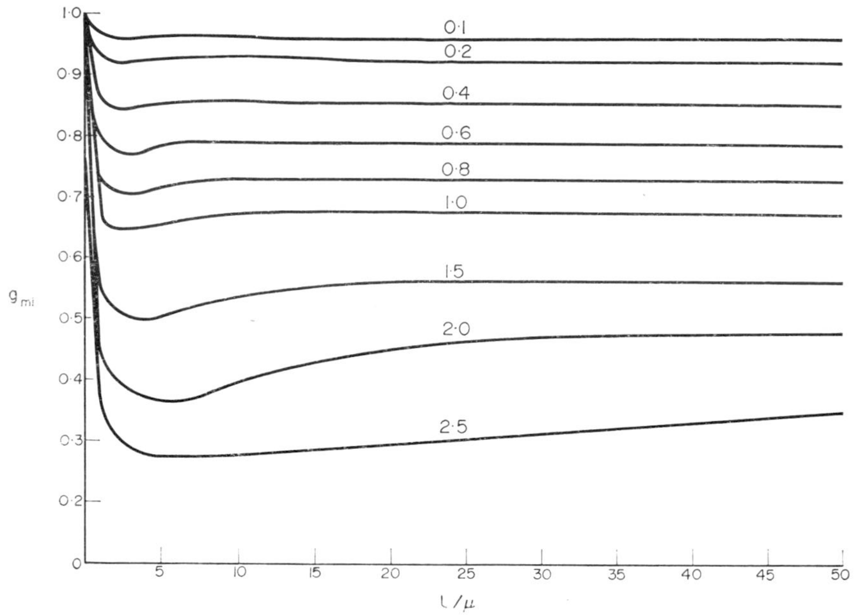

FIG. 8.3. The self-shielding factor g_{mi} as a function of l/μ for various values of $\mu\Delta\Sigma$. The value of $V = 0{\cdot}4$ and $(1+\sigma_\Delta^2/\mu^2) = 1{\cdot}125$. (From Randall (1960), USAEC, KAPL-M-CHR-2.)

these values the inequalities become $0{\cdot}4 \gg f_{mi}V$ for B_4C and $2 \gg f_{mi}V$ for UO_2, where $f_{mi} \simeq 0{\cdot}85$. Volume fractions must therefore be very much less than 0·5 for B_4C and unity for UO_2. Further examples are given by Randall who stresses that in any particular problem all contributing factors should be considered before applying the factors f_{mi} and g_{mi}: it is essential, for example, to make sure that the case under investigation falls into the region where f_{mi} and g_{mi} are relatively insensitive to l.

In connection with the general approach developed above by Randall, it is of interest to note that considerations of the effect of randomness on the first flight probability have been made by Barrett and Thompson (1969) in connection with boiling and the consequent two-phase nature of the moderating medium. Boiling is a problem which should in principle account for the simultaneous effect of space and time in determining its stochastic effect on neutron behaviour since the two phases are themselves time dependent, as opposed to the situation discussed above. Nevertheless, any progress, however limited, in this difficult subject is worth while. Barrett and Thompson are concerned with the effect of the vapour voids on the collision probabilities involved in the solution of the integral transport equation. They assume that scattering is isotropic in the laboratory system of coordinates and can therefore write an integral equation for the scalar flux. On application of stochastic averaging and use of the closure approximation of eqn. (8.22) (without much justification), they are faced with the calculation of statistical averages in the form

$$\left\langle \exp\left\{-\int_0^x \Sigma(y,u)\,dy\right\}\right\rangle, \tag{8.55}$$

$\Sigma(y,u)$ being a random variable (cross-section at position y, lethargy u). The statistical average has been rewritten as

$$\langle p(x,u)\rangle = \int_0^1 \exp[\sigma(u)\{M_L(1-\alpha)+M_V\alpha\}x]P(\alpha)\,d\alpha, \tag{8.56}$$

where M_L is the molecular concentration in the liquid phase and M_V in the vapour phase. α is the total void fraction along the path of length x and $P(\alpha)\,d\alpha$ is the probability that a neutron travelling a distance x will encounter a void fraction between α and $\alpha+d\alpha$.

In the absence of detailed knowledge of the statistics of the local void distribution a number of probability laws are tried, the most consistent being a Gaussian with mean value $\langle\alpha\rangle$ and variance $\sigma^2(\alpha)$ in the range $0 \leqslant \alpha \leqslant 1$. We shall not write down the expression for $\langle p(x,u)\rangle$ here but simply give the values of the factor

$$\epsilon = \left[\frac{\langle p(x,u)\rangle}{p(x,u)}-1\right]\times 100\%, \tag{8.57}$$

where

$$p(x,u) = \exp[-\sigma(u)\{M_L(1-\langle\alpha\rangle)+M_V\langle\alpha\rangle\}x] \tag{8.58}$$

is the "conventional" average. The results are given in Table 8.1 for 20 eV neutrons traversing a path of 4 cm through boiling H_2O.

TABLE 8.1

$\langle\alpha\rangle$	0·5	0·5	0·5	0·25	0·25	0·25	0·1	0·1	0·1
$\sigma(\alpha)/\langle\alpha\rangle$	0·1	0·15	0·25	0·1	1·15	0·25	0·1	0·15	0·25
ϵ	4·5	10·4	23·4	1·1	2·5	7·1	0·11	0·39	1·1

The comment is made that for slug flow where $\sigma(\alpha)/\langle\alpha\rangle$ may well be in excess of 0·25, the effect of randomness in the medium properties is significant in increasing the probability that a neutron will traverse all or part of the random medium without suffering a collision. One might add to this comment that the effect of randomly moving voids could have an equally important effect, but whether this would enhance or diminish the static effect is not yet clear. We complete this section by referring the reader to some work by Yamagishi and Sekiya (1968) which obtains an exact one-speed transport theory solution for neutrons diffusing from a plane source in a system containing random gaps, distributed in space according to a Poisson probability law.

8.7. *Localized noise sources*

We consider now a heterogeneous reactor and describe a method which will enable localized noise sources to be described. Such sources can arise, for example, from the random vibration of fuel elements or control rods, uncertainties in the fissile or absorbing concentration of each rod due to variations in manufacture (as described in the last section), random variations in the mean position of a fuel element or control rod and finally the variable composition of the moderator due either to manufacturing tolerances or the existence of boiling where this exists.

A mathematical formulation of the above problem is indeed formidable but, nevertheless, if we are to employ noise analysis as a method of continuous surveillance of reactor behaviour with the aim of incipient malfunction detection then such a complete treatment will, in principle, be necessary. To put it otherwise, it is not sufficient to know that a malfunction is present, as one might obtain from a homogeneous reactor model, but to pinpoint the source of the trouble. The only way in which this can be done is to build into the model the location of the important noise sources.

It is clear that a detailed heterogeneous transport theory calculation or even diffusion theory approximation will not be profitable except perhaps at the expense of substantial numerical work on a specific system which, in turn, would not lead to generalizations of any significance. It is therefore necessary to examine other methods of representing the situation and an obvious choice is the so-called source-sink method (Horning, 1954) or, as it is better known, the method of Feinberg and Galanin (Galanin, 1960). In this method, the fuel elements and control rods, and indeed any isolated component, are represented as a line sink of thermal neutrons and, in the case of fissile components, as a line source of fast neutrons. Such a description is naturally an approximation since it neglects the finite size of the element concerned and treats it as a filament. Nevertheless, there are considerable advantages to be gained by this method which compensate for the assumptions involved.

Let us consider, then, a reactor system with axially orientated fuel elements, each of differing composition and located at points in the system denoted by the two-dimensional vector $\boldsymbol{\xi}_{\mathbf{k}}(z,t)$, where $\boldsymbol{\xi}_{\mathbf{k}} = (x_{k_1}, y_{k_2})$, k_1 and k_2 being integers locating the element in the lattice. We note, however, that because an element may be bent or in vibration, the location in the (x,y) plane will depend on axial height z and time t. With this definition of the element's position, we can define mathematically the effect of any element on the thermal neutron behaviour by the absorbing term

$$\gamma_{\mathbf{k}}(z)\,N(\boldsymbol{\xi}_{\mathbf{k}}(z,t),z,t)\,\delta(\boldsymbol{\xi}-\boldsymbol{\xi}_{\mathbf{k}}(z,t)), \tag{8.59}$$

where $\gamma_{\mathbf{k}}(z)$ is the strength of the absorber at axial position z (Galanin's constant), $N(\boldsymbol{\xi},z,t)$ is the thermal neutron density at the point (x,y,z) at time t and the delta function shows that the element is a line filament located at $\boldsymbol{\xi}_{\mathbf{k}}(z,t)$. In the case of a fissile element we can represent the number of fast neutrons born at $\boldsymbol{\xi}_{\mathbf{k}}(z',t)$ leading to thermal neutrons at the point $(\boldsymbol{\xi},z)$ by the expression

$$\eta_{\mathbf{k}}(z')\,\gamma_{\mathbf{k}}(z')\,K(\boldsymbol{\xi}_{\mathbf{k}}(z',t),z' \to \boldsymbol{\xi},z)\,N(\boldsymbol{\xi}_{\mathbf{k}}(z',t),z',t), \tag{8.60}$$

where $K(\boldsymbol{\xi}',z' \to \boldsymbol{\xi},z)$ is the slowing-down kernel, assuming instantaneous slowing down, and $\eta_{\mathbf{k}}(z')$ is the thermal fission factor at $(\boldsymbol{\xi}_{\mathbf{k}}(z',t),z')$ (i.e. the number of fast neutrons produced in fission per thermal neutron captured).

With these definitions in mind we can write for the complete reactor the following balance equation,

$$\begin{aligned}\frac{\partial N(\boldsymbol{\xi},z,t)}{\partial t} = {} & \nabla . D_0(\boldsymbol{\xi},z,t)\,\nabla N(\boldsymbol{\xi},z,t) - \overline{v\Sigma_{am}(\boldsymbol{\xi},z,t)}\,N(\boldsymbol{\xi},z,t) \\ & - \sum_{\mathbf{k}} \gamma_{\mathbf{k}}(z)\,N(\boldsymbol{\xi}_{\mathbf{k}}(z,t),z,t)\,\delta(\boldsymbol{\xi}-\boldsymbol{\xi}_{\mathbf{k}}(z,t)) \\ & + \sum_{\mathbf{k}} \int_0^L dz'[1-\beta_{\mathbf{k}}(z')]\,\eta_{\mathbf{k}}(z')\,\gamma_{\mathbf{k}}(z')\,K(\boldsymbol{\xi}_{\mathbf{k}}(z',t),z' \to \boldsymbol{\xi},z)\,N(\boldsymbol{\xi}_{\mathbf{k}}(z,t),z',t) \\ & + \sum_{\mathbf{k},i} \int_0^L dz'\,\lambda_{\mathbf{k},i}(z')\,\beta_{\mathbf{k},i}(z')\,\eta_{\mathbf{k}}\,\gamma_{\mathbf{k}}(z') \int_0^t dt'\,K_i(\boldsymbol{\xi}_{\mathbf{k}}(z',t'),z' \to \boldsymbol{\xi},z) \\ & \times \exp[-\lambda_{\mathbf{k},i}(z')(t-t')]\,N(\boldsymbol{\xi}_{\mathbf{k}}(z',t'),z',t') + S(\boldsymbol{\xi},z,t) = 0,\end{aligned} \tag{8.61}$$

where $\beta_{k,i}(z)$ is the delayed neutron fraction of the ith group at the position (ξ_k, z), $\lambda_{k,i}(z)$ is the corresponding decay constant of the delayed neutron precursor and $L = L(\xi, t)$ is the height of the reactor which due to motion of the boiling boundary may be random. $K_i(\ldots)$ is the slowing-down kernel for neutrons born from the ith delayed neutron precursor.

We note in the above equation that ξ_k is a random variable in space and time because of possible random vibrations of the elements; however, even for no vibration the time-independent quantity $\xi_k(z)$ could be random, corresponding to manufacturing and/or assembly error. Apart from intrinsic fluctuations due to fission, η_k and γ_k can be random due to non-uniform burn-up or manufacturing tolerances or temperature changes. Similarly, if the moderator is boiling water, D_0 and $\overline{v\Sigma_{am}}$ are random functions of space and time due to bubble formation and collapse: even in solid moderators D_0 and $\overline{v\Sigma_{am}}$ could fluctuate due to irregularities in the physical composition. Coupled with these equations we have the corresponding heat-transfer and fluid-flow feedback equations.

Basically, there are two separate ways in which the parameters entering the above equations can be random. To understand this we must recall that the coefficients D_0, $\overline{v\Sigma_{am}}$, η, γ, etc., are all parameters averaged over the thermal neutron spectrum. The simplest case is that of $\overline{v\Sigma_{am}}$ which we may write as $N_{am}\,\overline{v\sigma_{am}}$ where N_{am} is the number density of absorber atoms and $\overline{v\sigma_{am}}$ is the microscopic reaction rate for capture averaged over the thermal neutron spectrum. $N_{am} = N_{am}(\mathbf{r}, t)$ will be a random function of position and time, in general, because of the changes in its value caused by boiling and other mechanical effects. On the other hand, $\overline{v\sigma_{am}}$ will depend upon the physical temperature of the moderator through its effect on the thermal neutron spectrum. D_0 is similarly related to $N\overline{v\sigma}$ for scattering, also η and γ whilst depending on self-shielding factors will be similarly affected, although for solid-fuel elements the difficult problem of boiling will not affect the associated number density in the particular element.

It is clear from this discussion that no further progress can be made without a knowledge of the temperature distribution in the system. This temperature distribution arises from power production within the fuel elements, its conduction to the surface and subsequent transfer to the coolant. In addition, the coolant velocity will be determined by the rate of heat removal necessary to balance the power production. The bubble formation in the coolant will depend on the surface nucleation sites on the fuel element surface, which will also affect the heat-transfer coefficient. Turbulent pressure fluctuations, whose amplitudes depend on coolant velocity, will cause fuel-element and control-rod vibrations which in turn affect reactivity as indeed do all of the random sources mentioned above. The complexity of the problem is clearly immense; nevertheless, under certain limited conditions a formalism for dealing with these problems can be developed even if considerable further development is required to incorporate it into a complete description. We begin our discussion of this formalism by discussing the heat-transfer and fluid-flow aspects of the problem; we shall later discuss the problems of random vibration, heat conduction and convection and also a statistical model of boiling.

8.8. *Heat transfer and fluid flow*

In principle, to describe the way in which heat is transported from the fuel elements in a reactor we should write down the complete set of Navier–Stokes equations for a compressible fluid in turbulent flow with heat sources determined by the power produced at the points (ξ_k, z). It is clearly an unrealistic hope to expect that any result of practical value will result from such a step and it is necessary to seek a less ambitious approach. One such method is to consider the heat balance in a single cell of the system (e.g. the cell surrounding the **k**th element) and to neglect any interaction of this cell with adjacent cells, at least as far as heat transfer is concerned. Consider, therefore, a simple cell in which there is a solid moderator surrounding an annular region containing coolant which, in turn, surrounds the fuel element with its associated

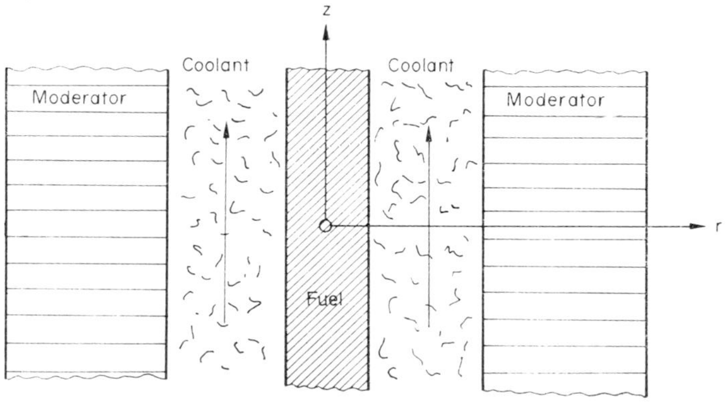

FIG. 8.4. Schematic diagram of a cell in a reactor used for studying the heat-transfer problem.

fins. Figure 8.4 is a schematic diagram of the situation. For completeness we assume that the coolant is a liquid. The neglect of interactions between cells might well have to be relaxed in practice since it might be of interest to study the effect of fluctuations in, say, the coolant flow of one channel on the heat-transfer rate in another. Similarly, in the case of a boiling-water reactor, where moderator and coolant are one, it may well be more profitable to consider the reactor as a whole, with localized heat sources but with suitable approximations for the Navier–Stokes equations in the fluid. Let us for the moment, however, simply consider the unit cell approach.

The fuel element in the neutronics eqn. (8.61) was, it will be recalled, treated as a line filament of infinitesimal radius. This is convenient in the context of neutron absorption and emission but not when the detailed properties of the heat generation within it have to be calculated. Thus in our Fig. 8.4 we show the filament as it really is and consider the heat balance within it. Neglecting the cladding and fins, we can write down the equation of heat conduction as

$$\nabla . k(\mathbf{r},t) \nabla T_f(\mathbf{r},t) + Q(\mathbf{r},t) = \rho C_p \frac{\partial T_f(\mathbf{r},t)}{\partial t}, \tag{8.62}$$

where $T_f(\mathbf{r},t)$ is the temperature in the fuel element, $Q(\mathbf{r},t)$ is the heat source term due to fission and $k(\mathbf{r},t)$ is the coefficient of thermal conductivity. There are, of course, boundary conditions for this equation which link the temperatures in adjacent media. Thus at a surface we would normally have continuity of temperature and heat flux $q = -k\nabla T$. In practice, because of the approximations that have to be made to describe the heat transport in the coolant and because of uncertaintics in the completeness of thermal contact between the two media, the boundary conditions that are often employed are somewhat modified.

The randomness in the above equation can arise in several distinct ways: (1) random heat generation, i.e. random Q; (2) random boundary conditions, due, for example, to variations in the surface temperature arising from the coolant; (3) random material properties, these affect $k(\mathbf{r}, t)$ and could be present in the ceramic fuels discussed in Section 8.4; (4) random geometry, this is due to surface roughness and leads to a form of random boundary condition. Clearly, solutions of this equation subject to one or more of conditions (1)–(4) will require the methods described in Section 8.2. Before attempting such a procedure, however, it is necessary to consider the form of the heat-transport equations in the surrounding coolant.

If the coolant were in laminar single-phase flow it could be described by the Navier–Stokes equations appropriate to annular geometry and supplemented by an appropriate equation of state. It might then be possible to obtain a solution, if not analytically, then possibly by numerical methods. However, such an ideal situation is not the case and even a single-phase medium in turbulent flow would present insuperable difficulties as far as any analytical solution is concerned (Hinze, 1959). Two-phase flow makes the problem even more difficult. The present state of knowledge regarding convective heat transfer, i.e. that involving the coolant, is strictly limited and the prediction of the rate at which heat is convected away from a solid surface by an ambient fluid involves a thorough understanding of the principles of heat conduction, fluid dynamics and boundary-layer theory. As we have explained, such an investigation is beyond the power of present analytical methods and the pragmatic approach is taken to lump together all of these effects in terms of a single parameter by the introduction of a law of cooling (à la Newton). Thus if $T_f(\mathbf{r}_s, t)$ is the surface temperature of the fuel element and T_c the temperature of the coolant, we write for the heat flux q from fuel to coolant

$$q = h(T_f(\mathbf{r}_s, t) - T_c) \tag{8.63}$$

which, using the relation $q = -k\nabla T_f$, leads to the boundary condition

$$\nabla T_f|_{\mathbf{r}_s} = -\frac{h}{k}(T_f(\mathbf{r}_s, t) - T_c). \tag{8.64}$$

h in eqn. (8.63) is called the *heat-transfer coefficient*. It is not an intrinsic property of the material as is k, but depends upon the composition of the fluid and on the nature of the flow and type of geometry of the surface. It is a macroscopic quantity, a "fudge" factor, which attempts to represent an effect without containing any explanation of it. Although theoretical methods exist for calculating h, they are not entirely satisfactory and it is usually found experimentally in terms of the well-known groups of non-dimensional numbers (see eqn. (6.95)). As we have seen already, h is generally very important in the generation of noise sources in reactors at power. However, in view of the fact that it enters the problem via a boundary condition, it will clearly be a surface noise effect and a spatial study of the problem will reveal some differences when compared with the corresponding point model calculation. An example of this will be given later. What now of the equation for the coolant in the channel? We cannot use the Navier–Stokes equations because of their complexity: we must therefore consider some semi-empirical model which accounts in an average way for turbulence but still retains some effect of the space–time behaviour of the fluid temperature. We shall consider single-phase flow only at this stage.

In eqn. (8.63) the exact definition of T_c, the coolant temperature, was not given: in fact it is a rather ambiguous quantity in view of our decision to abandon the use of the Navier–Stokes equations. However, let us consider the situation where the fluid flows through a passage bounded on all sides by surfaces through which heat passes into the fluid. The fluid temperature varies across the passage and we must characterize it by some appropriate weighted average. In this connection, the most *useful* average temperature is that which, when multiplied by the mass flow rate, W, and its specific heat, C_c, gives the net transport of heat along the passage. This has the advantage of preserving the correct heat flow.

Now if A is the area of the passage, $T(\mathbf{r},t)$ the temperature of the fluid at $(\mathbf{r},t)$, $u(\mathbf{r},t)$ the corresponding velocity, ρ the fluid density and $C(\mathbf{r},t)$ the specific heat at temperature T, we can write

$$WC_cT_c \equiv \int_A T(\mathbf{r},t)u(\mathbf{r},t)\rho(\mathbf{r},t)C(\mathbf{r},t)\,dA. \tag{8.65}$$

In general, it should be noted that T_c is a function of channel height z: T_c is referred to as the *bulk mean value*. Also of importance in this integral formulation of the coolant heat-transport properties is the mean fluid velocity, U, in the passage, which we define as

$$U(z,t) = \frac{1}{A\rho_c}\int_A \rho(\mathbf{r},t)u(\mathbf{r},t)\,dA \equiv \frac{W}{A\rho_c}, \tag{8.66}$$

where ρ_c is the density evaluated at the temperature T_c.

Any other physical quantities which enter into the calculation such as thermal conductivity k, viscosity μ, specific heat C are all assumed to be evaluated at T_c and to be given a subscript c. The shape and size of the channel and its heated surfaces are characterized by an effective diameter, d, defined by

$$d = \frac{4A}{S},$$

where S is the total perimeter of the channel including the perimeters of any separate fuel rods, etc., that it may contain.

In general, the experimental data which correlate all of these variables may be grouped as follows:

$$F\left(\frac{h}{\rho UC}, \frac{\rho Ud}{\mu}, \frac{C\mu}{k}\right) = 0, \tag{8.67}$$

where the Stanton number St $= h/\rho UC$, the Reynolds number Re $= \rho Ud/\mu$ and the Prandtl number Pr $= C\mu/k$.

It is the Stanton number which is concerned with heat transfer, whilst the Reynolds number determines the nature of the flow in the system. The Prandtl number is basically a physical property of the heat-transfer fluid and, apart from a temperature dependence, is not affected by the geometry or other engineering constraints. To complete the picture of this semi-empirical approach to heat transfer we define the Nusselt number $\mathrm{Nu} = hd/k = \mathrm{St.Re.Pr}$ and the Peclet number $\mathrm{Pe} = \mathrm{Re.Pr}$. We have already encountered various forms for St in our point model analysis in Section 6.7.

Whilst the spatial dependence of the coolant temperature has been averaged out in the plane perpendicular to its flow, it still has an important variation in the axial direction which can be described with some accuracy. Also, it should be added, the area-averaged values apply only to a particular cell and T_c, U, etc., will, in general, vary from channel to channel, i.e. will depend on the $\mathbf{k}(k_1, k_2)$ of eqn. (8.61).

In the fuel region, heat transport is by conduction. Thus if we neglect the axial flow of heat, the heat-balance equation at a height z from the centre of the reactor can be written

$$\frac{1}{r}\frac{\partial}{\partial r}\left(rk_f(r,z,t)\frac{\partial T_f(r,z,t)}{\partial r}\right) + Q_f(r,z,t) = \rho_f C_f \frac{\partial T_f(r,z,t)}{\partial t}, \tag{8.68}$$

where r is the radial coordinate relative to the centre of the cell. k_f, ρ_f and C_f are thermal conductivity, density and specific heat, respectively, of the fuel material. T_f is the fuel temperature and Q_f is the amount of heat generated by fission deposited directly in the fuel. Similar equations can be written for cladding and moderator with the appropriate values of k, ρ and C and also remembering that a certain amount of heat is generated there by radiation and neutron energy degradation.

The equation for the coolant temperature is less precise since we are unable to calculate its radial dependence (or at least have decided against such an attempt). We can therefore write down the following equation for unit length of coolant:

$$A_c C_c \rho_c \left(\frac{\partial T_c(z,t)}{\partial t} + U(z,t)\frac{\partial T_c(z,t)}{\partial z}\right) = q_s(z,t), \tag{8.69}$$

where C_c is the specific heat, $\rho_c A_c$ is the mass per unit length, $U(z,t)$ is the mean fluid velocity and $q_s(z,t)$ is the heat flux into the fluid from cladding and moderator plus any heat generated directly in the coolant. More specifically

$$q_s(z,t) = -2\pi(a+\Delta)k_{cl}\left.\frac{dT_{cl}}{dr}\right|_{a+\Delta} + 2\pi b k_m \left.\frac{dT_m}{dr}\right|_b + A_c \bar{Q}_c(z,t), \tag{8.70}$$

k_{cl} and k_m being the thermal conductivities of the cladding and moderator, respectively. Integration of the equations of thermal conduction in the three regions over their corresponding areas shows that

$$\begin{aligned} q_s(z,t) = {} & A_f \bar{Q}_f(z,t) + A_{cl}\bar{Q}_{cl}(z,t) + A_m \bar{Q}_m(z,t) + A_c \bar{Q}_c(z,t) \\ & - A_f \rho_f C_f \frac{\partial \bar{T}_f}{\partial t} - A_{cl}\rho_{cl} C_{cl}\frac{\partial \bar{T}_{cl}}{\partial t} - A_m \rho_m C_m \frac{\partial \bar{T}_m}{\partial t}, \end{aligned} \tag{8.71}$$

where a bar indicates an average over the corresponding domain in the radial coordinate.

We have then a set of four equations which, together with the boundary conditions of continuity of heat flux, enables the temperature at any (z,t) to be obtained in the cell components. The special case of steady-state temperatures with uniform radial heat generation in the fuel only and a cosine distributed heat source $\bar{Q}_f(z) = Q_0 \cos Bz$ is a standard problem in reactor heat transfer (El-Wakil, 1971). A number of interesting time-dependent problems are discussed by Stein (1957).

To be precise, the above equations should have associated with them the index $\mathbf{k}(k_1, k_2)$ to denote the cell to which they apply. This being the case we can relate the heat generation in the fuel, $\bar{Q}_f$, to the quantities appearing in the source-sink eqn. (8.61) as follows.

The heat generated in the fuel element in a slice dz is

$$dz\, 2\pi \int_0^a r\, dr\, Q_f(r,z,t) \equiv dz\, A_f \bar{Q}_f(z,t) \tag{8.72}$$

$$= \mu\, dz\, 2\pi \int_0^a \overline{v\Sigma_f(r)}\, N(r,z,t)\, r\, dr. \tag{8.73}$$

But in the spirit of source-sink theory we have for the kth element,

$$\overline{v\Sigma_f(r)} \to \frac{\eta_{\mathbf{k}}\gamma_{\mathbf{k}}}{\bar{\nu}_{\mathbf{k}}} \frac{\delta(r - r_{\mathbf{k}})}{2\pi r_{\mathbf{k}}}, \tag{8.74}$$

hence

$$A_f \bar{Q}_f(z,t) = \mu \frac{\eta_{\mathbf{k}}\gamma_{\mathbf{k}}}{\bar{\nu}_{\mathbf{k}}} N(\boldsymbol{\xi}_{\mathbf{k}}, z, t). \tag{8.75}$$

Since, in general, the coefficients D_0, $\overline{v\Sigma_{am}}$, $\eta_{\mathbf{k}}$ and $\gamma_{\mathbf{k}}$ in the neutron balance equation are temperature dependent it is necessary to write down additional relationships to couple the heat-transfer equations with the neutronics equation. This will require a knowledge of the thermal neutron-energy spectrum in the system so that average reaction rates may be calculated, e.g.

$$\overline{v\Sigma_{am}} = \frac{\int_0^\infty dE\, \Sigma_{am}(E)\phi_m(E)}{\int_0^\infty dE\, \frac{1}{v}\, \phi_m(E)}, \tag{8.76}$$

where $\phi_m(E)$ is the neutron spectrum in the moderator which will in turn depend upon T_f, T_c, T_{cl} and T_m (Williams, 1966). In practice we can introduce the idea of temperature coefficients and assuming only small changes from some equilibrium temperature, $\overline{v\Sigma_{am}}$ can be written

$$\overline{v\Sigma_{am}} \simeq (\overline{v\Sigma_{am}})_0 \{1 + \alpha_f(\bar{T}_f - \bar{T}_{f_0}) - \alpha_{Cl}(\bar{T}_{Cl} - \bar{T}_{Cl_0}) + \alpha_c(T_c - T_{c_0}) + \alpha_m(\bar{T}_m - \bar{T}_{m_0})\}. \tag{8.77}$$

The temperature coefficients α_l are obtained from separate calculations. In practice it is likely that $\overline{v\Sigma\alpha_m}$ is dominated by the term involving α_m and it is acceptable to write

$$\overline{v\Sigma_{am}} = (\overline{v\Sigma_{am}})_0 \{1 + \alpha_m(\bar{T}_m - \bar{T}_{m_0})\}. \tag{8.78}$$

Similarly,

$$\gamma_{\mathbf{k}} = (\gamma_{\mathbf{k}})_0 \{1 + \alpha_{f\mathbf{k}}(\bar{T}_f - \bar{T}_{f_0})\} \tag{8.79}$$

and

$$\eta_{\mathbf{k}}\gamma_{\mathbf{k}} = (\eta_{\mathbf{k}}\gamma_{\mathbf{k}})_0 \{1 + \tilde{\alpha}_{f\mathbf{k}}(\bar{T}_f - \bar{T}_{f_0})\}. \tag{8.80}$$

Thus as far as the temperature coupling is concerned we have a closed set of equations which *in principle* can be solved either deterministically or, in the case that $U_{\mathbf{k}}(z,t)$ or $T_c(z,t)$ is a random variable, in a statistical manner.

If we consider the situation studied by Greef (1971), discussed in Chapter 6, it is not difficult to see that we now have a method for extending his point model approach to include the effect of spatial dependence. All that is necessary is to assume a constant thermal conductivity in fuel and cladding, to set $U(z,t) = \langle U \rangle + \tilde{u}(t)$ and linearize the equations about some reference values. With appropriate linearization of the kinetics equation and use of the feedback terms (8.79) and (8.80) above, the spatially dependent transfer function for any channel can be obtained. This task will be straightforward but rather tedious since the heat-transfer coefficient, h, is a function of coolant-flow fluctuation $\tilde{u}(t)$ and this variable will enter as a random boundary condition via eqn. (8.69).

8.9. *Rod vibration*

We have not yet considered the problem of rod vibration, i.e. the connection between $\xi_k(z,t)$ and the operational conditions of the reactor. It is clear that vibration is caused by turbulent pressure fluctuations in the coolant and is therefore related to coolant velocity and thereby to reactor power. A comprehensive survey of this problem has been given by Wambsganss (1967). It is pointed out there that, with the exception of control rods, the in-core components of a reactor are stationary elements and not designed for motion. However, because every structural component possesses mass and elasticity it is clear that there is a potential for mechanical vibration provided an appropriate source of excitation is present. This source can only be the coolant, either directly from turbulence or indirectly if the source of energy comes from pump oscillations which will be transmitted by the coolant to the structure. A survey of vibrational problems in reactor cores shows that the majority have been flow-induced. Typical vibration frequencies are in the range 5–10 Hz. The importance of suppressing such oscillations cannot be overestimated since they can lead to reduced clearance between rods in closely packed clusters and hence to hot spots. Also sustained vibration increases the failure rate due to fatigue and cracking defects. It is of some concern that in practice many vibration problems have not been allowed for in original designs but have been eliminated only after reactor operation has shown them to be a hazard.

The basic problem is that the various mechanisms by which the coolant transmits energy to a reactor component are not completely understood. However, some progress has been made and three significant areas of qualitative understanding exist, (1) vibration of rods in cross-flow, (2) vibration of rods in parallel flow and (3) flat-plate vibration. Vibrations of rods in cross-flow was recognized quite early in connection with flow through heat exchanges. Whilst this is an important topic it is not of direct interest in the present context and we shall consider parallel flows only. In fact the study of such problems is relatively new, being pioneered by Burgreen *et al.* (1958) in their studies of rod bundles. Experiments were performed in air and water and it was observed that, when resonance occurred, it was at a lower frequency in water. The vibration frequencies themselves remained fairly constant over a range of coolant flow velocities and hence led to the conclusion that the vibrations were self-excited rather than forced. There was a general trend for the amplitude of vibration to increase with coolant velocity.

A definitive study of this problem has been carried out by Paidoussis (1965, 1966, 1969) who sets up a force balance equation and obtains a differential equation for the displacement of the beam, ϵ, from its mean position as a function of position, z, along the beam at a given time t. The equation takes the following form:

$$EI\frac{\partial^4 \epsilon}{\partial z^4}-\frac{\partial}{\partial z}\left(T\frac{\partial \epsilon}{\partial z}\right)+M\left(\frac{\partial}{\partial t}+U\frac{\partial}{\partial z}\right)^2 \epsilon+m\frac{\partial^2 \epsilon}{\partial t^2}+\tfrac{1}{2}\rho DU^2\left[C_1\left(\frac{V}{U}\right)^2+C_2\frac{V}{U}\right] = 0, \tag{8.81}$$

where C_1 and C_2 are coefficients of form drag and skin friction drag, D is the diameter of the rod, EI is the flexural rigidity of the rod, M, m are the virtual masses of fluid and cylinder, respectively, per unit length, T is the longitudinal tension, U is the mean flow velocity parallel to the axis of the cylinder, ρ is the fluid density and V is the flow velocity perpendicular to the rod.

Physically, the terms arise as follows: the first term is due to flexural restoring forces and the second to tension forces. The third term is due to the change of fluid momentum as a result of lateral motion and the fourth term is the inertial force. Finally, the last term arises from transverse viscous forces on the cylinder (Taylor, 1952). This term is in effect the random source excitation since the lateral velocity V is determined mainly by the turbulent nature of the fluid.

No attempt was made by Paidoussis to solve eqn. (8.81); instead he used an argument based on non-dimensional groups to obtain a value for the maximum displacement ϵ_{max} and, by comparison with experimental work, has obtained a useful working formula for predicting the displacement. Later studies by Paidoussis (1966) are of an analytical nature and involve a stability analysis of eqn. (8.81). It should be mentioned, however, that there still remains some controversy as to whether the basic excitation source is forced or self-excited (Burgreen *et al.*, 1958; Quinn, 1962; Paidoussis, 1965). Results pertinent to flat-plate vibration may be found in Wambsganss (1967) and in Schlösser (1962).

Two important comments can be made at this stage regarding the above discussion. The first is that the equation deduced by Paidoussis for the deflection is only one of several possible mathematical models. An equally important model is the Timoshenko beam which results in a pair of coupled differential equations for the deflection and corresponding bending angle. By means of a well-established approximation, the bending angle can be eliminated and we again obtain an equation for the deflection in terms of the transverse forces acting on the beam (Samuels and Eringen, 1958). Indeed all models can eventually be written in the form

$$\hat{L}\epsilon(z,t) = F(z,t), \tag{8.82}$$

where $\hat{L}$ denotes a partial differential operator in (z,t) and $F(z,t)$ is the transverse force acting at (z,t). The second comment concerns the nature of $F(z,t)$ about which it must be admitted very little is known. All that can be said with some certainty is that it arises from random pressure fluctuations acting on the lateral surface resulting from turbulence in the coolant boundary layer. It will be impossible to obtain theoretically any useful statistical information about F other than $\langle F\rangle = 0$. It is therefore usual to attempt to obtain experimentally the lowest order measure of statistical behaviour of $F(z,t)$, namely the covariance

$$Q_{FF}(z_1,z_2;\, t_2-t_1) = \langle F(z_1,t_1)F(z_2,t_2)\rangle. \tag{8.83}$$

Such a programme of research has been conducted at the Chalk River Laboratories by Gorman (1969). We shall return to this problem in the next chapter but it is useful to state now that a fairly useful analytical form has been found for $\phi_{FF}(z_1, z_2; \tau)$ in terms of three arbitrary constants which can be found by experiment to depend on the mean coolant velocity U. With a knowledge of ϕ_{FF}, and on the basis that the operator $\hat{L}$ is non-random and appropriate boundary conditions for end fixings of the rod are known, we can calculate the auto-correlation function of $\epsilon(z, t)$. This information will be of value in specifying, partially, the statistical behaviour of $\boldsymbol{\xi}_k(z, t)$ if we note that

$$\boldsymbol{\xi}_k(z, t) = \langle \boldsymbol{\xi}_k \rangle + \boldsymbol{\epsilon}_k(z, t). \tag{8.84}$$

If the vibrations are isotropically distributed about the mean value, then

$$\boldsymbol{\epsilon}_k(z, t) = \mathbf{i}\epsilon_k(z, t) + \mathbf{j}\epsilon_k(z, t), \tag{8.85}$$

where $\mathbf{i}$ and $\mathbf{j}$ are unit vectors in the $(x-y)$ plane and $\epsilon_k(z, t)$ is the random variable discussed above.

Finally, then, we now have a completely self-contained system. Given a fixed power output for each channel and prescribed inlet coolant temperature the corresponding coolant velocity is known, also the heat-transfer coefficient and the corresponding temperatures. From these parameters we can calculate rod vibration effects and the way in which they feed back on the neutronic behaviour. In addition, if appropriate neutron density detectors and thermocouples are inserted in the core it should be possible by methods of triangulation to detect the location of any channel which is behaving in an abnormal fashion. This Utopian situation has, of course, not been achieved but the above theory lays a basis for such a scheme. We shall discuss the details of some of the mechanical-statistical problems mentioned above in more detail later.

8.10. *Boiling moderator and coolant*

When the coolant or moderator boils, the number density in the terms involving the diffusion coefficient and absorption cross-sections themselves become random functions of space and time. Thus in addition to the treatment of localized noise sources by the source-sink method, it will be necessary to treat the distributed noise sources due to random absorbing and moderating properties. Even without power-reactivity coupling, vibration of fuel elements, or even nuclear noise from the fuel rods, the solution of the general Feinberg–Galanin eqn. (8.61) for such a case will present formidable difficulties. Indeed, we have seen in Section 8.2 in a "homogeneous" system with a fuel density which is *spatially* random only that the problem has no easy solution. When the diffusion coefficient is random as well, and both space and time coordinates are involved, we are faced with a considerable problem.

In order to obtain some idea of the approximations that it might be possible to introduce it will be useful to study the mechanism of boiling and the information that is available regarding the statistical behaviour of bubble motion. We have, therefore, to examine the problem of two-phase flow and its causes.

Boiling can take several forms, the most important of which are pool boiling, bulk boiling, nucleate boiling and film boiling. Pool boiling arises when vapour is formed by adding heat to the liquid by a surface in contact with or immersed in the liquid.

Bulk boiling arises when heat is generated within the fluid itself, as, for example, by chemical or nuclear reactions: a homogeneous fluid-fuelled water boiler is an example (Wilkins, 1945). In nucleate boiling, bubbles form around irregularities on a heated surface or around a small vapour nucleus. Film boiling arises when the nucleate boiling near a surface becomes so vigorous that the bubbles coalesce to form a continuous vapour film between liquid and surface: this is clearly a situation to be avoided in practice since it reduces heat transfer markedly.

Nucleate boiling is the most important of the above processes as far as normal reactor operation is concerned since it is associated with maximum heat transfer; however, since we are also interested in abnormal conditions the study of the transition from nucleate to film boiling is of importance (departure from nucleate boiling: DNB), as indeed is the heat flux necessary to initiate nucleate boiling. A detailed and satisfactory physical and mathematical study of the causes of the nucleate boiling is not available although there do exist a number of conflicting theories which lead to formulae containing empirical constants that can usually be adjusted to agree with experiment. The main problem it seems is the precise specification of the nucleation sites, which can only then be known in a statistical sense, and the development of a theory which will predict, again statistically, the subsequent position and size of a bubble at a given time. A review of the situation to date has been given by Sandervag (1971) who describes the various theories of bubble formation and their conflicting assumptions (the work by Greenfield *et al.* (1954) is also of considerable interest regarding the dynamics of bubble formation). Without suitable statistical information on bubble formation, the appropriate statistics for $D_0(\mathbf{r},t)$ and $\overline{v\Sigma_{am}(\mathbf{r},t)}$ cannot be specified and so cruder methods must be used: such methods are acceptable for basic heat-transfer design studies but are not entirely satisfactory from the point of view of noise analysis. A further problem in a boiling-reactor system is the random motion of the boundary between the coolant and the vapour above it and also the position of the boiling boundary. In effect this leads to a two-region reactor with randomly varying axial lengths: we shall have more to say regarding this phenomenon later.

As we have said, in practice, an experimental correlation exists for the heat flux from a fuel element and the corresponding temperature difference between the surface temperature and the saturation temperature, i.e. the temperature at which the liquid boils when its vapour pressure is equal to the pressure on the liquid surface. In the case of boiling in the nucleate regime a form which has been found to fit the experimental facts can be written as follows (Rohsenow, 1952):

$$\frac{C_{pl}(T_{fs}-T_{\text{sat}})}{H_{lv}P_r^{1\cdot 7}} = C\left[\frac{q}{\mu_l H_{lv}}\left\{\frac{\sigma_{lv}}{g(\rho_l-\rho_v)}\right\}^{\frac{1}{2}}\right]^{0\cdot 33}, \tag{8.86}$$

where C_{pl} is the specific heat of the saturated liquid, T_{fs} is the temperature of the heated surface, T_{sat} is the saturation temperature, H_{lv} is latent heat of vaporization at system pressure, C is an empirical constant that varies for different liquids, q is the required heat flux, μ_l is the viscosity of the liquid, σ_{lv} is the surface tension at the liquid–vapour interface, g is the acceleration due to gravity, and ρ_l and ρ_v are the densities of saturated liquid and vapour phases, respectively. In general, the values of H_{lv}, μ_l, etc., all depend on T_{sat} so that calculations of q as a function of T_{fs} can be tedious. As well as normal boiling noise whose effects on the neutron noise we shall discuss in more detail below, there exists another important effect which can be influenced by statistical perturbations: we refer to the problem of burn-out. This phenomenon can best be explained by reference to the classical curve of heat flux versus temperature difference ($T_{fs}-T_{sat}$) which is shown in Fig. 8.5. The various boiling regimes are as follows: below A there is natural convection, from B to C nucleate boiling occurs, between D and E we have film boiling and from E to F there is film boiling and radiation. Now cooling by a saturated liquid is very different from cooling by a gaseous coolant since, for the liquid, the temperature T_{sat} remains constant. Thus for a given value of surface temperature T_s (as determined by metallurgical considerations) it is the heat flux

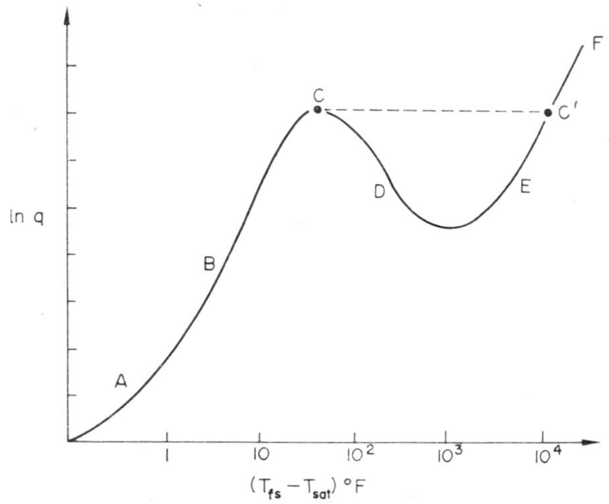

FIG. 8.5. The Nukiyama curve for surface boiling heat flux, q, versus temperature difference ($T_{fs}-T_{sat}$).

which becomes the independent variable. In this case, when the critical heat flux and the corresponding temperature at C in Fig. 8.5 are reached a further increase in the heat flux q leads to a sudden jump from C to C'. Now C' is in the film-boiling region where the thermal resistance is high, thus, as shown in the figure, the value of T_s-T_{sat} increases enormously and for a fixed T_{sat}, the surface temperature T_s is likely to exceed safe limits, and rupture or melting of the fuel element may result. The heat flux necessary to cause T_s to rise sufficiently to cause damage is known as the burn-out heat flux; however, it is also referred to as DNB (departure from nucleate boiling) or the "boiling crisis". Several experimental studies have been performed on the burn-out heat flux, which is shown to increase with pressure (Rohsenow, 1952), with the inclusion of certain additives and with ultrasonic and electrostatic fields (Berger and Derian, 1965). However, it decreases in the presence of dissolved gases and agents which reduce the surface tension at the liquid–vapour interface. It is true, however, that a great deal of uncertainty remains as to the precise causes that promote or inhibit burn-out. This is unfortunate since it is necessary in reactor design to be able to predict burn-out to avoid accidents. A number of experimental correlations exist for predicting the burn-out heat flux which are valid according to whether the liquid is saturated or sub-cooled and whether it is pool or flow boiling (El-Wakil, 1971). It should be noted, however, that from a statistical point of view, small fluctuations in the heat flux in the neighbourhood of the burn-out value could lead to large statistical fluctuations in fuel temperature.

Whilst the above correlations for heat transfer give a satisfactory method of predicting net power output very little is known about the detailed spatial behaviour of two-phase flow itself. There are in fact a number of ways in which bubbles can disperse themselves in a liquid, ranging from bubble flow which is the independent motion of bubbles up the tube, through slug flow where the bubbles coalesce to form large vapour voids filling the tube, and also annular flow which is virtually a continuous vapour phase with a few dispersed liquid droplets moving up the centre of the tube and leaving an annulus of superheated liquid on the walls. Finally there is fog flow which is analogous to bubble flow but with liquid and vapour exchanged.

In the absence of any precise or even useful mathematical representation of the bubble distribution (see, for example, Hulburt and Katz, 1964) it is usual to resort to the expedient of the *void fraction* α which is the ratio of volume of vapour to volume of vapour and liquid. The physics of bubble growth and collapse are omitted in this concept and it is based on the formulation of balance equations which are set up for conservation of mass, momentum and energy for the two phases. These equations contain the void fraction, the enthalpies and the velocities of the vapour and liquid phases as well as certain physical properties. The void fraction itself is clearly dependent on position and is also a random function of time despite the fact that the detailed bubble mechanics are omitted. These variations will clearly influence the reactivity of the system and we shall therefore examine the simple methods used for calculating α and attempt to assess the probable causes of any fluctuation in it in the sense of the Langevin technique.

8.11. *The void fraction*

The *quality*, x, of a vapour–liquid mixture where there is relative motion between the phases, i.e. where the vapour because of its buoyancy tends to slip past the liquid, is defined as

$$x = \frac{\text{mass flow rate of vapour}}{\text{mass flow rate of mixture}}. \tag{8.87}$$

The void fraction α, on the other hand, is defined as

$$\alpha = \frac{\text{volume of vapour in the mixture}}{\text{total volume of liquid–vapour mixture}}. \tag{8.88}$$

To prevent large excursions in the moderating properties of the reactor coolant it is desirable to keep x small (a few percent). If the average velocity of the vapour is U_v and that of the liquid U_l then we define the slip ratio $S = U_v/U_l$. If in addition A_v and A_l are the cross-sectional areas of the two phases perpendicular to the flow direction and v_v and v_l are the specific volumes of vapour and liquid we can write α in terms of x as follows:

$$\frac{1}{\alpha} = 1+\left(\frac{1-x}{x}\right)\left(\frac{v_l}{v_v}\right)\left(\frac{U_v}{U_l}\right) \equiv 1+\left(\frac{1-x}{x}\right)\psi. \tag{8.89}$$

v_l and v_g are obtained from steam tables since they depend on pressure. The slip ratio S must be obtained experimentally and is found to depend on pressure, volumetric flow rate, power density, steam quality and inlet coolant velocity: appropriate correlations are available (El-Wakil, 1971).

As an example, if we take $S = 1$ and a pressure of 1000 p.s.i.a. a mixture quality of 2% leads to a void fraction $\alpha = 30\%$. Thus in general a small steam fraction by mass corresponds to a large fraction by volume.

We are also interested in the variation of α with height and must therefore examine the way in which the boiling varies up the channel. Basically, there are two distinct heights in a boiling core: the non-boiling height L_0 and the boiling height L_B. The total core is of height $L = L_0 + L_B$. The non-boiling height starts at the core base where the incoming sub-cooled fluid enters and in this region the heat generated and transported up to the point z is $W_0(z)$. As we move up the core the coolant temperature increases until at $z = L_0$ it reaches saturation temperature and boiling begins; the total heat added is then $W_0(L_0)$. The distance from the point at which boiling starts, to the core top, is the boiling height L_B. The total heat flux from the fuel element over the complete height of the reactor is W_T. If H_i is the enthalpy at the inlet of the channel, H_l is the enthalpy of the saturated liquid and H_{lv} is the latent heat of vaporization at system pressure, then clearly the value of $W_0(L_0) = H_l - H_i$. On the other hand, in the boiling region $z > L_0$, we have for the total heat generated and transported to the point z

$$W(z) = (H_l + x(z) H_{lv}) - H_i, \tag{8.90}$$

where $x(z)$ is the quality of the saturated liquid at z. Naturally $x(z) = 0$ for $z \leqslant L_0$. Thus the total heat generated and transferred to the top of the core is

$$W_T = W(L) = (H_l + x(L) H_{lv}) - H_i. \tag{8.91}$$

We also have the relation between $x(z)$ and $\alpha(z)$ given by eqn. (8.89). However, to obtain an explicit relation for $\alpha(z)$ we will assume a sinusoidal flux variation and hence that the heat flux at z is

$$q(z) = q_0 \sin\left(\frac{\pi z}{L}\right). \tag{8.92}$$

Thus

$$\frac{W(z)}{W_T} = \frac{\int_0^z q(z')\,dz'}{\int_0^L q(z')\,dz'} = \frac{\pi}{2L}\int_0^z \sin\left(\frac{\pi z}{L}\right) dz, \tag{8.93}$$

$$= \frac{1}{2}\left(1 - \cos\frac{\pi z}{L}\right). \tag{8.94}$$

Combining eqns. (8.90), (8.91) and (8.94) we find that

$$x(z) = \left(\frac{W_T}{2H_{lv}} - \frac{H_l - H_i}{H_{lv}}\right) - \frac{W_T}{2H_{lv}} \cos\left(\frac{\pi z}{L}\right) \tag{8.95}$$

or more simply

$$x(z) = A + B\cos\left(\frac{\pi z}{L}\right). \tag{8.96}$$

Using this in eqn. (8.89) gives for $\alpha(z)$ the following expression:

$$\frac{1}{\alpha(z)} = 1+\psi\left\{\frac{1+A-B\cos\left(\frac{\pi z}{L}\right)}{A+B\cos\left(\frac{\pi z}{L}\right)}\right\} \tag{8.97}$$

which is valid in the range $L_0 \leqslant z \leqslant L$.

L_0 is obtained by noting that, at $z = L_0$, $x(0) = 0$ and W_T is given by eqn. (8.91), thus

$$\frac{1}{2}\left(1-\cos\frac{\pi L_0}{L}\right) = \frac{H_l-H_i}{H_l-H_i+x(L)H_{lv}}. \tag{8.98}$$

We note from eqn. (8.97) that the void fraction increases smoothly from zero at L_0 to a maximum at the core top L.

With knowledge of $\alpha(z)$ we can calculate the density ρ_B of the coolant (moderator) in the boiling region at height z from the relation

$$\rho_B(z) = (1-\alpha(z))\rho_l+\alpha(z)\rho_v. \tag{8.99}$$

In the non-boiling region it is reasonable to assume that the density varies linearly and we can write

$$\rho_0(z) = \rho_i+(\rho_l-\rho_i)\frac{z}{L_0}, \tag{8.100}$$

where ρ_i is the inlet density.

At this point we might rightly ask where the randomness enters the situation. An answer to this question is difficult since we have essentially removed any statistical information about bubble formation. The best that can be done at this level is to assume that fluctuations in the fuel temperature T_f or the saturation temperature T_{sat} due to pressure fluctuations will cause U_l and U_v to vary through the correlation of eqn. (8.86). This in turn makes the constants A and B random and hence $\alpha(z)$. It must be recalled, however, that time dependence as such has not entered the derivation of $\alpha(z)$ and quite clearly additional relationships are required. Before discussing this aspect of the problem, however, it is necessary to appreciate the full implications of introducing a void fraction. As we have said, ideally one would like to know how the voids are distributed in space and time, e.g. $P_N(R,\Delta,t)$, where $P_N(R,\Delta,t)$ is the probability that there are N voids of volume Δ in the region of space R. We must consider the void fraction $\alpha(\xi,z,t)$ as some average over this distribution, viz.

$$\alpha(R,t) = \sum_N N\int d\Delta\, P_N(R,\Delta,t). \tag{8.101}$$

Thus $\alpha(R,t)$ is the total void volume in R at time t. This remains a statistical quantity but nevertheless contains no information about individual bubbles and indeed will require further arguments to deduce its value. In many cases the variation of α in the direction perpendicular to the flow is also averaged out so that we have

$$\alpha(z,t) = \frac{1}{A}\int_A d\xi\,\alpha(R,t), \tag{8.102}$$

where A is the area of the flow channel.

The value of $\alpha(z)$ discussed above was in the context of a stationary system so that in practice it is a statistical time average

$$\alpha(z) = \lim_{T\to\infty} \frac{1}{2T}\int_{-T}^{T} \alpha(z,t)\,dt. \tag{8.103}$$

To make the void fraction a useful concept in the study of boiling reactor noise it will be necessary to include, specifically, the time dependence. Investigations of boiling reactor stability by a number of authors have done this by means of a mass, momentum and energy balance of the liquid–vapour mixture and an evaluation of the sources of vapour production (Sandervag, 1971; Zuber, 1967; Beckjord, 1958; Pederson, 1966; Holt and Rasmussen, 1972; Siegmann, 1971; Fleck, 1960; De Shong, 1959).

In the notation used above, these balance equations may be written as follows:

Equation of continuity of the mixture:

$$\frac{\partial}{\partial t}\{(1-\alpha)\rho_l+\alpha\rho_v\}+\frac{\partial}{\partial z}\{(1-\alpha)\rho_l U_l+\alpha\rho_v U_v\} = 0. \tag{8.104}$$

Equation of continuity of the vapour phase:

$$\frac{\partial}{\partial t}(\alpha\rho_v)+\frac{\partial}{\partial z}(\alpha\rho_v U_v) = \Psi, \tag{8.105}$$

where Ψ is the rate of vapour formation per unit volume.

The conservation of momentum:

$$\frac{\partial}{\partial t}\{(1-\alpha)\rho_l U_l+\alpha\rho_v U_v\}+\frac{\partial}{\partial z}\{(1-\alpha)\rho_l U_l^2+\alpha\rho_v U_v^2\} = -\frac{\partial P}{\partial z}+g\{(1-\alpha)\rho_l+\alpha\rho_v\}-\lambda, \tag{8.106}$$

where P is the local pressure, g is the acceleration due to gravity and λ is the two-phase friction.

Finally, we have conservation of energy:

$$\frac{\partial}{\partial t}\{(1-\alpha)\rho_l H_l+\alpha\rho_v H_v\}+\frac{\partial}{\partial z}\{(1-\alpha)\rho_l U_l H_l+\alpha\rho_v U_v H_v\} = \frac{qP_h}{A}, \tag{8.107}$$

where q is the heat flux, P_h the heated perimeter and A the cross-sectional area of the channel.

There are ten variables in these equations and only four equations. Sandervag shows how by additional assumptions and experimental correlations the additional six unknowns can be specified. However, the most important point concerns Ψ, the rate of vapour production, for this surely is an important noise source that will drive the fluctuations in α. If eqns. (8.104)–(8.107) are interpreted in the Langevin sense then the task remains to find the noise source Ψ. Sandervag considers Ψ to be composed of two distinct terms Ψ_{SF} due to surface formation and Ψ_B due to formation in the bulk liquid. We shall not describe the analysis used to obtain these terms since it is quite detailed. However, suffice to say that one considers the equation of growth for the bubble on the surface, as developed by Plesset (1952), and obtains an expression for Ψ_{SF} which depends in a relatively simple manner on the bubble radius at detachment, the thickness of the saturated liquid layer on the surface T_f, the liquid temperature T_l, the boiling heat flux q_B and a number of readily available constants.

To obtain Ψ_B, the rate of bubble growth, Rayleigh's equation (Rayleigh, 1917) is used together with the theory of Plesset and Zwick (1952, 1954, 1955). The collapse of bubbles as calculated by Wittke and Chao (1967) was also employed together with considerations of bubble coalescence. Whilst a correlation was used for the final form of Ψ_B, as a by-product it was possible to predict the bubble diameter distribution function in a reasonable fashion, and also the average void fraction in the radial direction of a heated tube was obtained experimentally.

Whilst the above discussion is rather superficial, a detailed study of Sandervag's work will show that it might be possible to study the effect on $\alpha(z,t)$ of a number of basic noise sources. This can then be used as input to eqn. (8.61) for the specification of $D_0(z,t)$ and $\overline{v\Sigma_{am}(z,t)}$.

8.12. *A simplified model of boiling*

The above discussion, whilst reasonably fundamental, is too complicated to extract information analytically. For this reason a number of attempts have been made to reduce the complexity of the mass, energy, momentum relations by means of rational approximations. In this way we can study the effect of small disturbances in various parameters on the void fraction, the fuel temperatures and the boiling boundary. Typical of these models is one due to Fleck and Huseby (1958) and Fleck (1960) for a natural circulation boiling-water system. This is a semi-point model approximation because it assumes the void fraction and water velocity to depend linearly on position. However, it results in a very useful set of equations for the time dependence of the outlet void fraction, the fuel temperature and the position of the boiling boundary. These equations when linked to the point model neutron kinetics constitute a non-linear set for studying boiling reactor transients. They could be used for noise-analysis studies by linearizing about mean values; however, it is unlikely that this method would predict the oscillatory packets studied by Akcasu (1961) and discussed in Section 7.8. For that type of noise study we noted that a parametric excitation was needed. A study of the full non-linear model, however, with various parameters considered random might well reveal some useful facts in this respect, even in the point model approximation.

Despite the fact that the Fleck–Huseby model is limited, its study under various sources of noise excitation would be a valuable tool for understanding many aspects of boiling reactors. Nevertheless, as we have said it is not strictly a space-dependent model and we therefore discuss in some detail another model for describing the hydrodynamic behaviour in boiling reactors as developed by Akcasu (1960).

We first consider a particular boiling channel in the reactor, containing fuel element and coolant. This is further divided axially into the non-boiling region $0 < z < L_0(t)$ and the boiling region $L_0(t) < z < L$, where we have explicitly noted the dependence of the height of the non-boiling region on time. Figures 8.6*a* and 8.6*b* illustrate the situation and show that L_0 and L can vary from channel to channel.

In the fuel element the usual heat-conduction equation applies from which the temperatures in the fuel and cladding can be obtained in terms of the heat generated by fission and the surface temperature of the fuel element. If, therefore, for simplicity

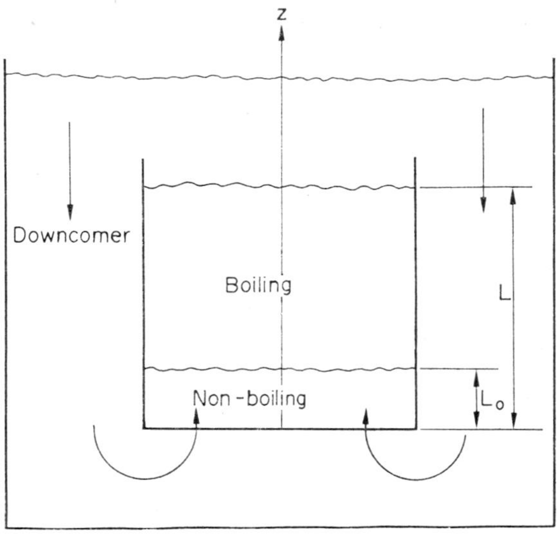

FIG. 8.6*a*. Schematic diagram of boiling-water reactor with L the boiling height and L_0 the non-boiling height.

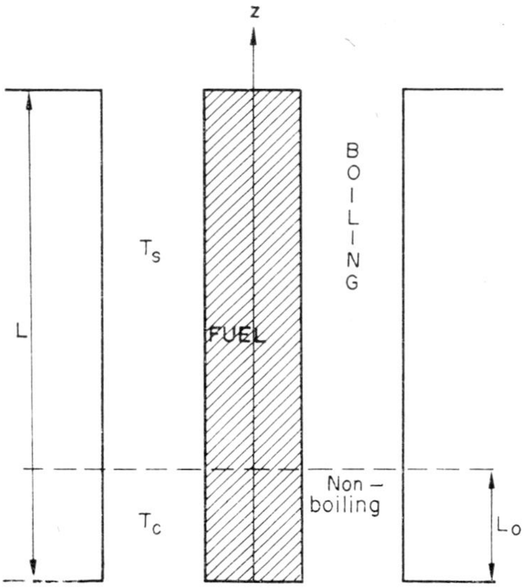

FIG. 8.6*b*. Typical fuel channel in the reactor of Fig. 8.6*a*.

we consider a plate-type fuel element and neglect the cladding material, we can write for the heat-conduction equation

$$k\frac{\partial^2 T_f(x,t)}{\partial x^2}+Q_f(x,t) = P_f C_f\frac{\partial T_f(x,t)}{\partial t}, \tag{8.108}$$

where we have neglected heat conduction in the z-direction.

In the non-boiling region $z < L_0$ the equation for the bulk coolant temperature is

$$A_c C_c \rho_c\left(\frac{\partial T_c(z,t)}{\partial t}+U(z,t)\frac{\partial T_c(z,t)}{\partial z}\right) = q_{nb}(z,t), \tag{8.109}$$

where q_{nb} is the heat flux into the coolant, which may be written

$$q_{nb}(z,t) = -A_f k_f\left.\frac{\partial T_f}{\partial x}\right|_s, \tag{8.110}$$

A_f is the area of one side of the fuel element per unit length and the subscript "s" indicates the surface value. Integration of the equation for T_f over x leads to

$$q_{nb}(z,t) = A_f\bar{Q}_f(z,t)-A_f\rho_f C_f\frac{\partial \bar{T}_f}{\partial t}. \tag{8.111}$$

Thus from eqn. (8.108) we find $T_{fs}(z,t)$ (i.e. surface temperature of fuel) and from (8.109) and (8.111) we get $T_c(z,t)$. These two temperatures are coupled through the boundary condition

$$-k_f\left.\frac{\partial T_f}{\partial x}\right|_s = h_{nb}(T_{fs}-T_c) \tag{8.112}$$

where h_{nb} is the non-boiling heat-transfer coefficient.

In the boiling region $L_0 < z < L$, the coolant temperature remains sensibly constant at its saturation value T_{sat}. In this case the fuel temperature is determined by the boiling heat flux q_B where

$$\begin{aligned} q_B(z,t) &= -A_f k_f\left.\frac{\partial T_f}{\partial x}\right|_s \\ &\equiv A_f\bar{Q}_f(z,t)-A_f\rho_f C_f\frac{\partial \bar{T}_f}{\partial t} \end{aligned} \tag{8.113}$$

with q_B related to T_{fs} and T_{sat} by the boiling heat-transfer correlation of equation (8.86). Sometimes the relationship is written simply as

$$q_B(z,t) = A_f h_B(T_{fs}-T_{\text{sat}}), \tag{8.114}$$

where T_{fs} is the fuel surface temperature in the boiling region and h_B is a boiling heat-transfer coefficient.

The temperatures and heat fluxes obtained from these calculations will affect crucially the void fraction in the boiling region and also the position of the boiling boundary. Thus any fluctuations in the above quantities will give rise to fluctuations in the void volume and steam-production rate. In practice, it is usual to assume that random variations of the heat input into the coolant, resulting from random steam-bubble formation at the surface of the fuel, are responsible for the observed noise and we shall adopt this assumption below. It is important to appreciate, however, that even in an ideal situation where steam bubbles are produced uniformly at a constant rate, there remains a random fluctuation, not in terms of total volume of steam produced, but from the spatial migration, collapse and coalescence of the bubbles themselves. The present theory is unable to treat such an effect since by definition of the void fraction these effects have been averaged out [see eqn. (8.101)].

If the noise source is assumed to arise from fluctuations in h_B, then clearly it will result in fluctuations in T_{fs} and q_B, the input heat source to the coolant. We shall assume that this is the main source of noise exciting the coolant void fraction although there may well be additional noise sources which are indirectly caused by variations in heat generation.

8.13. *Hydrodynamics of the moderator*

Let us consider now the model used by Akcasu for the boiling region. Consider eqns. (8.104) and (8.107) and make the following assumptions: (1) neglect pressure changes and the variation of potential and kinetic energies along the channel; (2) saturation enthalpies depend only on pressure; (3) steam density and latent heat H_{lv} are constant; (4) steam and liquid phases are in equilibrium at all times; (5) the axial distribution of power does not change with time; (6) the slip between water and steam is ignored so that $U_l = U_v = U$. Then the equations of continuity of mixture and of conservation of energy can be combined to give the following equation:

$$\frac{\partial \alpha}{\partial t}+\frac{\partial}{\partial z}(U\alpha) = \frac{Q}{\rho_v H_{vl}}-\frac{1}{\rho_v H_{vl}}\left\{\alpha\rho_v \frac{dH_v}{dP}+(1-\alpha)\rho_l \frac{dH_l}{dP}\right\}\frac{dP}{dt}, \tag{8.115}$$

where $Q = q_B P_h/A \simeq A_f P_h \bar{Q}_f/A$ and we have rearranged H_v and H_l to indicate their dependence on pressure (note that $H_{vl} = H_v - H_l$).

Before proceeding further with this equation let us note the possible variables whose dependence on each other we should like to study. Some obvious ones are:

1. Powervoid fraction.
2. Powerboiling boundary position.
3. Pressurevoid fraction.
4. Pressureboiling boundary position.
5. Boiling boundary positionvoid fraction.
6. Non-boiling coolant temperaturepower.
7. Pressure–saturation temperature in boiling region.

To illustrate the procedure we shall discuss items 1 and 2. The calculation it is to be remembered will be done in the small perturbation approximation since the noise level is assumed to be small compared with the average value. This being the case the various effects described in items 1–7 can be dealt with by linearizing the appropriate equations and holding constant those parameters not being studied. The final effect on say the void fraction will be obtained by summing all of the individual effects. However, we shall explain this point further below.

In the power–void calculation it is assumed that pressure is constant and that we may write $\alpha = \alpha_0+\delta\alpha$, $U = U_0+\delta U$ and $Q = Q_0+\delta Q$, where α_0, U_0 and Q_0 are the steady, average values. Inserting these expressions into eqn. (8.115) and linearizing we find

$$\frac{\partial\,\delta\alpha}{\partial t}+\frac{\partial}{\partial z}(U_0\delta\alpha) = \frac{\delta Q}{\rho_v H_{vl}}. \tag{8.116}$$

It should be noted at this point that we do not have any knowledge of $U_0(z)$. Thus we need an additional equation relating U and α which is provided by the momentum balance (8.106). However, this equation is analytically difficult and so in the spirit of the present inquiry it is postulated that $U_0(z)$ is a prescribed function which depends only through a constant multiplicative factor on pressure. The sensitivity of U_0 to pressure, for high pressures, is slight and we shall therefore ignore it altogether. Some justification for this has been given by Akcasu (1960).

Returning to eqn. (8.116) we can write the initial and boundary conditions as

$$\delta\alpha(z,0) = 0 \tag{8.117}$$

and

$$\delta\alpha(L_0,t) = 0. \tag{8.118}$$

Taking the Laplace transform of eqn. (8.116), such that

$$\overline{\delta\alpha(z,s)} = \int_0^\infty e^{-st}\,\delta\alpha(z,t)\,dt, \tag{8.119}$$

we find

$$\frac{d}{dz}[U_0(z)\overline{\delta\alpha(z,s)}]+s\,\overline{\delta\alpha(z,s)} = \frac{1}{\rho_v H_{vl}}\,\overline{\delta Q(z,s)}. \tag{8.120}$$

Solving this equation we find

$$\overline{\delta\alpha(z,s)} = \frac{1}{U_0(z)\rho_v H_{vl}}\int_{L_0}^{z} dz'\,\overline{\delta Q(z',s)}\exp\left\{-s\int_{z'}^{z}\frac{dz''}{U_0(z'')}\right\} \tag{8.121}$$

and inverting the Laplace transform leads to

$$\delta\alpha(z,t) = \frac{1}{U_0(z)\rho_v H_{vl}}\int_{L_0}^{z} dz'\,\delta Q(z',t-\tau(z',z)), \tag{8.122}$$

where

$$\tau(z',z) = \int_{z'}^{z}\frac{dz''}{U_0(z'')} \tag{8.123}$$

is the transit time of a disturbance to travel between the points z' and z. From eqn. (8.122) the correlation function

$$\langle\delta\alpha(z_1,t_1)\,\delta\alpha(z_2,t_2)\rangle \tag{8.124}$$

can be evaluated in terms of the noise source δQ. For the sake of example let U_0 be independent of z and let $\delta Q(z,t) = \eta(t)F(z)$ where $\eta(t)$ is a random variable and the variation with z follows the axial power distribution. Then we find

$$\delta\alpha(z,t) = \frac{1}{U_0\rho_0 H_{vl}}\int_{L_0}^{z} dz'\,F(z')\,\eta\left[t-\frac{1}{U_0}(z'-z)\right]. \tag{8.125}$$

The correlation function is then given by

$$\langle\delta\alpha(z_1,t_1)\,\delta\alpha(z_2,t_2)\rangle = \frac{1}{U_0^2\rho_v^2 H_{vl}^2}\int_{L_0}^{z_1}dz'\int_{L_0}^{z_2}dz''\,F(z')F(z'')\,\phi_\eta\left(t_2-t_1-\frac{1}{U_0}\{z''-z'-z_2+z_1\}\right), \tag{8.126}$$

where
$$\phi_\eta(t_2-t_1) = \langle \eta(t_1)\eta(t_2)\rangle. \tag{8.127}$$

For the special case of $F = 1$ and white noise η, such that $\phi_\eta(\tau) = \sigma^2\delta(\tau)$, we find

$$\phi_\alpha(z_1, z_2; \tau) = \frac{\sigma^2}{U_0^2\rho_v^2 H_{vl}^2} G(z_1, z_2; \tau), \tag{8.128}$$

where, for $z_2 > z_1$,

$$\begin{aligned} G(z_1, z_2; \tau) &= 0 &&; \quad \tau > (z_1-L_0)/U_0 \\ &= z_1-L_0-U_0\tau; && 0 < \tau < (z_1-L_0)/U_0 \\ &= z_1-L_0 &&; \quad -\frac{(z_2-z_1)}{U_0} < \tau < 0 \\ &= z_2-L_0+U_0\tau; && -\frac{(z_2-L_0)}{U_0} < \tau < -\frac{(z_2-z_1)}{U_0} \\ &= 0 &&; \quad \tau < -(z_2-L_0)/U_0. \end{aligned} \tag{8.129}$$

Thus the correlation function is not symmetrical about $\tau = 0$ as indeed we expect since it is a measure of the correlation between voids at different space points and will clearly depend upon the sign of U_0. In practice, a more realistic form for $F(z)$ would be $\sin(\pi z/L)$; however, the additional algebra involved does not lead to any significant new qualitative information. It is of interest though to consider the p.s.d. of the above correlation function, which for the special case of $z_1 = z_2 = z$ becomes

$$\Phi_\alpha(z; \omega) = \frac{\sigma^2}{U_0\rho_v^2 H_{vl}^2}\frac{1}{\omega^2}\left[1-\cos\left(\frac{\omega(z-L_0)}{U_0}\right)\right]. \tag{8.130}$$

We observe that this vanishes whenever

$$f_n = \frac{\omega_n}{2\pi} = \frac{nU_0}{z-L_0} \quad (n = 1, 2, 3, \ldots), \tag{8.131}$$

which indicates a cyclic effect which can be attributed to the rate of bubble flow between L_0 and z. If z is set equal to the core height L and we use the typical values $U_0 = 260$ cm sec^{-1}, $L = 62$ cm, $L_0 = 36$ cm, then $f_0 = 10$ Hz which is certainly in the frequency range where structure in the p.s.d. is to be expected. Naturally, however, due to the superposition of this effect from all channels, the sink will not be as sharp as suggested by the above formula.

As far as the influence of $\delta\alpha$ on the neutronics is concerned we must return to eqn. (8.61) and note that D_0 and $\overline{v\Sigma_{am}}$ depend upon the physical density ρ. If we write this for the voided coolant as

$$\rho_m(z, t) = (1-\alpha(z, t))\rho_l + \alpha(z, t)\rho_v \tag{8.132}$$

we find that ρ_m is given in terms of $\delta\alpha$ by

$$\rho_m(z, t) = \bar{\rho}_m - (\rho_l - \rho_v)\delta\alpha(z, t) \tag{8.133}$$

and hence the fluctuation can be incorporated into the neutron density equation.

The other effect that we shall consider is that of motion of the boiling boundary, since this will inevitably affect the void fraction. The heat supplied to the channel in the non-boiling region brings the subcooled water to the saturation temperature at the boiling boundary. The variation of feedwater enthalpy along the channel is given by the balance eqn. (8.107) with $\alpha = 0$, thus we have

$$\frac{\partial}{\partial z}(\rho_l U_l H_l)+\frac{\partial}{\partial t}(\rho_l H_l) = Q(z,t). \tag{8.134}$$

Assuming constant flow velocity U_l and writing $H_l = H_{l_0}+\delta h$ and $Q = Q_0+\delta Q$, we find that

$$\rho_l U_0 \frac{d}{dz} H_{l_0}(z) = Q_0(z) \tag{8.135}$$

and

$$\rho_l U_0 \frac{\partial \delta h}{\partial z}+\rho_l \frac{\partial \delta h}{\partial t} = \delta Q. \tag{8.136}$$

The definition of the boiling boundary in the steady state is obtained by integrating (8.135), from which we find

$$\rho_l U_0(H_{l_0}(L_0)-H_{l_0}(0)) = \int_0^{L_0} Q_0(z)\,dz, \tag{8.137}$$

where $H_{l_0}(0)$ is the feedwater enthalpy and $H_{l_0}(L_0) = H_l$ that of the liquid at saturation temperature.

The solution of (8.136) with $\delta h(z,0) = 0$ and $\delta h(0,t) = 0$ leads to

$$\overline{\delta h(z,s)} = \frac{1}{\rho_l U_0}\int_0^z dz'\,\overline{\delta Q(z',s)}\exp\left\{-\frac{s}{U_0}(z-z')\right\}. \tag{8.138}$$

If we now consider a small movement of the boiling boundary, $\Delta(t)$ say, we have

$$H_l = H_{l_0}(L_0+\Delta)+\delta h(L_0+\Delta,t) \tag{8.139}$$

$$\simeq H_{l_0}(L_0+\Delta)+\delta h(L_0,t), \tag{8.140}$$

since an increase of Δ in δh is a second-order effect.

Now if eqn. (8.135) is integrated over z in the range $L_0 < z < L_0+\Delta$ and $Q(z)$ is assumed to remain sensibly constant over Δ, we find that

$$H_{l_0}(L_0+\Delta)-H_{l_0}(L_0) = \frac{Q_0(L_0)\Delta(t)}{U_0\rho_l}. \tag{8.141}$$

Combining (8.140) and (8.141) we get

$$\Delta(t) = -\frac{U_0\rho_l}{Q_0(L_0)}\,\delta h(L_0,t). \tag{8.142}$$

Thus from (8.138)

$$\overline{\Delta(s)} = -\frac{1}{Q_0(L_0)}\int_0^{L_0} dz'\,\overline{\delta Q}(z',s)\,e^{-s(L_0-z')/u_0} \tag{8.143}$$

or

$$\Delta(t) = -\frac{1}{Q_0(L_0)}\int_0^{L_0} dz'\,\delta Q\left(z',t-\frac{1}{U_0}(L_0-z')\right). \tag{8.144}$$

From this we can obtain $\langle\Delta(t_1)\Delta(t_2)\rangle$ in the usual way; indeed apart from the limits and a constant factor the equation for $\Delta(t)$ is the same as that for $\delta\alpha$.

If the form of $\delta Q(z,t)$ is

$$\eta(t)\frac{\pi}{2L}\sin\left(\frac{\pi z}{L}\right)$$

then $\overline{\Delta(s)}$ may be evaluated and leads to

$$\overline{\Delta(s)} = -\frac{\overline{\eta(s)}}{2Q_0(L_0)} \frac{e^{-\beta x}+x\sin\beta-\cos\beta}{1+x^2}, \tag{8.145}$$

where $x = s/\omega_0$, $\omega_0 = \beta/T_0$, $\beta = \pi L_0/L$ and $T_0 = L_0/U_0$.

The p.s.d. can therefore be obtained directly as

$$\Phi_\Delta(\omega) = \langle \Delta(i\omega)\Delta^*(i\omega)\rangle$$

$$= \frac{\Phi_\eta(\omega)}{4Q_0^2(L_0)} \left|\frac{e^{-\beta x}+x\sin\beta-\cos\beta}{1+x^2}\right|^2_{x=i\omega/\omega_0} \tag{8.146}$$

which we note has zero phase angle. It is also interesting to note that the numerator of this expression is very sensitive to β especially in the range $\pi/3 < \beta < \pi$. $\beta = \pi$ corresponds to the boiling boundary, $L_0 = L$ being at the top of the core, whereas $\beta = \pi/3$ is characteristic of, say, the EBWR (ANL-5607, 1957). Numerical investigation shows that Φ_Δ can exhibit sharp sinks in the range $\omega/\omega_0 \simeq 3$ and 5, etc., although, again, the depths of these sinks may in practice be smoothed out when the effect of other channels is superimposed. Some physical insight into the reason for these sinks can be obtained by use of the simpler model where δQ is independent of z. Then we find

$$\Delta(s) = -\overline{\eta(s)}\frac{L_0}{L}\frac{1}{Q_0(L_0)}\frac{1-e^{-sT_0}}{sT_0} \tag{8.147}$$

or

$$\Phi_\Delta(\omega) = \frac{\Phi_\eta(\omega)L_0^2}{L^2Q_0^2(L_0)}\left|\frac{1-e^{-i\omega T_0}}{i\omega T_0}\right|^2 \tag{8.148}$$

$$= \frac{\Phi_\eta(\omega)L_0^2}{L^2T_0^2}\frac{2}{\omega^2T_0^2}(1-\cos\omega T_0) \tag{8.149}$$

which vanishes when

$$f_n = \frac{\omega_n}{2\pi} = \frac{n}{T_0} \quad (n = 1,2,3,\ldots). \tag{8.150}$$

At the sink frequency f_1, water "packets" experience a complete cycle of the power generation during their transit through the non-boiling region. Thus the net change in water enthalpy is zero at the boiling boundary and no change in its position will occur.

Finally, we mention the effect on the void fraction of the boiling boundary variations. That this will have an effect is clear since it will cause the total steam volume in the channel to fluctuate. We return therefore to eqn. (8.105) for the continuity of steam mass and note that Ψ, the steam production rate, can be written as the sum of the normal production rate at steady-state Ψ_0, plus the term $\delta\Psi$ due to the additional steam produced by movement of the boiling boundary. Thus if we note that $Q_0(L_0)\Delta(t)$ is the heat generated in the length $\Delta(t)$, we can write for the amount of steam produced at L_0 the expression

$$\delta\Psi = -\frac{Q_0(L_0)}{H_{vl}}\Delta(t)\delta(z-L_0), \tag{8.151}$$

where the delta function indicates that it is a localized effect.

With a constant value of $U_v = U_0$ (which is not essential) and writing $\alpha = \alpha_0+\delta\alpha$, we find that the perturbation in $\delta\alpha$ due to the perturbation $\delta\Psi$ is given by

$$\rho_v \frac{\partial}{\partial t}\delta\alpha+\rho_v U_0 \frac{\partial}{\partial z}\delta\alpha = -\frac{Q_0(L_0)}{H_{vl}}\Delta(t)\delta(z-L_0). \tag{8.152}$$

Laplace transform in time leads to

$$\overline{\delta\alpha(z,s)} = -\frac{Q_0(L_0)}{H_{vl}\rho_v U_0}\overline{\Delta(s)}\exp\{-s(z-L_0)/U_0\} \tag{8.153}$$

or

$$\delta\alpha(z,t) = -\frac{Q_0(L_0)}{H_{vl}\rho_v U_0}\Delta\left(t-\frac{1}{U_0}\{z-L_0\}\right). \tag{8.154}$$

We therefore find that

$$\phi_\alpha(z_1,z_2;\tau) = \frac{Q_0^2(L_0)}{H_{vl}^2\rho_v^2 U_0^2}\phi_\Delta\left(\tau-\frac{1}{U_0}(z_2-z_1)\right). \tag{8.155}$$

The connection between $\delta\alpha$ and δQ is obtained by using (8.143) in (8.153) to eliminate $\overline{\Delta(s)}$, whence

$$\overline{\delta\alpha(z,s)} = \frac{1}{H_{vl}\rho_v U_0}e^{-s(z-L_0)/U_0}\int_0^{L_0} dz'\,\overline{\delta Q(z',s)}\,e^{-s(z'-L_0)/U_0}$$

$$= \frac{1}{H_{vl}\rho_v U_0}\int_0^{L_0} dz'\,\overline{\delta Q(z',s)}\,e^{-s(z-z')/U_0}. \tag{8.156}$$

It should be noted that this equation for $\overline{\delta\alpha}$ is formally equivalent to that for void-power coupling as given by (8.121). However, the statistics of δQ for $z < L_0$ are different from those for δQ in $z > L_0$, due to the differences in the heat-transfer characteristics of the boiling and non-boiling regions.

When all of the effects that cause α to fluctuate have been evaluated the total effect is obtained by summing, viz.

$$\rho_m(z,t) = \bar{\rho}_m-(\rho_l-\rho_v)\sum_i \delta\alpha_i(z,t). \tag{8.157}$$

There will also be temperature feedback effects but these are obtained in a standard manner from the equations of heat transfer discussed earlier. Such effects will influence the energy averaged nuclear reaction rates but may also affect the densities of steam and water to some degree.

Other authors have made studies of the effect of voidage on reactivity and a very straightforward method is given by Nomura (1968) who considers the process of heat transfer to the coolant as a shot-noise process: a shot being the transfer of a quantity of heat leading to a bubble. The randomness arises therefore from the random time series generated by the successive "shots". The shot noise is related to void fraction, pressure and temperature which are subsequently, either directly or indirectly, related to reactivity of the system. This is then fed into the linearized neutronics equations. Nomura compares his results for the p.s.d. of the neutron fluctuations with some experiments performed on the JPDR reactor, and whilst agreement is not excellent, at least the theory and experiment are in qualitative agreement as far as the general structure of the p.s.d. is concerned.

Finally we note the work of Robinson (1970) who sets up a theoretical model for the Molten Salt Reactor Experiment using the hydraulic model described above and calculates the effect, on the voidage entrained in the salt, of pressure variations. Comparison of the theory with experiment was made by Robinson and Fry (1970) who introduced pressure fluctuations into the pump bowl of the reactor and measured the resulting neutron fluctuations. Theory and experiment were in substantial agreement over the frequency range 0·025–0·125 Hz.

Further discussion of the work done and inherent weaknesses of the various techniques used in boiling-water noise may be found in the review article by Kosály (1973).

It is clearly obvious, but nevertheless worth mentioning, that the void fractions obtained by any of these methods must be fed back into the source-sink equations for heterogeneous systems. In this way the neutron noise at any position may be obtained. Needless to say, such an ambitious programme has not yet been performed.

8.14. *Application of the source-sink method to a zero-power problem*

An interesting application of the heterogeneous formalism described above is to zero-power reactor noise. In Chapter 7, this was dealt with by homogenizing the core and use of the appropriate eigenfunctions of the Helmholtz equation to find the Green's function and subsequently the covariance. We now have an opportunity to test the validity of the homogenization procedure. Let us return to the exprcssion for the count rate correlation function $\Gamma_c(\mathbf{r}_1,\mathbf{r}_2;\tau)$ as defined by eqn. (7.35), viz.

$$\begin{aligned}\Gamma_c(\mathbf{r}_1^*,\mathbf{r}_2^*;\tau) &= \tilde{\epsilon}(\mathbf{r}_1^*)f(\mathbf{r}_1^*)\delta(\mathbf{r}_1^*-\mathbf{r}_2^*)\delta(\tau)\\ &\quad+\tilde{\epsilon}(\mathbf{r}_1^*)\tilde{\epsilon}(\mathbf{r}_2^*)\int d\mathbf{r}_0\overline{\nu(\nu-1)f(\mathbf{r}_0)v\Sigma_f(\mathbf{r}_0)}\\ &\quad\times\int_0^\infty d\tau_0F(\mathbf{r}_0,\mathbf{r}_1^*;\tau_0)F(\mathbf{r}_0,\mathbf{r}_2^*;\tau+\tau_0),\end{aligned}\tag{8.158}$$

where

$$\begin{aligned}F(\mathbf{r}_0,\mathbf{r};\tau) &= (1-\beta)\int d\mathbf{r}'K(\mathbf{r}_0\to\mathbf{r}')G(\mathbf{r}'\to\mathbf{r};\tau)\\ &\quad+\sum_{i=1}^{l}\beta_i\lambda_i\int_0^\tau dt_0e^{-\lambda_it_0}\int d\mathbf{r}'K(\mathbf{r}_0\to\mathbf{r}')G(\mathbf{r}'\to\mathbf{r};\tau-t_0)\end{aligned}\tag{8.159}$$

and the Green's function G is defined by

$$\begin{aligned}&\left[-\frac{\partial}{\partial t}+D_0\nabla^2-\overline{v\Sigma_a(\mathbf{r})}\right]G(\mathbf{r}'\to\mathbf{r};t)+(1-\beta)\int d\mathbf{r}''\overline{\nu v\Sigma_f(\mathbf{r}'')}K(\mathbf{r}''\to\mathbf{r})G(\mathbf{r}'\to\mathbf{r}'';t)\\ &\quad+\sum_{i=1}^{l}\beta_i\lambda_i\int_0^t dt_0e^{-\lambda_it_0}\int d\mathbf{r}''\overline{\nu v\Sigma_f(\mathbf{r}_0)}K(\mathbf{r}''\to\mathbf{r})G(\mathbf{r}'\to\mathbf{r}'';t-t_0)+\delta(t)\delta(\mathbf{r}-\mathbf{r}')=0,\end{aligned}\tag{8.160}$$

where it has been assumed that prompt and delayed neutrons have the same slowing-down kernels.

The p.s.d. is obtained as the Fourier transform of Γ_c, viz.

$$\begin{aligned}\Phi_{12}(\omega) &= \tilde{\epsilon}(\mathbf{r}_1^*)\delta(\mathbf{r}_1^*-\mathbf{r}_2^*)f(r_1^*)+\tilde{\epsilon}(\mathbf{r}_1^*)\tilde{\epsilon}(\mathbf{r}_2^*)|\Theta(i\omega)|^2\\ &\quad\times\int d\mathbf{r}_0\overline{\nu(\nu-1)}\overline{v\Sigma_f(\mathbf{r}_0)f(\mathbf{r}_0)}\bar{F}_0(\mathbf{r}_0,\mathbf{r}_1^*;-i\omega)\bar{F}_0(\mathbf{r}_0,\mathbf{r}_2^*;i\omega),\end{aligned}\tag{8.161}$$

where
$$\bar{F}_0(\mathbf{r}_0, \mathbf{r}; s) = \int d\mathbf{r}' K(\mathbf{r}_0 \to \mathbf{r}') \bar{G}(\mathbf{r}' \to \mathbf{r}; s), \tag{8.162}$$

$\bar{G}$ being the Laplace transform of G and
$$\Theta(s) = 1 - \sum_{i=1}^{l} \frac{s\beta_i}{s+\lambda_i}. \tag{8.163}$$

Now, in the spirit of source-sink, we set
$$\bar{\nu} \overline{v \Sigma_f(\mathbf{r}_0)} \to \sum_{\mathbf{k}} (\eta\gamma)_{\mathbf{k}} \delta(\boldsymbol{\xi}_0 - \boldsymbol{\xi}_{\mathbf{k}}), \tag{8.164}$$

where $\eta\gamma$ can still depend upon the axial coordinate z_0. Similarly
$$v\Sigma_a(\mathbf{r}) \to \sum_{\mathbf{k}} \gamma_{\mathbf{k}} \delta(\boldsymbol{\xi} - \boldsymbol{\xi}_{\mathbf{k}}) + \overline{v\Sigma_{am}}. \tag{8.165}$$

The correlation function now becomes, with $\mathbf{r} \equiv (\boldsymbol{\xi}, z)$,
$$\begin{aligned}\Gamma_c(\boldsymbol{\xi}_1^*, \boldsymbol{\xi}_2^*, z_1^*, z_2^*; \tau) &= \tilde{\epsilon}(\boldsymbol{\xi}_1^*, z_1^*)\delta(\boldsymbol{\xi}_1^* - \boldsymbol{\xi}_2^*)\delta(z_1^* - z_2^*) f(\boldsymbol{\xi}_1^*, z_1^*)\delta(\tau) \\ &+ \tilde{\epsilon}(\boldsymbol{\xi}_1^*, z_1^*)\tilde{\epsilon}(\boldsymbol{\xi}_2^*, z_2^*) \int dz_0 \sum_{\mathbf{k}} \left[\frac{\overline{\nu(\nu-1)}}{\bar{\nu}}\right]_{\mathbf{k}} \eta_{\mathbf{k}} \gamma_{\mathbf{k}} f(\boldsymbol{\xi}_{\mathbf{k}}, z_0) \\ &\times \int_0^\infty d\tau_0 F(\boldsymbol{\xi}_{\mathbf{k}}, \boldsymbol{\xi}_1^*, z_0, z_1^*; \tau_0) F(\boldsymbol{\xi}_{\mathbf{k}}, \boldsymbol{\xi}_2^*, z_0, z_2^*; \tau + \tau_0).\end{aligned} \tag{8.166}$$

An analogous expression for Φ_{12} may also be written down. Note that the integration over z_0 extends over the height of the reactor. The corresponding equation for the Green's function now becomes
$$\begin{aligned}&\left[-\frac{\partial}{\partial t} + D_0 \nabla^2 + \overline{v\Sigma_{am}}\right] G(\boldsymbol{\xi}' \to \boldsymbol{\xi}, z' \to z; t) - \sum_{\mathbf{k}} \gamma_{\mathbf{k}} \delta(\boldsymbol{\xi} - \boldsymbol{\xi}_{\mathbf{k}}) G(\boldsymbol{\xi}' \to \boldsymbol{\xi}_{\mathbf{k}}, z' \to z; t) \\ &+ \int dz_0 \sum_{\mathbf{k}} (1 - \beta_{\mathbf{k}}) \eta_{\mathbf{k}} \gamma_{\mathbf{k}} K(\boldsymbol{\xi}_{\mathbf{k}} \to \boldsymbol{\xi}, z_0 \to z) G(\boldsymbol{\xi}' \to \boldsymbol{\xi}_{\mathbf{k}}, z' \to z_0; t) \\ &+ \int dz_0 \sum_{\mathbf{k}} \sum_{i=1}^{l} \beta_{\mathbf{k},i} \lambda_{\mathbf{k},i} \int_0^t dt_0 \exp\{-\lambda_{\mathbf{k},i} t_0\} \eta_{\mathbf{k}} \gamma_{\mathbf{k}} K(\boldsymbol{\xi}_{\mathbf{k}} \to \boldsymbol{\xi}; z_0 \to z) \\ &\times G(\boldsymbol{\xi}' \to \boldsymbol{\xi}_{\mathbf{k}}, z' \to z_0; t - t_0) + \delta(t)\delta(\boldsymbol{\xi} - \boldsymbol{\xi}')\delta(z - z') = 0.\end{aligned} \tag{8.167}$$

To obtain a simple result which may be compared directly with the homogeneous results of Chapter 7, we make the following assumptions: (1) the medium is infinite in the axial direction; (2) the volume of fuel is very much less than that of the moderator so that $K(\mathbf{r}' \to \mathbf{r}) \simeq K(|\mathbf{r}' - \mathbf{r}|)$; (3) the fuel elements all have the same properties which do not vary with z; (4) detectors are infinite in length and their efficiency is constant with z. In this case we can define a new correlation function by
$$\tilde{\Gamma}_c(\boldsymbol{\xi}_1^*, \boldsymbol{\xi}_2^*; \tau) = \lim_{H\to\infty} \frac{1}{H} \int_{-H/2}^{H/2} dz_1^* \int_{-H/2}^{H/2} dz_2^* \, \Gamma_c(\boldsymbol{\xi}_1^*, \boldsymbol{\xi}_2^*, z_1^*, z_2^*; \tau), \tag{8.168}$$

where
$$\begin{aligned}\tilde{\Gamma}_c(\boldsymbol{\xi}_1^*, \boldsymbol{\xi}_2^*; \tau) &= \tilde{\epsilon}(\boldsymbol{\xi}_1^*)\delta(\boldsymbol{\xi}_1^* - \boldsymbol{\xi}_2^*) f(\boldsymbol{\xi}_1^*)\delta(\tau) + \tilde{\epsilon}(\boldsymbol{\xi}_1^*)\tilde{\epsilon}(\boldsymbol{\xi}_2^*) \frac{\overline{\nu(\nu-1)}}{\bar{\nu}} \eta\gamma \sum_{\mathbf{k}} f(\boldsymbol{\xi}_{\mathbf{k}}) \\ &\times \int_0^\infty d\tau_0 \tilde{F}(\boldsymbol{\xi}_{\mathbf{k}}, \boldsymbol{\xi}_1^*; \tau_0) \tilde{F}(\boldsymbol{\xi}_{\mathbf{k}}, \boldsymbol{\xi}_2^*; \tau + \tau_0),\end{aligned} \tag{8.169}$$

where

$$\tilde{F}(\xi_0, \xi; \tau) = (1-\beta)\int d\xi' K(|\xi_0-\xi'|)G(\xi' \to \xi; \tau)$$
$$+\sum_{i=1}^{l} \beta_i \lambda_i \int_0^{\tau} dt_0 e^{-\lambda_i t_0} \int d\xi' K(|\xi_0-\xi'|)G(\xi' \to \xi; \tau-t_0) \quad (8.170)$$

with $K(\ldots)$ the slowing-down kernel from a line source and G the two-dimensional Green's function given by the following equation:

$$\left[-\frac{\partial}{\partial t}+D_0\nabla_\xi^2-\overline{v\Sigma_{am}}\right]G(\xi' \to \xi; t)-\gamma\sum_{\mathbf{k}}\delta(\xi-\xi_{\mathbf{k}})G(\xi' \to \xi_{\mathbf{k}}; t)$$
$$+(1-\beta)\eta\gamma\sum_{\mathbf{k}}K(|\xi_{\mathbf{k}}-\xi|)G(\xi' \to \xi_{\mathbf{k}}; t)+\eta\gamma\sum_{i=1}^{l}\beta_i\lambda_i\sum_{\mathbf{k}}\int_0^t dt_0$$
$$\times e^{-\lambda_i t_0}K(|\xi_{\mathbf{k}}-\xi|)G(\xi' \to \xi_{\mathbf{k}}; t-t_0)+\delta(t)\delta(\xi-\xi') = 0. \quad (8.171)$$

The corresponding p.s.d. is

$$\tilde{\Phi}_{12}(\omega) = \tilde{\epsilon}(\xi_1^*)\delta(\xi_1^*-\xi_2^*)f(\xi_1^*)-\tilde{\epsilon}(\xi_1^*)\tilde{\epsilon}(\xi_2^*)|\Theta(i\omega)|^2\frac{\overline{\nu(\nu-1)}}{\overline{\nu}}$$
$$\times\sum_{\mathbf{k}}f(\xi_{\mathbf{k}})\bar{\bar{F}}_0(\xi_{\mathbf{k}}, \xi_1^*; -i\omega)\bar{\bar{F}}_0(\xi_{\mathbf{k}}, \xi_2^*; i\omega) \quad (8.172)$$

with

$$\bar{\bar{F}}_0(\xi_0, \xi; s) = \int d\xi' K(|\xi_0-\xi'|)\bar{G}(\xi' \to \xi; s). \quad (8.173)$$

Whilst we could apply the methods for treating finite reactors developed by Feinberg and Galanin, for the sake of simplicity we shall consider a plane lattice, infinite in the lateral dimensions. Thus by integrating over the y-coordinates the p.s.d. reduces once more to the simpler form

$$\Phi_{12}(\omega) = \tilde{\epsilon}(x_1^*)f(x_1^*)\delta(x_1^*-x_2^*)+\tilde{\epsilon}(x_1^*)\tilde{\epsilon}(x_2^*)\frac{\overline{\nu(\nu-1)}}{\overline{\nu}}\eta\gamma|\Theta(i\omega)|^2$$
$$\times\sum_{k=-\infty}^{\infty}f(x_k)\bar{F}_0(x_k, x_1^*; -i\omega)\bar{F}_0(x_k, x_2^*; i\omega), \quad (8.174)$$

where we assume that the detectors are infinite plates in the planes $x = x_1^*$ and $x = x_2^*$.

$\bar{F}_0$ is given by

$$\bar{F}_0(x_0, x; s) = \int_{-\infty}^{\infty} dx' K(|x_0-x'|)\bar{G}(x' \to x; s). \quad (8.175)$$

The equation for the Green's function is obtained by integrating (8.171) over the y-coordinate. Taking the Laplace transform with respect to time and the Fourier transform with respect to x leads after rearrangement and Fourier inversion to

$$\bar{G}(x' \to x; s) = \frac{1}{2\alpha D_0}\exp(-\alpha|x-x'|)+\frac{\gamma}{2D_0}\sum_{k=-\infty}^{\infty}G(x' \to x_k; s)$$
$$\times\frac{1}{2\pi}\int_{-\infty}^{\infty}\frac{[\eta\Theta(s)\bar{K}(B^2)-1]}{B^2+\alpha^2}\exp[iB(x_k-x)]\,dB, \quad (8.176)$$

where $\bar{K}(B^2)$ is the Fourier transform of the slowing-down kernel and

$$\alpha = L^{-1}(1+sl)^{\frac{1}{2}}$$

with $l = (\overline{v\Sigma_{am}})^{-1}$ and $L^2 = D_0/\overline{v\Sigma_{am}}$.

$\bar{F}_0$ can be obtained directly by multiplying eqn. (8.176) by $K(|x_l - x'|)$ and integrating over x'. Then we find

$$\bar{F}_0(x_l, x; s) = \frac{1}{2D_0} S(|x_l - x|) + \frac{\gamma}{2D_0} \sum_{k=-\infty}^{\infty} H(|x_k - x|; s) \bar{F}_0(x_l, x_k; s) \quad (8.177)$$

with
$$S(|x_l - x|; s) = \frac{1}{\pi} \int_{-\infty}^{\infty} K(|x_l - x|) \exp(-\alpha|x - x'|) dx'$$

$$\equiv \frac{1}{\pi} \int_{-\infty}^{\infty} \frac{\exp[iB(x - x_l)]}{B^2 + \alpha^2} \bar{K}(B^2) dB \quad (8.178)$$

and
$$H(|x|; s) = \frac{1}{\pi} \int_{-\infty}^{\infty} \frac{[\eta \Theta(s) \bar{K}(B^2) - 1]}{B^2 + \alpha^2} e^{iBx} dB. \quad (8.179)$$

We now recall that in an infinite lattice $f(x_k) = f(0) =$ constant. Further, to solve eqn. (8.177) we set $x = x_q = aq$ where a is the lattice pitch. Then we find in an obvious notation that

$$\bar{F}_0(l, q; s) = \frac{1}{2D_0} S(a|l - q|; s) + \mu_0 \sum_{k=-\infty}^{\infty} H(a|k - q|; s) \bar{F}_0(l, k; s) \quad (8.180)$$

with $\mu_0 = \gamma/2D_0$.

This equation may be solved by a finite Fourier series expansion which leads readily to the following result:

$$\bar{F}_0(l, q; s) \equiv \bar{F}_0(|l - q|; s) = \frac{1}{2\pi} \int_{-\infty}^{\infty} \exp[-i\nu(q - l)] I(\nu, s) d\nu, \quad (8.181)$$

where
$$I(\nu, s) = \frac{1}{2D_0} \cdot \frac{\mathscr{S}(\nu, s)}{1 - \mu_0 h(\nu, s)} \quad (8.182)$$

and
$$\mathscr{S}(\nu, s) = \sum_{p=-\infty}^{\infty} \exp(-ip\nu) S(a|p|; s), \quad (8.183)$$

$$h(\nu, s) = \sum_{p=-\infty}^{\infty} \exp(-ip\nu) H(a|p|; s). \quad (8.184)$$

Knowing $\bar{F}_0(l, q; s)$ we can substitute into eqn. (8.177) to find $\bar{F}_0(ak, x; s)$ and hence the p.s.d.

A very simple case of this problem arises if we set $x_1^* = 0$ and $x_2^* = am$ ($m = 0, 1, 2, \ldots$), then the cross-correlation between any two fuel plates a distance am apart is given by

$$\Phi_{0m}(\omega) = \tilde{\varepsilon} d f(0) \delta_{0,m} + (\tilde{\varepsilon} d)^2 \frac{\overline{\nu(\nu - 1)}}{\bar{\nu}} \eta \gamma f(0) |\Theta(i\omega)|^2 \cdot \frac{1}{2\pi} \int_{-\pi}^{\pi} d\nu \exp(i\nu m) |I(\nu, i\omega)|^2, \quad (8.185)$$

where to include the detector volume we have integrated the p.s.d. over a thin detector of thickness d.

If the system considered above is critical, the steady-state flux is easily shown to be

$$f(x) = f(0) \mu_0 \sum_{k=-\infty}^{\infty} H(|x - ak|; 0) \quad (8.186)$$

and the critical equation is $\quad 1 = \mu_0 \sum_{k=-\infty}^{\infty} H(a|k|; 0). \quad (8.187)$

In the paper by Williams (1967*a*) some general models of slowing down are considered and the form of $\Phi_{12}(\omega)$ and $\Phi_{0m}(\omega)$ is given. It is of special interest, however, to consider the case of one-speed theory in which $\bar{K}(B^2) = 1$. It then follows that the sums in eqns. (8.183) and (8.184) can be performed, resulting in

$$I(\nu, s) = \frac{1}{2D_0\alpha}\frac{\sinh\alpha a}{C-\cos\nu}, \tag{8.188}$$

where

$$C = \cosh\alpha a + \mu_0 a(1-\eta\Theta)\frac{\sinh\alpha a}{\alpha a}. \tag{8.189}$$

Using the integral

$$\frac{1}{\pi}\int_0^{\pi}\frac{\cos\nu m}{C-\cos\nu}\,d\nu = \frac{[C-(C^2-1)^{\frac{1}{2}}]^m}{(C^2-1)^{\frac{1}{2}}} \tag{8.190}$$

we find that

$$\Phi_{0m}(\omega) = \tilde{\varepsilon}df(0) + (\tilde{\varepsilon}d)^2\,\frac{\overline{\nu(\nu-1)}}{\bar{\nu}}\,\frac{\eta\gamma f(0)}{4D_0^2}\,\frac{|\sinh\alpha a|^2}{|\alpha|^2}\,|\Theta|^2$$
$$\times\frac{1}{(C^*-C)}\left\{\frac{[C-(C^2-1)^{\frac{1}{2}}]^m}{(C^2-1)^{\frac{1}{2}}} - \frac{[C^*-(C^{*2}-1)^{\frac{1}{2}}]^m}{(C^{*2}-1)^{\frac{1}{2}}}\right\}, \tag{8.191}$$

where C^* is the complex conjugate of C.

This is an exact expression for an infinite plate reactor in the one-speed approximation. To see how the expression reduces to the homogeneous limit we write $\alpha = \xi + i\mu$, where

$$\xi = \frac{1}{\sqrt{2}L}\,[1+(1+\omega^2 l^2)^{\frac{1}{2}}]^{\frac{1}{2}} \tag{8.192}$$

and

$$\mu = \frac{1}{\sqrt{2}L}\,[(1+\omega^2 l^2)^{\frac{1}{2}}-1]^{\frac{1}{2}}. \tag{8.193}$$

$1/\xi$ is the attenuation length, and $2\pi/\mu$ is the wavelength of a neutron wave in the pure moderator of the lattice (Weinberg and Wigner, 1958).

If we now insert the expression for α in eqn. (8.189) and set $\sin\mu a \simeq \mu a$, $\cos\mu a = 1$, $\sinh\xi a \simeq \xi a$ and $\cosh\xi a \simeq 1$, we find that

$$C \simeq 1 + \frac{1}{2}\frac{a^2}{L^2} + \mu_0 a[1-\eta(1-\beta)] + \frac{ia^2}{2L^2}\,\omega l. \tag{8.194}$$

With this approximation we can write

$$[C-(C^2-1)^{\frac{1}{2}}]^m \equiv \exp\{m\log[C-(C^2-1)^{\frac{1}{2}}]\}, \tag{8.195}$$

$$\simeq \exp\left\{-ma\left(\frac{\omega_0+i\omega}{D_0}\right)^{\frac{1}{2}}\right\} \tag{8.196}$$

and obtain

$$\Phi_{0m}(\omega) \simeq \tilde{\varepsilon}df(0) + (\tilde{\varepsilon}d)^2\,\frac{\overline{\nu(\nu-1)}}{\bar{\nu}}\,\frac{\eta\gamma f(0)}{4D_0\omega}\,\frac{i(D_0)^{\frac{1}{2}}}{a}$$
$$\times\left\{\frac{\exp\left[-ma\left(\frac{\omega_0+i\omega}{D_0}\right)^{\frac{1}{2}}\right]}{(\omega_0+i\omega)^{\frac{1}{2}}} - \frac{\exp\left[-ma\left(\frac{\omega_0-i\omega}{D_0}\right)^{\frac{1}{2}}\right]}{(\omega_0-i\omega)^{\frac{1}{2}}}\right\} \tag{8.197}$$

with

$$\omega_0 = \frac{1}{l} + \frac{\gamma}{a}\,[1-\eta(1-\beta)]. \tag{8.198}$$

Now $\gamma = \overline{v\Sigma_{au}}t$, where $\overline{v\Sigma_{au}}$ is the fuel-absorption rate and t its thickness, thus ω_0 can also be written

$$\omega_0 = \overline{v\Sigma_{am}}+\frac{t}{a}\,\overline{v\Sigma_{au}}-\frac{t}{a}\,\overline{v\Sigma_{au}}\,\eta(1-\beta). \tag{8.199}$$

In the homogeneous limit we can also write k_∞ as

$$k_\infty = \frac{\overline{v\Sigma_{au}}\,t\eta}{\overline{v\Sigma_{am}}\,a+\overline{v\Sigma_{au}}\,t} \tag{8.200}$$

and the average lifetime l^* as

$$\frac{1}{l^*} = \overline{v\Sigma_{am}}+\frac{t}{a}\,\overline{v\Sigma_{au}}. \tag{8.201}$$

Thus

$$\omega_0 = \frac{1}{l^*}\,[1-k_\infty(1-\beta)]. \tag{8.202}$$

With this definition, eqn. (8.197) can be cast into exactly the same form as the cross-p.s.d. between two points a distance $x = ma$ apart obtained by Natelson *et al.* (1966) and Williams (1967*b*) using homogeneous diffusion theory. Indeed, if we set $m = 0$ and calculate the auto-p.s.d., the result is

$$\Phi_{00}(\omega) = \tilde{\epsilon}df(0)+\frac{(\tilde{\epsilon}d)^2\,\overline{\nu(\nu-1)}f(0)\,\overline{v\Sigma_f}/2(2D_0)^{\frac{1}{2}}}{(\omega^2+\omega_0^2)^{\frac{1}{2}}\,[\omega_0+(\omega_0^2+\omega^2)^{\frac{1}{2}}]^{\frac{1}{2}}} \tag{8.203}$$

which is identical with eqn. (3.76) and which in turn can be obtained from eqn. (7.58) if we set

$$g(\mathbf{k}, V_1, V_2) = (2\pi)^2(\tilde{\epsilon}d)^2\,\delta(k_y)\,\delta(k_z). \tag{8.204}$$

The approximations made in going to the homogeneous limit enable the limitations of using that theory to describe heterogeneous reactors to be defined. Thus, we assumed that $\mu a \ll 1$ and $\xi a \ll 1$. Since, for a given ω, $\xi > \mu$ this leads to the following restriction,

$$\frac{a}{\sqrt{2}L}\,[(1+\omega^2 l^2)^{\frac{1}{2}}+1] \ll 1. \tag{8.205}$$

In other words, instead of the plate spacing being very much less than a diffusion length as we require for steady-state problems, we now have the added restriction that the plate spacing must be very much less than the attenuation length of a neutron wave in the moderator. Thus the homogeneous approximation becomes poorer for the higher frequencies. Whilst the present theory cannot be compared directly with any experiments, it is of interest to note that for some cross-correlation measurements made by Kylstra and Uhrig (1965) in H_2O and D_2O heterogeneous systems, it was found that deviations from homogeneous theory set in above a certain frequency. These deviations are consistent with the inequality (8.205); however, there are a number of other factors involved and it cannot be stated without doubt that failure of homogeneous theory is the cause of the discrepancy. Nevertheless, it is clear that the non-uniformity of the noise sources are of significance and estimates should be made of their importance.

Whilst we do not intend to give the complete theory here (see Williams, 1967*b*), it is worth noting that the source-sink technique has been employed to assess the effect of detector perturbation, i.e. flux depression, on the noise spectrum. It is noted that in

H_2O moderated systems significant effects on the p.s.d. could result and for highly absorbing detectors an underestimate of β/l would result when compared with the usual theory which neglects the perturbation.

8.15. *Vibrating absorber in an infinite sub-critical medium*

In the previous section we considered heterogeneous noise sources arising only from nuclear origin. Thus the mechanical feedback from temperature and vibrations were neglected. By the same token, in this section we shall neglect nuclear noise sources and study the effect of an independently driven, vibrating control element on the time dependence of the neutron flux and its statistical averages (Williams, 1970).

For simplicity, then, we consider an infinite sub-critical medium in which there exists a uniformly distributed source S_0. Embedded in the medium is an absorbing plate which executes random vibrations in the direction normal to its surface. We assume one-speed diffusion theory and write the neutron-balance equation (neglecting delayed neutrons) as

$$\frac{\partial N(x,t)}{\partial t} = D_0 \frac{\partial^2 N(x,t)}{\partial x^2} - \alpha N(x,t) - \gamma N(x,t)\,\delta(x-\epsilon(t)) + S_0, \tag{8.206}$$

where D_0 is the diffusion coefficient, $1/\alpha$ is the mean neutron lifetime ($\alpha > 0$), γ is Galanin's constant for the plate which we take to be vibrating about the plane $x = 0$ with a random, time-dependent amplitude $\epsilon(t)$.

We will further assume that $\epsilon(t)$ is a Gaussian variable with zero mean and that it is a stationary random process. Statistically, then, $\epsilon(t)$ is completely specified by its correlation function

$$R(t-t') = \langle \epsilon(t)\epsilon(t')\rangle \equiv \sigma^2 \rho(t-t'), \tag{8.207}$$

where $\rho(0) = 1$ and σ^2 is the mean square amplitude of vibration.

On the assumption that the plate has been in vibration for a long time such that any initial transients have decayed, the differential equation may be converted to integral form by the usual Green's function method (Morse and Feshbach, 1954). We obtain then

$$N(x,t) = \frac{S_0}{\alpha} - \frac{\gamma}{\sqrt{\pi}} \int_{-\infty}^{t} dt'\, N(\epsilon(t'),t').[4D_0(t-t')]^{-\frac{1}{2}} \exp\left\{-\frac{(x-\epsilon(t'))^2}{4D_0(t-t')} - \alpha(t-t')\right\}. \tag{8.208}$$

Setting $x = \epsilon(t)$ leads to a stochastic integral equation for $N(\epsilon(t),t)$, viz.

$$N(\epsilon(t),t) = \frac{S_0}{\alpha} - \frac{\gamma}{\sqrt{\pi}} \int_{-\infty}^{t} dt'\, N(\epsilon(t'),t').[4D_0(t-t')]^{-\frac{1}{2}} \exp\left\{-\frac{(\epsilon(t)-\epsilon(t'))^2}{4D_0(t-t')} - \alpha(t-t')\right\}. \tag{8.209}$$

Clearly to obtain the "exact" solution of this equation is a formidable task despite the fact that the statistics of ϵ are fully given. We apply an approximate method which we justify qualitatively but which should be subject to further tests.

Let us first seek the mean value of $N(\epsilon(t),t)$, i.e. $\langle N(\epsilon(t),t)\rangle$, which contrary to expectations is not the same as solving eqn. (8.209) with $\epsilon = \langle\epsilon\rangle = 0$. We proceed by statistically averaging eqn. (8.209) as follows:

$$\langle N(\epsilon(t),t)\rangle = \frac{S_0}{\alpha}-\frac{\gamma}{\sqrt{\pi}}\int_{-\infty}^{t} dt'\,\frac{\exp[-\alpha(t-t')]}{[4D_0(t-t')]^{\frac{1}{2}}}\left\langle N(\epsilon(t'),t')\exp\left\{-\frac{(\epsilon(t)-\epsilon(t'))^2}{4D_0(t-t')}\right\}\right\rangle. \tag{8.210}$$

To proceed further, we either have to derive a further equation for the expectation under the integral sign or make a closure assumption. We choose the latter technique for simplicity and on the grounds that the correlation between $N(\epsilon(t'),t')$ and the exponential term will not be strong. Thus we write

$$\langle N(\epsilon(t'),t')\exp\{\ldots\}\rangle \simeq \langle N(\epsilon(t'),t')\rangle\langle\exp\{\ldots\}\rangle. \tag{8.211}$$

Since the average $\langle N(\epsilon(t'),t')\rangle$ will be time independent, we can set it equal to $\tilde{N}(0)$ and from eqn. (8.210) obtain its value as

$$\tilde{N}(0) = \frac{S_0}{\alpha}\left(1+\frac{\gamma}{\sqrt{\pi}}\int_{-\infty}^{t} dt'\,\frac{\exp[-\alpha(t-t')]}{[4D_0(t-t')]^{\frac{1}{2}}}\left\langle\exp\left\{\frac{-(\epsilon(t)-\epsilon(t'))^2}{4D_0(t-t')}\right\}\right\rangle\right)^{-1}. \tag{8.212}$$

To evaluate the statistical average we note that it is a function of two correlated random variables $\epsilon(t)$ and $\epsilon(t')$. Thus with our Gaussian assumption we have

$$\langle\exp\{\ldots\}\rangle = \int_{-\infty}^{\infty} d\epsilon\int_{-\infty}^{\infty} d\epsilon'\, W_2(\epsilon,\epsilon')\exp\left\{-\frac{(\epsilon-\epsilon')^2}{4D_0(t-t')}\right\}, \tag{8.213}$$

where we have abbreviated $\epsilon(t)$ by ϵ and $\epsilon(t')$ by ϵ'. $W_2(\epsilon,\epsilon')$ is a two-dimensional Gaussian distribution (Middleton, 1960) and is given by

$$W_2(\epsilon,\epsilon') = \{2\pi\sigma^2[1-\rho^2(t-t')]^{\frac{1}{2}}\}^{-1}\exp\left\{-\frac{\epsilon^2+\epsilon'^2-2\epsilon\epsilon'\rho(t-t')}{2\sigma^2[1-\rho^2(t-t')]^{\frac{1}{2}}}\right\} \tag{8.214}$$

which after some lengthy algebra leads to

$$\langle\exp\{\ldots\}\rangle = \left\{1+\frac{\sigma^2[1-\rho(t-t')]}{D_0(t-t')}\right\}^{-\frac{1}{2}} \tag{8.215}$$

and hence $\tilde{N}(0)$ may be written

$$\tilde{N}(0) = \frac{S_0}{\alpha}\left[1+\frac{\gamma L}{\sqrt{\pi}\,.\,D_0}\int_0^{\infty}\frac{ds\exp(-s)}{\left\{s+\frac{\sigma^2}{L^2}[1-\rho(s/\alpha)]\right\}^{\frac{1}{2}}}\right]^{-1}. \tag{8.216}$$

Using the same statistical averaging on $N(x,t)$ and the same closure approximation, we find

$$\langle N(x,t)\rangle \equiv \tilde{N}(x) = \frac{S_0}{\alpha}-\frac{\gamma}{\sqrt{\pi}}\int_{-\infty}^{t} dt'\,\frac{\exp[-\alpha(t-t')]}{[4D_0(t-t')]^{\frac{1}{2}}}$$

$$\times\langle N(\epsilon(t'),t')\rangle\left\langle\exp\left\{-\frac{(x-\epsilon(t'))^2}{4D_0(t-t')}\right\}\right\rangle. \tag{8.217}$$

The exponential average now only involves the one-dimensional Gaussian, viz.

$$\langle\exp\{\ldots\}\rangle = \int_{-\infty}^{\infty} d\epsilon'\, W_1(\epsilon')\exp\left\{-\frac{(x-\epsilon')^2}{4D_0(t-t')}\right\}, \tag{8.218}$$

where

$$W_1(\epsilon) = \frac{1}{\sqrt{(2\pi\sigma)}}\exp\left\{-\frac{\epsilon^2}{2\sigma^2}\right\}. \tag{8.219}$$

Evaluating this integral, using $\langle N(\epsilon',t')\rangle = \tilde{N}(0)$ and inserting into eqn. (8.217), we find

$$\tilde{N}(x) = \frac{S_0}{\alpha}\left[1-\frac{\gamma L}{\sqrt{\pi}.D_0}\frac{\displaystyle\int_0^\infty \frac{dq\exp(-q)}{\left(q+\frac{\sigma^2}{2L^2}\right)^{\frac{1}{2}}}\exp\left(-\frac{x^2}{4qL^2+2\sigma^2}\right)}{\left\{1+\frac{\gamma L}{\sqrt{\pi}.D_0}\displaystyle\int_0^\infty \frac{dq\exp(-q)}{\left\{q+\frac{\sigma^2}{L^2}[1-\rho(q/\alpha)]\right\}^{\frac{1}{2}}}\right\}}\right]. \tag{8.220}$$

In the case of no vibration, $\sigma^2 = 0$, and the expression for $\tilde{N}(x)$ reduces to the deterministic value of

$$N(x) = \frac{S_0}{\alpha}\left[1-\frac{\gamma L}{\gamma L+2D_0}\exp\left(-\frac{x}{L}\right)\right]. \tag{8.221}$$

A number of general comments may be made about the results obtained. Firstly, we note from eqn. (8.220) that the flux drops off more rapidly than in the $\sigma^2 = 0$ case. A plausible reason for this can be obtained from eqn. (8.216) for $\tilde{N}(0)$ which, by comparison with eqn. (8.221) for $N(0)$, shows that $\tilde{N}(0) > N(0)$. Thus the reaction rate in the moving plate is greater than that for a stationary one and causes a greater flux depression and therefore an increased attenuation rate of the neutron flux from the source. The actual magnitude of the effect will depend on the correlation function of the vibrations $\epsilon(t)$ and this will be considered further in the next chapter. It should be pointed out at this stage that Barrett (1971) using a mixed diffusion theory-collision probability technique has concluded that vibration increases the thermal flux-disadvantage factor through an increase in the fuel to fuel collision probability. This is in contrast to the results of an extension of the theory described above to an infinite plate lattice of independently vibrating fuel elements (Williams, 1970). In that theory, we show that if a system is just critical with stationary plates, then when the plates are set into random vibration the reactor develops a positive period. Quite clearly the discrepancies between these two approaches deserves further study since it has important implications for reactor safety.

An important point that emerges from the above work, whether it be that of Barrett or of Williams, is the effect of parametric excitation and the fact that it changes the mean value. We have, of course, seen examples of this in previous chapters.

We ask now what is the correlation function resulting from the vibrations. This theory has not been worked out in any detail, but following the simple method outlined above for the control plate in a sub-critical infinite medium, we can immediately from eqn. (8.208) write down an expression for $\langle N(x_1,t_1)N(x_2,t_2)\rangle$. Thus with $N(x,t)$ expressed in the obvious notation

$$N(x,t) = \frac{S_0}{\alpha}-\int_{-\infty}^{t} dt' N(\epsilon(t'),t')G(x,\epsilon(t');t-t') \tag{8.222}$$

we have

$$\left\langle\left(N(x_1,t_1)-\frac{S_0}{\alpha}\right)\left(N(x_2,t_2)-\frac{S_0}{\alpha}\right)\right\rangle = \int_{-\infty}^{t_1} dt'\int_{-\infty}^{t_2} dt''\langle N(\epsilon(t'),t')N(\epsilon(t''),t'')$$

$$\times G(x_1,\epsilon(t');t_1-t')G(x_2,\epsilon(t'');t_2-t'')\rangle. \tag{8.223}$$

With the plausible closure approximation

$$\langle NN'G_1G_2\rangle \simeq \langle NN'\rangle\langle G_1G_2\rangle. \tag{8.224}$$

We have an integral equation for $\langle NN'\rangle$. However, we shall not pursue this topic further here. All of the work described in this section can be repeated for rods by the introduction of the appropriate mathematical functions.

8.16. *The variable strength absorber*

An alternative technique for the mathematical representation of a vibrating absorber has been suggested by Schwalm (1972). In the case of a homogeneous reactor, the effect of a vibrating rod, instead of taking the form

$$\gamma_k \delta(\mathbf{r}-\mathbf{r}_k-\epsilon_k(t))N(\mathbf{r}_k+\epsilon_k(t),t), \tag{8.225}$$

is denoted by

$$\tilde{\gamma}_k(t)\delta(\mathbf{r}-\mathbf{r}_k)N(\mathbf{r}_k,t), \tag{8.226}$$

where $\tilde{\gamma}_k(t)$ is a random function. That is, Schwalm assumes that the effect of random vibration about the point $\mathbf{r}_k$ can be simulated by a fixed absorber whose strength is a random function of time but which does not execute any motion. This assumption is open to some criticism since, whilst there is undoubtedly some connection between the two cases, it is clear that the statistical properties of $\epsilon_k(t)$ cannot be related linearly to those of $\tilde{\gamma}_k(t)$, which in fact is what Schwalm has done. As a matter of physical interest the formalism proposed by Schwalm would be more appropriate for the simulation of random fuel temperature changes or possibly, if the absorber were a pipe containing circulating fissile material in the form of a molten salt, it could represent fluctuations in the concentration.

To demonstrate the differences that can be expected to arise between such cases we might look at the linearized equations involving the terms (8.225) and (8.226). The linearization might arise, for example, in the Langevin method for small perturbations. Then we would have

$$N(\mathbf{r},t) = N_0(\mathbf{r})+\delta N(\mathbf{r},t) \tag{8.227}$$

and if $\tilde{\gamma}_k(t) = \langle\tilde{\gamma}\rangle_k(1+\Delta_k(t))$, where $\Delta_k(t)$ is considered a small effect and $\langle\gamma_k\rangle$ is the mean value, the expression (8.226) will become

$$\langle\tilde{\gamma}_k\rangle\delta(\mathbf{r}-\mathbf{r}_k)(N_0(\mathbf{r})+\delta N(\mathbf{r},t)+\Delta_k(t)N_0(\mathbf{r})), \tag{8.228}$$

where $\langle\tilde{\gamma}_k\rangle\delta(\mathbf{r}-\mathbf{r}_k)\Delta_k N_0$ will enter as the random source function in the equation for δN.

On the other hand, to linearize the expression (8.225) requires us to set not only $N = N_0+\delta N$, but also

$$\delta(\mathbf{r}-\mathbf{r}_k-\epsilon_k(t)) \simeq \delta(\mathbf{r}-\mathbf{r}_k)-\epsilon_k(t).\nabla\delta(\mathbf{r}-\mathbf{r}_k). \tag{8.229}$$

The linearized product can then be written as

$$\gamma_k\delta(\mathbf{r}-\mathbf{r}_k)N_0(r)+\gamma_k\delta(\mathbf{r}-\mathbf{r}_k)\delta N(r,t)-\gamma_k\epsilon_k.\nabla\delta(\mathbf{r}-\mathbf{r}_k)N_0(\mathbf{r}), \tag{8.230}$$

where now $-\gamma_k\epsilon_k.\nabla\delta(\mathbf{r}-\mathbf{r}_k)N_0(\mathbf{r})$ is the random source term. Not only is this entirely different in character from the previous expression, but the validity of the expansion of the delta function requires further justification.

To illustrate the point further, we can treat the plate problem discussed in Section 8.15 by Schwalm's proposal. Then the equation becomes

$$\frac{\partial N(x,t)}{\partial t} = D_0 \frac{\partial^2 N(x,t)}{\partial x^2} - \alpha N(x,t) - \gamma(t)\delta(x)N(x,t) + S_0. \tag{8.231}$$

Conversion to integral form follows as before and we find

$$N(x,t) = \frac{S_0}{\alpha} - \frac{1}{\sqrt{\pi}} \int_0^\infty dt_0 \gamma(t-t_0) N(0,t-t_0) \frac{\exp\left\{-\frac{x^2}{4D_0 t_0} - \alpha t_0\right\}}{(4D_0 t_0)^{\frac{1}{2}}}. \tag{8.232}$$

Now setting $x = 0$ and averaging we get

$$\langle N(0,t)\rangle = \frac{S_0}{\alpha} - \frac{1}{\sqrt{\pi}} \int_0^\infty dt_0 \langle \gamma(t-t_0) N(0,t-t_0)\rangle \frac{e^{-\alpha t_0}}{(4D_0 t_0)^{\frac{1}{2}}}. \tag{8.233}$$

Setting $\langle \gamma N\rangle \simeq \langle \gamma\rangle\langle N\rangle$ is not profitable since we would arrive at the stationary plate solution. Instead we use the technique discussed in Section 8.2 and after setting $x = 0$ in (8.232) we multiply by $\gamma(t)$ and again average. This leads to

$$\langle \gamma(t) N(0,t)\rangle = \frac{S_0}{\alpha}\gamma_0 - \frac{1}{\sqrt{\pi}} \int_0^\infty dt_0 \langle \gamma(t)\gamma(t-t_0) N(0,t-t_0)\rangle \frac{e^{-\alpha t_0}}{(4D_0 t_0)^{\frac{1}{2}}}, \tag{8.234}$$

where $\gamma_0 = \langle \gamma\rangle$.

Now we use the closure approximation $\langle \gamma\gamma N\rangle \simeq \langle \gamma\gamma\rangle\langle N\rangle$, where

$$\langle \gamma(t)\gamma(t-t_0)\rangle = R_\gamma(t_0) \tag{8.235}$$

is the auto-correlation function of $\gamma(t)$.

Inserting this expression for $\langle \gamma N\rangle$ into eqn. (8.233) for $\langle N(0,t)\rangle$, we finally get

$$\langle N(0,t)\rangle \equiv \langle N(0)\rangle = \frac{\frac{S_0}{\alpha}\left\{1 - \frac{\gamma_0}{2(D_0\alpha)^{\frac{1}{2}}}\right\}}{1 - \frac{1}{4D_0(\pi\alpha)^{\frac{1}{2}}} \int_0^\infty dt_0 R_\gamma(t_0) \frac{e^{-\alpha t_0}}{\sqrt{t_0}}}. \tag{8.236}$$

Remembering that

$$\begin{aligned} R_\gamma(t_2 - t_1) &= \gamma_0^2 \langle (1+\Delta(t_1))(1+\Delta(t_2))\rangle \\ &= \gamma_0^2\{1 + R_\Delta(t_2 - t_1)\} \end{aligned} \tag{8.237}$$

we find eventually

$$\langle N(0)\rangle = \frac{\frac{S_0}{\alpha}\left\{1 - \frac{\gamma_0}{2(D_0\alpha)^{\frac{1}{2}}}\right\}}{1 - \frac{\gamma_0^2}{4\alpha D_0} - \frac{\gamma_0^2}{4D_0(\pi\alpha)^{\frac{1}{2}}} \int_0^\infty \frac{dt_0}{\sqrt{t_0}} R_\Delta(t_0) e^{-\alpha t_0}}. \tag{8.238}$$

$\langle N(x,t)\rangle$ can be obtained by averaging eqn. (8.232) and inserting the expression for $\langle \gamma N\rangle$.

Comparison of eqn. (8.238) with the corresponding one for the delta function case, i.e. eqn. (8.216), indicates that differences can be expected to arise. However, we note that for the case of stationary plates where both R_Δ and σ^2 are zero, both methods reduce to the correct value. It might be argued, therefore, that for small perturbations the methods might not be too different in numerical value. Nevertheless, this remains to be verified and, moreover, the covariances $\langle N_1 N_2\rangle$ will undoubtedly be more sensitive to the method used. Finally, it should be noted that Schwalm's method is a great deal more easily dealt with mathematically so that an investigation into the precise numerical discrepancies would be of some value from a practical point of view.

8.17. *Critical mass for fuel elements in a random array*

Lloyd (1959) has reported a series of buckling measurements on random arrays of fuel elements to see if such configurations would have bucklings significantly different from those resulting from a uniform array. H_2O/uranium lattices were used and whilst the experimental errors were large it was concluded that the critical mass of a given random rod assembly would be larger than that of a similar but uniform array.

Whilst insufficient data were given by Lloyd to make a numerical comparison possible a method for treating such a problem results from the general theory of Section 8.7. Thus if we consider a system of fuel rods with random but fixed positions in the $x-y$ plane, $\boldsymbol{\xi}_{\mathbf{k}}$, and characterized by an axial buckling B_z^2, we can write the critical equation as follows:

$$D_0 \nabla_{\xi}^2 N(\boldsymbol{\xi}) - (D_0 B_z^2 + v\overline{\Sigma_{am}}) N(\boldsymbol{\xi}) - \gamma \sum_{\mathbf{k}} \delta(\boldsymbol{\xi}-\boldsymbol{\xi}_{\mathbf{k}}) N(\boldsymbol{\xi}_{\mathbf{k}}) + \eta\gamma \sum_{\mathbf{k}} K(|\boldsymbol{\xi}_{\mathbf{k}}-\boldsymbol{\xi}|;\, B_z^2) N(\boldsymbol{\xi}_{\mathbf{k}}) = 0. \quad (8.239)$$

If we use a slowing-down kernel of the form

$$K(|\mathbf{r}-\mathbf{r}'|) = \frac{e^{-|\mathbf{r}-\mathbf{r}'|/L_s}}{8\pi L_s^3} \quad (8.240)$$

then we have defined $K(|\boldsymbol{\xi}-\boldsymbol{\xi}'|;\, B_z^2)$ as

$$K(|\boldsymbol{\xi}-\boldsymbol{\xi}'|;\, B_z^2) = \int_{-\infty}^{\infty} e^{iB_z z} K(\sqrt{[(\boldsymbol{\xi}-\boldsymbol{\xi}')^2+z^2]})\, dz. \quad (8.241)$$

Assuming that the displacement of the fuel rods from a uniform position $\langle\boldsymbol{\xi}_{\mathbf{k}}\rangle$ is such that

$$\boldsymbol{\xi}_{\mathbf{k}} = \langle\boldsymbol{\xi}_{\mathbf{k}}\rangle + \Delta_{\mathbf{k}}, \quad (8.242)$$

where the $\Delta_{\mathbf{k}}$ are uncorrelated Gaussian distributed random variables with zero mean and variance σ^2, we get, after using the standard source-sink procedure and the simplest non-trivial closure approximation, the following critical equation:

$$1 = \frac{\gamma}{2\pi D_0} \sum_{\mathbf{k}} \int_0^{\infty} \frac{ds s[\eta \overline{K}(s^2+B_z^2)-1]}{s^2+B_z^2+\kappa^2}\, e^{-\sigma^2 s^2} J_0(s|\boldsymbol{\rho}-\langle\boldsymbol{\xi}_{\mathbf{k}}\rangle|), \quad (8.243)$$

where κ is the reciprocal of the diffusion length in the moderator, $\overline{K}(B^2)$ is the Fourier transform of the slowing-down kernel and $|\boldsymbol{\rho}|$ is the radius of a fuel rod.

Whilst we have not put numbers into eqn. (8.243) it is clear that, for given material properties, it will be necessary to increase η as σ^2 increases. Thus we have proved, at least qualitatively and in accord with experiment, that a random array requires a larger critical mass than does the corresponding ordered lattice.

References

Adomian, G. (1963) Linear stochastic operators, *Rev. Mod. Phys.* **35**, 185.
Adomian, G. (1964) Stochastic Green's functions, *Proc. Symp. Appl. Math.* xvi, 1, Am. Math. Soc., Prov., R.I.
Adomian, G. (1970) Random operator equations in mathematical physics I, *J. Math. Phys.* **11**, 1069.
Adomian, G. (1971 *a*) Linear random operator equations in mathematical physics II, *J. Math. Phys.* **12**, 1944.
Adomian, G. (1971 *b*) Linear random operator equations in mathematical physics III, *J. Math. Phys.* **12**, 1948.
Akcasu, Z. (1960) Theoretical feedback analysis in boiling-water reactors, USAEC, ANL-6221.
Akcasu, Z. (1961) Mean square instability in boiling reactors, *Nucl. Sci. Engng*, **10**, 337.
Allio, R. J. and Randall, C. H. (1962) The quantitative analysis of microstructure with densitometer data, *Trans. Metallurgical Soc. of AIME*, **24**, 221.
Barrett, P. R. (1971) The effect of random amplitudes and phases of vibrating fuel elements upon multigroup neutron lattice parameters, *Atomkernenergie*, **16**, 285.
Barrett, P. R. and Thompson, J. J. (1969) The effect of random media upon neutron transport, *Nukleonik*, **12**, 159.
Beckjord, E. S. (1958) Dynamic analysis of natural circulation boiling-water power reactors, USAEC, ANL-5799.
Berger, F. P. and Derian, L. (1965) The influence of an electric field on the heat transfer to gaseous coolants in a nuclear reactor, *Acta Technica Csav*, No. 3, 312.
Bharucha-Reid, A. T. (1968) *Probabilistic Methods in Applied Mathematics*, Academic Press, N.Y.
Bourret, R. C. (1962) Stochastically perturbed fields with applications to wave propagation in random media, *Il Nuovo Cimento*, xxvi, 1.
Bourret, R. C. (1965) Ficton theory of dynamical systems with noisy parameters, *Can. J. Phys.* **43**, 619.
Burgreen, D. *et al.* (1958) Vibration of rods induced by water in parallel flow, *Trans. ASME*, **80**, 991.
Burrus, W. R. (1960) Radiation transmission through boral and similar heterogeneous materials consisting of randomly distributed absorbing chunks, ORNL 2528.
Case, K., de Hoffmann, F. and Placzek, G. (1953) *Introduction to the Theory of Neutron Diffusion*, USAEC.
De Shong, J. A. (1959) Flux, reactivity, steam void, and steam-water interlayer noise spectrums in the EBWR, *J. Nucl. Energy*, **10**, 147.
El-Wakil, M. M. (1971) *Nuclear Heat Transport*, International Textbook Co.
Feller, W. (1957) *An Introduction to Probability Theory and its Applications*, i, 2nd ed., Wiley, N.Y.
Fleck, J. A. (1960) The dynamic behaviour of boiling-water reactors, *J. Nucl. Energy*, **11**, 114.
Fleck, J. A. and Huseby, J. (1958) A mathematical model for calculating boiling reactor transients, *Proc. Int. Conf. Geneva*, **11**, 457.
Galanin, A. D. (1960) *Thermal Reactor Theory*, Pergamon Press.
Gorman, D. J. (1969) The role of turbulence in the vibration of reactor fuel elements in liquid flow, AECL 3371.
Greef, C. (1971) A Study of Fluctuations in a Nuclear Reactor at Power, Ph.D. Thesis (Council for Nat. Acad. Awards).
Greenfield, M. L. *et al.* (1954) Studies on density transients in volume heated boiling systems, USAEC Rep. AECU-2950.
Halperin, M. (1959) Some asymptotic results for a coverage problem, KAPL-M-MH-2, USAEC.
Hinze, J. O. (1959) *Turbulence, an Introduction to its Mechanism and Theory*, McGraw-Hill.
Holt, R. and Rasmussen, J. (1972) *Ramona II, a Fortran Code for Transient Analysis of Boiling-water Reactors*, Kjeller rep. (Norway), KR-147.
Horning, W. A. (1954) *A Summary of Small Source Theory Applied to Thermal Reactors*, HW-34021, General Electric Co. Rep., Washington.
Hulburt, H. M. and Katz, S. (1964) Some problems in particle technology, *Chem. Engng Sci.* **19**, 555.
Kosály, G. (1973) Remarks on a few problems in the theory of power reactor noise, *J. Inst. Nucl. Engrs*, **14**, 67.
Kylstra, C. D. and Uhrig, R. E. (1965) Spatially dependent transfer functions for nuclear systems, *Nucl. Sci. Engng*, **22**, 191.
Lloyd, R. C. (1959) Buckling measurements for fuel elements in a random array versus a uniform array, USAEC, HW-57919.

MIDDLETON, D. (1960) *An Introduction to Statistical Communication Theory*, McGraw-Hill, N.Y.
MORSE, P. M. and FESHBACH, H. (1954) *Methods of Mathematical Physics*, McGraw-Hill, N.Y.
NATELSON, M. *et al.* (1966) Space and energy effects in reactor fluctuation experiments, *J. Nucl. Energy*, **20**, 557.
NOMURA, T. (1968) Noise analysis of boiling water reactor, *Japan–United States Seminar on Nuclear Reactor Noise Analysis*, p. 197.
PAIDOUSSIS, M. P. (1965) The amplitude of fluid induced vibration of cylinders in axial flow, AECL-2225.
PAIDOUSSIS, M. P. (1966*a*) Dynamics of flexible slender cylinders in axial flow. Part I, Theory, *J. Fluid Mech.* **26**, 717.
PAIDOUSSIS, M. P. (1966*b*) Dynamics of flexible slender cylinders in axial flow. Part II, Experiment, *J. Fluid Mech.* **26**, 737.
PAIDOUSSIS, M. P. (1969) An experimental study of vibration of flexible cylinders induced by nominally axial flow, *Nucl. Sci. Engng*, **35**, 127.
PEDERSON, N. K. (1966) *Dynamic Aspects of Boiling-heavy-water Nuclear Reactors, Parts I and II*, Danish Atomic Energy Com. Reps. RISO 128, 129.
PLESSET, M. S. (1952) Note on the flow of vapour between liquid surfaces, *J. Chem. Phys.* **20**, 790.
PLESSET, M. S. and ZWICK, S. A. (1952) A non-steady heat diffusion problem with spherical symmetry, *J. Appl. Phys.* **23**, 95.
PLESSET, M. S. and ZWICK, S. A. (1954) The growth of vapour bubbles in superheated liquids, *J. Appl. Phys.* **25**, 493.
PLESSET, M. S. and ZWICK, S. A. (1955) On the dynamics of small vapour bubbles in liquids, *J. Math. Phys.* **33**, 308.
QUINN, E. P. (1962) *Vibration of Fuel Rods in Parallel Flow*, USAEC rep. GEAP-4059, General Electric Co.
RANDALL, C. H. (1960) Microscopic effects in multiphase mediums (neutron self-shielding), KAPL-M-CHR-2, USAEC.
RAYLEIGH, LORD (1917) On the pressure developed in a liquid during the collapse of a spherical cavity, *Phil. Mag.* **34**, 94.
ROBINSON, J. C. (1970) Analytical determination of the neutron flux to pressure frequency response: application to the Molten Salt Reactor Experiment, *Nucl. Sci. Engng*, **42**, 382.
ROBINSON, J. C. and FRY, D. N. (1970) Experimental neutron flux to pressure frequency response for the Molten Salt Reactor Experiment: determination of void fraction in fuel salt, *Nucl. Sci. Engng*, **42**, 397.
ROHSENOW, W. M. A. (1952) A method of correlating heat transfer data for surface boiling liquids, *Trans. ASME*, **74**, 969.
SAMUELS, J. C. and ERINGEN, A. C. (1958) Response of a simply supported Timoshenko beam to a purely random Gaussian process, *Trans. ASME* (*J. Appl. Mechanics*), December, 496.
SANDERVAG, ODDBJORN (1971) *Thermal Non-equilibrium and Bubble Size Distributions in an Upward Steam Water Flow*, Kjeller report (Norway) KR-144.
SCHLÖSSER, J. (1962) On the stability of flat plates under the influence of coolant flow, *J. Nucl. Energy*, **16**, 351.
SCHWALM, VON D. (1972) Some remarks on failure detection in nuclear power reactors by noise measurements, *Atomkernenergie*, **19**, 263.
SIEGMANN, E. R. (1971) Theoretical study of transient sodium boiling in reactor coolant channels utilizing a compressible flow model, USAEC, ANL-7842.
STEIN, R. P. (1957) Transient heat transfer in reactor coolant channels, USAEC. AECU-3600.
STEPANOV, A. B. (1965) *Symposium on Pulsed Neutron Research*, Vol. 1, p. 339, IAEA, Vienna.
STEPANOV, A. B. (1968) *Symposium on Neutron Thermalization and Reactor Spectra*, Vol. 1, p. 193, IAEA, Vienna.
TAYLOR, G. I. (1952) Analysis of the swimming of long and narrow animals, *Proc. Roy. Soc.* A **214**, 158.
THIE, J. (1958) Dynamic behaviour of boiling reactors, ANL-5894.
WAMBSGANSS, M. W. (1967) Vibration of reactor core components, *Reactor and Fuel Processing Technology*, **10**, 208.
WEINBERG, A. M. and WIGNER, E. P. (1958) *The Physical Theory of Neutron Chain Reactors*, Chicago University Press.
WILKINS, J. ERNEST (1945) The effect of bubbling on the multiplication constant I, II, USAEC, CP-2709, 3067.
WILLIAMS, M. M. R. (1966) *The Slowing Down and Thermalization of Neutrons*, North-Holland.
WILLIAMS, M. M. R. (1967*a*) An application of slowing down kernels to thermal neutron density fluctuations in nuclear reactors, *J. Nucl. Energy*, **21**, 321.

WILLIAMS, M. M. R. (1967*b*) Reactor noise in heterogeneous systems: 1. Plate-type elements, *Nucl. Sci. Engng*, **30**, 188.

WILLIAMS, M. M. R. (1970) Reactivity changes due to the random vibration of control rods and fuel elements, *Nucl. Sci. Engng*, **40**, 144.

WILLIAMS, M. M. R. (1971) The kinetic behaviour of simple neutronic systems with randomly fluctuating parameters, *J. Nucl. Energy*, **25**, 563.

WITTKE, D. D. and CHAO, B. T. (1967) Collapse of vapour bubbles with translatory motion, *J. Heat Transfer*, Feb. 17.

YAMAGISHI, T. and SEKIYA, T. (1968) Calculations of neutron distributions by statistical method, *J. Nucl. Sci. Technol.* **5**, 614.

ZUBER, N. (1967) Flow excursions and oscillations in boiling two phase flow systems with heat addition, *Symposium on Two-phase Flow Dynamics*, Eindhoven.

CHAPTER 9

Associated Fluctuation Problems

9.1. *Introduction*

A number of problems, not directly of a nuclear nature, but which affect reactor behaviour in a random fashion, can arise. In this final chapter we shall consider a few of these aspects; in particular, (1) vibration of rods, beams and plates; (2) fatigue due to random stresses; (3) variable heat conduction; (4) the mechanism of void formation. Each of these topics has been mentioned in earlier chapters and its effect on reactor response noted. Here we shall examine the origins of these noise sources in more detail.

9.2. *Random vibration of mechanical components*

We have mentioned the fact that the energy arising from coolant passing through the core can excite and sustain vibratory motion of control rods and fuel elements and of any other structural component. Let us consider now in more detail the equation of motion of the vibrating system.

In Section 8.9 we noted that the transverse displacement from equilibrium $\epsilon(z,t)$ of a point on a rod distance z from the lower end, at time t, could be represented formally by the equation

$$\hat{L}(z,t)\epsilon(z,t) = F(z,t), \tag{9.1}$$

where $\hat{L}$ is an operator defined by the model use for the rod motion and F is the random transverse loading per unit length generated by the turbulent flow of the coolant or some other specified effect.

To proceed further it is necessary to decide upon the form of $\hat{L}$, the boundary conditions and the nature of the random term F.

The simplest beam model of any engineering value is the damped Bernoulli–Euler model in which the load per unit length is equated to the rate of change of shear. The shear in turn is related to the bending moment and hence to the displacement. The equation for the displacement therefore becomes (Burgreen *et al.*, 1958; Thomson, 1965; Bogdanoff and Goldberg, 1960):

$$EI\frac{\partial^4\epsilon}{\partial z^4}+\mu\frac{\partial\epsilon}{\partial t}+\rho A\frac{\partial^2\epsilon}{\partial t^2} = F(z,t), \tag{9.2}$$

where E is Young's modulus, I is the cross-sectional moment of inertia, μ is the viscosity of water (or of the fluid in which the rod vibrates), ρ is the density of the rod and A its cross-sectional area.

A more complete description of the motion of a rod or beam is given by Timoshenko and Goodier (1951) who take into account the rotatory inertia as well as the shear deformation. This leads to a pair of coupled equations for displacement and rotation. External viscous damping is accounted for in both the transverse and rotatory senses. In addition, Crandall and Yildiz (1962) have included an internal resistance based upon the Voight type of visco-elasticity in the force–deformation relations.

The resulting equations may be written as follows:

$$\frac{F(z,t)}{\rho A}+a_2^2\left(1+\epsilon_2\frac{\partial}{\partial t}\right)\left(\frac{\partial^2\epsilon}{\partial z^2}-\frac{\partial\psi}{\partial z}\right)=\frac{\partial^2\epsilon}{\partial t^2}+\beta_1\frac{\partial\epsilon}{\partial t} \tag{9.3}$$

and

$$\frac{a_2^2}{r^2}\left(1+\epsilon^2\frac{\partial}{\partial t}\right)\left(\frac{\partial\epsilon}{\partial z}-\psi\right)+a_1^2\left(1+\epsilon_1\frac{\partial}{\partial t}\right)\frac{\partial^2\psi}{\partial z^2}=\frac{\partial^2\psi}{\partial t^2}+\beta_2\frac{\partial\psi}{\partial t}, \tag{9.4}$$

where ψ is the angular displacement at (z,t).

The parameters in the above equation are defined as follows: $\beta_1 = C_1/\rho A$, $\beta_2 = C_2/\rho I \equiv C_2/\rho Ar^2$, where C_1 and C_2 are external viscous damping coefficients for transverse and rotatory motion, respectively, and depend on the fluid in which the rod vibrates. ϵ_1 and ϵ_2 are the Voight visco-elastic damping coefficients for bending moment and shear, respectively, and are physical properties of the rod. Finally, $a_1^2 = E/\rho \equiv EI/\rho Ar^2$ and $a_1^2 = kG/\rho$ with G the shear modulus, r the radius of the rod and k is a shape factor which is about $\frac{2}{3}$ (Timoshenko, 1957, 1966). This is clearly a very comprehensive model and a number of attempts have been made to simplify it. One approximation is to neglect the rotatory inertia term $\partial^2\psi/\partial t^2$ in eqn. (9.4) and is known as the "shear beam" approximation. This does not appear to lead to any appreciable reduction in complexity and another approach has been adopted by Rayleigh which includes rotatory inertia but neglects shear deformation. Thus $\psi = \partial\epsilon/\partial z$ and the shear force is defined by

$$V+\frac{\partial M}{\partial z}=\rho I\left(\frac{\partial^2\psi}{\partial t^2}+\beta_2\frac{\partial\psi}{\partial t}\right), \tag{9.5}$$

M being the bending moment. V, M and ψ can then be eliminated from the basic Timoshenko equations and a single equation for ϵ obtained, viz.

$$r^2a_1^2\left(1+\epsilon_1\frac{\partial}{\partial t}\right)\frac{\partial^4\epsilon}{\partial z^4}+\left(\beta_1-\beta_2r^2\frac{\partial^2}{\partial z^2}\right)\frac{\partial\epsilon}{\partial t}+\left(1-r^2\frac{\partial^2}{\partial z^2}\right)\frac{\partial^2\epsilon}{\partial t^2}=\frac{F}{\rho A}. \tag{9.6}$$

This is the Rayleigh model. A modified Bernoulli–Euler equation can be obtained by dropping the rotatory inertia effect $r^2\partial^4\epsilon/\partial z^2\partial t^2$, but which retains the visco-elastic forces and also the rotatory external forces. Setting $\epsilon_1 = \beta_2 = 0$ leads to the simple Bernoulli–Euler equation given earlier. Further work on these problems may be found in Samuels and Eringen (1958).

The most recent work on rod vibration is due to Paidoussis (1965, 1966) which, however, neglects the internal resistance of the rod but accounts more accurately for the axial motion of the fluid past the rod. As far as the author is aware this particular equation has not been employed in any depth for noise studies although it has proved useful in the interpretation of experimental results for the maximum displacement of a fuel rod from its equilibrium position (Paidoussis, 1969, 1974).

9.3. *Statistical considerations*

Whilst the experimental correlations of Burgreen *et al.* (1958) and Paidoussis (1965) are valuable from the design point of view they are based upon correlations which have only an indirect connection with the underlying theory of rod motion. Let us consider therefore the general analysis of eqn. (9.1) for any arbitrary model $\hat{L}$ and forcing function F. We will assume for the moment arbitrary boundary conditions at the endpoints and later specialize to a particular model (Bourgaine, 1970).

Define the eigenfunctions $\psi_n(z)$ by the equation

$$\hat{L}(z,t)\psi_n(z) = \hat{T}_n(t)\psi_n(z), \tag{9.7}$$

where $\hat{T}_n(t)$ is a time-dependent operator and the $\psi_n(z)$ satisfy the boundary conditions. Now we expand $\epsilon(z,t)$ and $F(z,t)$ as follows:

$$\epsilon(z,t) = \sum_n a_n(t)\psi_n(z), \tag{9.8}$$

$$F(z,t) = \sum_n f_n(t)\psi_n(z), \tag{9.9}$$

insert into eqn. (9.1), multiply by $\psi_m(z)$ and integrate over the domain of z. Using the fact that the eigenfunctions $\psi_n(z)$ are orthonormal, i.e.

$$\int_l dz\,\psi_m(z)\psi_n(z) = \delta_{n,m} \tag{9.10}$$

we obtain

$$\hat{T}_n(t)a_n(t) = f_n(t). \tag{9.11}$$

Note that $f_n(t)$ is a random function of time which we shall take to be stationary and that if $F(z,t)$ is a random function of space as well as time, the $f_n(t)$ are themselves randomly distributed for a given value of t.

We may formally write the solution of (9.11) in terms of the Green's function (impulse response function) $h_n(t)$ as

$$a_n(t) = \int_{-\infty}^{t} dt'\,h_n(t-t')f_n(t'), \tag{9.12}$$

where we have set the lower limit equal to minus infinity on the assumption that the initial transients have died away.

Writing $a_n(t)$ as

$$a_n(t) = \int_0^{\infty} dt'\,h_n(t')f_n(t-t') \tag{9.13}$$

we can form the complete solution, viz.

$$\epsilon(z,t) = \sum_n \psi_n(z)\int_0^{\infty} dt'\,h_n(t')f_n(t-t')$$

$$\equiv \sum_n \psi_n(z)\int_0^{\infty} dt'\,h_n(t')\int_0^{L} dz'\,\psi_n(z')F(z',t-t'). \tag{9.14}$$

In general, the mean value of F, i.e. $\langle F\rangle$, is zero, so it follows that $\langle\epsilon\rangle = 0$. The covariance function $\langle \epsilon(z_1,t_1)\epsilon(z_2,t_2)\rangle$ is given by

$$\langle \epsilon(z_1,t_1)\epsilon(z_2,t_2)\rangle = \sum_{nm} \psi_n(z_1)\psi_m(z_2)\int_0^\infty dt' \int_0^\infty dt''\, h_n(t')h_m(t'')\langle f_n(t_1-t')f_m(t_2-t'')\rangle$$

$$\equiv \sum_{nm} \psi_n(z_1)\psi_m(z_2)\int_0^\infty dt' \int_0^\infty dt''\, h_n(t')h_m(t'')\int_0^L dz' \int_0^L dz''\, \psi_n(z')$$
$$\times\, \psi_m(z'')\langle F(z',t_1-t')\times F(z'',t_2-t'')\rangle. \quad (9.15)$$

But for a stationary process

$$\langle F(z_1,t_1)F(z_2,t_2)\rangle = \phi_{FF}(z_1,z_2;\, t_2-t_1). \quad (9.16)$$

With an obvious abbreviation for the covariance of ϵ we can now write

$$\phi_{\epsilon\epsilon}(z_1,z_2;\,\tau) = \sum_{nm} \psi_n(z_1)\psi_m(z_2)\int_0^\infty dt' \int_0^\infty dt''\, h_n(t')h_m(t'')\int_0^L dz' \int_0^L dz''\, \psi_n(z')$$
$$\times\, \psi_m(z'')\phi_{FF}(z',z'';\,\tau-t''+t'). \quad (9.17)$$

If we define the Fourier transform of $h(t)$, viz.

$$h(t) = \frac{1}{2\pi}\int_{-\infty}^{\infty} d\omega\, e^{i\omega t} H(\omega), \quad (9.18)$$

then we can write (9.17) as

$$\phi_{\epsilon\epsilon}(z_1,z_2;\,\tau) = \frac{1}{2\pi}\sum_{nm} \psi_n(z_1)\psi_m(z_2)\int_{-\infty}^{\infty} d\omega'\, H_n(\omega')H_m^*(\omega')e^{-i\omega'\tau}$$
$$\times \int_0^L dz' \int_0^L dz''\psi_n(z')\psi_m(z'')\Phi_{FF}(z',z'';\,\omega'), \quad (9.19)$$

where $\Phi_{FF}(z',z'';\,\omega)$ is the Fourier transform of $\langle F(z_1,t_1)F(z_2,t_2)\rangle$. We have therefore the cross-correlation function of the amplitudes of vibration in terms of the p.s.d. of the transverse force fluctuations. Clearly, the p.s.d. of $\epsilon(z,t)$ is

$$\Phi_{\epsilon\epsilon}(z_1,z_2;\,\omega) = \sum_{nm} \psi_n(z_1)\psi_m(z_2)H_n(\omega)H_m^*(\omega)\int_0^L dz' \int_0^L dz''\, \psi_n(z')$$
$$\times\, \psi_m(z'')\Phi_{FF}(z',z'';\,\omega). \quad (9.20)$$

To make further progress we must investigate Φ_{FF} more closely. Following Reavis (1969) and Gorman (1969*a*, *b*) we use the following simplifying assumptions: (1) effects on wall pressure due to surface boiling and heat-transfer fin mixing turbulence are neglected; (2) there is no correlation between the pressure fields on different rods; (3) rod motion is assumed to have no effect on the pressure field in its neighbourhood; (4) the pressure field is homogeneous over the rod length; (5) the pressure field for turbulent air in a duct is identical to that for water over a fuel rod; (6) the lateral shear stress of the fluctuating pressure is normal to the wall: lateral shear stress can therefore be ignored as a significant source of excitation.

With these restrictions in mind we relate the forcing function $\langle F(z_1,t_2)F(z_2,t_2)\rangle$ to the turbulent pressure correlation function $\langle p(z_1,t_1)p(z_2,t_2)\rangle$. The connection can be seen by studying Fig. 9.1 which shows a cross-sectional area of the rod of diameter d. It is

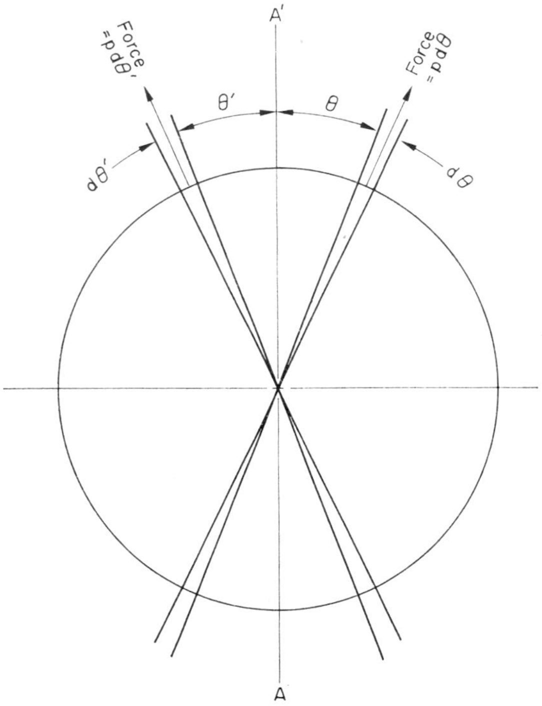

FIG. 9.1. Lateral differential force elements per unit length acting on a vibrating rod.

assumed that the pressure difference between a point on the surface of the rod and its value diametrically opposite can be measured; this is called $p(z,\theta,t)$. Now the upward force on the rod in the section $d\theta$ in the direction $A \to A'$ is $\frac{1}{2}pd\cos\theta\, d\theta$. The total force is therefore

$$F(z,t) = \frac{d}{2}\int_{-\pi/2}^{\pi/2} p(z,\theta,t)\cos\theta\, d\theta \tag{9.21}$$

or the correlation function is

$$\langle F(z_1,t_1)F(z_2,t_2)\rangle = \frac{d^2}{4}\int_{-\pi/2}^{\pi/2}\int_{-\pi/2}^{\pi/2}\langle p(z_1,\theta,t_1)p(z_2,\theta',t_2)\rangle \cos\theta\cos\theta'\, d\theta\, d\theta'. \tag{9.22}$$

Similarly we can write the p.s.d. as

$$\Phi_{FF}(z_1,z_2;\omega) = \frac{d^2}{4}\int_{-\pi/2}^{\pi/2}\int_{-\pi/2}^{\pi/2}\Phi_{pp}(z_1,z_2;\theta,\theta';\omega)\cos\theta\cos\theta'\, d\theta\, d\theta'. \tag{9.23}$$

But it has been shown by Bakewell (1962, 1964) from experimental measurements that

$$\Phi_{pp}(z_1,z_2;\theta,\theta';\omega) = \Phi(\omega)\cos\left\{\frac{\omega}{U_c}(z_1-z_2)\right\}\exp\left\{-\frac{0{\cdot}11\omega}{U_c}|z_1-z_2|\right\} \times\exp\left\{-\frac{3}{2\pi}\frac{\omega d}{U}|\theta-\theta'|\right\}, \tag{9.24}$$

where $\Phi(\omega)$ is the p.s.d. of the fluctuating pressure and may be regarded as originating from turbulent eddies: its precise calculation depends therefore on a solution of a very difficult problem but has been given as a correlation by Corcos (1962) and Baroudi (1963) in terms of the hydraulic radius D and the coolant flow velocity U. $U_c = 0{\cdot}8U$ and d is the rod diameter.

Inserting (9.24) into (9.23) and integrating over θ and θ' we obtain:

$$\Phi_{FF}(z_1, z_2; \omega) = \Phi(\omega)\cos\left\{\frac{\omega}{U_c}(z_1 - z_2)\right\}\exp\left\{-\frac{0{\cdot}11\,\omega}{U_c}|z_1 - z_2|\right\} d^2\psi_d^2, \tag{9.25}$$

where, with $S = \omega d/U$,

$$\psi_d^2 = \frac{3{\cdot}26S}{S^2+4{\cdot}34}\left\{1 - \frac{0{\cdot}69S[1-\exp\{-3{\cdot}02S\}]}{S^2+4{\cdot}34}\right\}. \tag{9.26}$$

We are now in a position to choose the appropriate model for the rod. Using for simplicity the Bernoulli–Euler model of equation (9.2), in the manner of Bogdanoff and Goldberg (1960), and a simply supported structure such that the boundary conditions are

$$\epsilon(0, t) = \epsilon(L, t) = 0 \tag{9.27}$$

and

$$\left.\frac{\partial^2 \epsilon}{\partial z^2}\right|_{z=0} = \left.\frac{\partial^2 \epsilon}{\partial z^2}\right|_{z=L} = 0, \tag{9.28}$$

where L is the length of the rod, we can insert eqns. (9.8) and (9.9) for $\epsilon(z, t)$ and $F(z, t)$ and with $\psi_n(z) = (2/L)^{\frac{1}{2}}\sin(n\pi z/L)$, $n = 1, 2, 3, \ldots$, we get†

$$\rho A\ddot{a}_n(t) + \mu\dot{a}_n(t) + EI\left(\frac{n\pi}{L}\right)^4 a_n(t) = f_n(t). \tag{9.29}$$

The impulse response function $h_n(t)$ is obtained by replacing $f_n(t)$ by $\delta(t)$, whence

$$\overline{h_n(s)} = \frac{1}{\rho A}\{s^2 + 2\mu_0 s + p_n^2\}^{-1}, \tag{9.30}$$

where $\quad p_n^2 = EI(n\pi/L)^4/\rho A, \quad 2\,\mu_0 = \mu/\rho A \quad$ and $\quad H_n(\omega) = \overline{h_n(i\omega)}$.

Inversion of $\overline{h_n(s)}$ gives $h_n(t)$, viz.

$$h_n(t) = \frac{1}{\rho A\alpha_n}\,e^{-\mu_0 t}\sin\alpha_n t \quad (t > 0), \tag{9.31}$$

where $\alpha_n^2 = p_n^2 - \mu_0^2$ and we have assumed $\alpha_n^2 > 0$.

We note at this point that should the influence of a defective support on the p.s.d. wish to be investigated then the appropriate "defective" boundary condition would have to be used to calculate $\psi_n(z)$ and $h_n(t)$.

Insertion of $h_n(t)$ or $H_n(\omega)$ and Φ_{FF} into (9.19) or (9.20) leads to the correlation function or p.s.d. for the vibrational displacement. The only unknown is $\Phi(\omega)$ and in practical calculations this is taken to be independent of ω since it varies very slowly over the region of ω where $H_1(\omega)$ varies rapidly. In fact the frequency region of

† In some approximate treatments, the damping term proportional to $\dot{a}_n(t)$ is replaced by introducing a small imaginary term in the coefficient multiplying $a_n(t)$. This procedure has the same qualitative effect in that it introduces damping but it is clearly not a fundamental approach (Robson, 1963).

interest is usually dominated by the first resonance frequency corresponding to $|H_1(\omega)|^2$, viz.

$$|H_1(\omega)|^2 = \frac{1}{\rho^2 A^2}\{(p_1^2-\omega^2)^2+4\mu_0^2\omega^2\}^{-1}, \tag{9.32}$$

i.e. the p.s.d. is very peaked for $\omega \approx p_1$. Reavis has gone so far as to neglect all higher mode contributions to the mid-span ($z_1 = z_2 = L/2$) mean square deflection $\langle\epsilon^2\rangle$, such that

$$\langle\epsilon^2\rangle = \frac{2L}{\pi}\int_0^\infty d\omega |H_1(\omega)|^2 \int_0^L dz' \int_0^L dz'' \psi_1(z')\psi_1(z'')\Phi_{FF}(z',z'';\omega). \tag{9.33}$$

Similarly the average fundamental p.s.d. which we define as

$$\Phi_v(\omega) = \int_0^L dz_1 \int_0^L dz_2 \Phi_{\epsilon\epsilon}(z_1,z_2;\omega)\psi_1(z_1)\psi_1(z_2) \tag{9.34}$$

$$= |H_1(\omega)|^2 \int_0^L dz' \int_0^L dz'' \psi_1(z')\psi_1(z'')\Phi_{FF}(z',z'';\omega) \tag{9.35}$$

can be written for this simple model as

$$\Phi_v(\omega) = \frac{B(\omega)}{(\omega^2-\omega_0^2)^2+4\mu_0^2\omega^2}, \tag{9.36}$$

where ω_0 is the fundamental frequency and $B(\omega)$ is the spatial average of $\Phi_{FF}(\ldots)$.

Equation (9.36) should be compared with eqn. (6.85) used in the point model approximation of vibration. In fact the results are not quite the same because the earlier result employed the synthetic damping approach of introducing a term $(1+i\eta_0)\epsilon$ instead of $\mu_0\dot{\epsilon}$. However, in the neighbourhood of ω_0 the term $4\mu_0^2\omega^2 \simeq 4\mu_0^2\omega_0^2$ and then with $\omega = 2\pi f$, $\eta_0 = \mu_0/\pi$ and $A(f) = B(\omega)/(2\pi)^4$, we arrive at the same result. The present analysis highlights the approximations inherent in the point model approach.

Insertion of the various analytical forms for $H_1(\omega)$, $\psi_1(z)$ and Φ_{FF} lead to integrals which can be evaluated. However, the result is rather cumbersome from an algebraic point of view and so we will not write it down: the interested reader can go to Reavis (1969). In Reavis's paper his formula is compared with a number of experimental results. The basic result is not in very good agreement in absolute terms but it does give a formula from which the important physical parameters influencing vibration can be deduced. Also by means of a simple correlation factor a working formula for root mean square deflection can be obtained which is in excellent agreement with experiment. An example of the expected deflection is given by Reavis using the following data: rod diameter = 0·422 in., length = 20 in. The support conditions are unknown but the natural frequency is found to be $p_1/2\pi$ = 41 Hz and have 1% of critical damping. The rod is mounted in a channel with hydraulic diameter 0·534 in. and is subjected to a parallel flow of water at 193 in. sec^{-1} at a temperature of 575 °F. The value of $\sqrt{\langle\epsilon^2\rangle}$ is 0·00026 in. or about 0·006 mm, which is very small but could, over a long period of time, lead to fatigue of a component.

Closely related to the displacement, ϵ, and of considerable importance as regards fatigue studies, is the bending moment $M(z,t)$ or the bending stress $S = M(z,t)/Z$

where Z is the section modulus of the rod. The bending moment is given according to Bernoulli–Euler beam theory as

$$M(z,t) = EI\frac{\partial^2\epsilon(z,t)}{\partial z^2}. \tag{9.37}$$

Thus, returning to eqn. (9.14), we find that

$$M(z,t) = EI\sum_n \psi_n''(z)\int_0^\infty dt'\, h_n(t')\int_0^L dz'\, \psi_n(z')F(z',t-t') \tag{9.38}$$

and hence the cross-correlation function, viz.

$$\langle M(z_1,t_1)\,M(z_2,t_2)\rangle = (EI)^2\sum_{nm}\psi_n''(z_1)\psi_m''(z_2)\int_0^\infty dt'\int_0^\infty dt''$$
$$\times h_n(t')h_m(t'')\int_0^L dz'\int_0^L dz''\,\psi_n(z')\psi_m(z'')\langle F(z',t_1-t')F(z'',t_2-t'')\rangle. \tag{9.39}$$

If we define the cross-correlation of the bending stress as $\Phi_{ss}(z_1,z_2;\tau)$ then we can write it as

$$\Phi_{ss}(z_1,z_2;\tau) = \left(\frac{EI}{Z}\right)^2\sum_{nm}\psi_n''(z_1)\psi_m''(z_2)\int_0^\infty dt'\int_0^\infty dt''\,h_n(t')h_m(t'')$$
$$\times\int_0^L dz'\int_0^L dz''\,\psi_n(z')\psi_m(z'')\,\Phi_{FF}(z',z'';\tau-t''+t'). \tag{9.40}$$

The maximum mean square bending stress, which we will later use, is defined as

$$\langle\sigma^2\rangle = \Phi_{ss}\left(\frac{L}{2},\frac{L}{2};0\right). \tag{9.41}$$

Equation (9.40) can, of course, be written in terms of the p.s.d. $\Phi_{FF}(z_1,z_2;\omega)$ when this is more convenient.

Some numerical calculations of mean square displacement and mean square stress along a beam, together with the associated correlation functions, have been made by Samuels and Eringen (1958). Bernoulli–Euler and Timoshenko models were used and it was concluded that for displacement calculations the Bernoulli–Euler method was satisfactory. However, for bending stress, the Bernoulli–Euler model failed completely and the Timoshenko result was required. This being the case, the appropriate form relating M to transverse displacement, ϵ, and angular distortion, ψ, is

$$M(z,t) = EI\left(1+\epsilon_1\frac{\partial}{\partial t}\right)\frac{\partial\psi(z,t)}{\partial z}. \tag{9.42}$$

The reader is referred to the source paper for more details.

Occasionally, longitudinal motion of a rod can arise in a reactor. A typical situation has been discussed by Nicholson (1958) in connection with a possible reactor-instability mechanism due to interaction of longitudinal vibrations, heat generation, thermal expansion and neutron kinetics. Nicholson's analysis shows that under certain conditions an instability will exist in a reactor of the type that has solid fuel elements that run continuously the length of the core. Self-sustained stresses can build up, causing the fuel elements to exceed their yield-point. Nicholson also shows, however, that in the presence of sufficiently high frictional damping the instability can be prevented. Calculations using data appropriate to the Enrico Fermi Fast Reactor show that the system is stable against this particular effect.

The mathematical model that represents longitudinal oscillations of a rod can be obtained by treating it as an elastic body subject to damping and excitation and setting up a force balance (Thomson, 1965). We find therefore that if $u(z,t)$ is the displacement from equilibrium, it is given by the equation,

$$\frac{m}{g}\, dz\, \frac{\partial^2 u(z,t)}{\partial t^2} = dP_E + dP_R + dP_A, \tag{9.43}$$

where $$dP_E \text{ (elastic restoring force)} = AE \frac{\partial^2 u}{\partial z^2}\, dz,$$

$$dP_R \text{ (resistive force)} = A\gamma \frac{\partial u}{\partial t}\, dz$$

(P_R assumed proportional to velocity) and $dP_A = F_A dz$ is an external force. Thus the equation for longitudinal motion is

$$E\frac{\partial^2 u}{\partial z^2} + \gamma \frac{\partial u}{\partial t} + \frac{\rho}{g}\frac{\partial^2 u}{\partial t^2} = F_A, \tag{9.44}$$

γ being a damping coefficient dependent on the external medium and the material of the rod which gives rise to internal friction.

The corresponding stress $S_{11}(z,t)$ is found from

$$S_{11}(z,t) = E\frac{\partial u(z,t)}{\partial z}. \tag{9.45}$$

If F_A is a random excitation, which is (for example) in Nicholson's treatment due to thermal expansion, the corresponding correlation functions for u and S_{11} can be found. Knowledge of $\langle \sigma_s^2 \rangle$ is useful for fatigue studies and some calculations are given by Thomson and Barton (1957).

We note in passing that a point model approach to the above problem was considered by Nicholson (1958) who set up an equation of motion for the endpoint of the rod by treating it as a loaded spring. Thus with $L(t) = L_0 + x(t)$, where $L(t)$ is the length of the rod at time t and L_0 is the value at equilibrium, it is found that

$$\ddot{x} + \omega_0^2(x - \alpha L_0 \theta) + \omega_0 \delta \dot{x} = 0, \tag{9.46}$$

where ω_0 is the natural frequency for free longitudinal vibrations, α is the coefficient of thermal expansion, δ is a damping coefficient and θ is the temperature difference $T(t) - T_0$ at time t. The equation for $\theta(t)$ is given by

$$H\dot{\theta} = P_1 = (P(t) - P_0), \tag{9.47}$$

where $P(t)$ is the power and H the heat capacity of the rod. The kinetic equation gives

$$\dot{P} = \frac{1}{l}(k_{ex}(t) - \beta)P(t) + \frac{\beta}{l}P_0 \tag{9.48}$$

with $k_{ex} = -\gamma x(t)$, where γ is a reactivity coefficient.

This set of equations when linearized can be solved for a sinusoidal input $P_1(t) \propto \exp(i\omega t)$ and the values of ω for stability obtained. It is found that to prevent instability the damping coefficient δ must be greater than δ^*, where

$$\delta^* = \frac{P_0 \alpha L \gamma}{\omega_0 \beta H[1+(\omega_0 l/\beta)^2]}. \tag{9.49}$$

Considering all possible mechanisms of damping it was shown that δ for the Fermi reactor could never be less than $10\delta^*$ and therefore that the reactor was stable against oscillations of this type. It would be useful to repeat this analysis using a space-dependent model. Also instead of exciting the system with a sinusoid a random source $P_1(t)$ could be used and the various p.s.d. obtained for $x(t)$ and $\theta(t)$.

9.4. *Fatigue and component failure*

The structural members inside a power reactor must be designed against failure for at least 30 years. This is due to radiation hazards which prevent maintenance being carried out inside the core. Clearly, due to irradiation creep, wear, corrosion and fatigue, the mechanical strength and other physical properties of the core components will change over the reactor life. Thus the prediction of the lifetime of a component is of considerable practical importance. Because the phenomena which cause the defects listed above are of a random nature it is necessary to apply statistical methods for lifetime estimates. In addition, the components themselves, even though outwardly identical, will differ in their internal structural composition, thereby introducing a further degree of uncertainty into the calculation.

In this section wc shall discuss one of the above topics, namely fatigue, since the underlying theory has been fairly well established in other engineering areas, in particular aeronautics (Miles, 1954). Failure due to corrosion, creep, etc., requires much further fundamental investigation before a reliable criterion for life can be given. Fatigue is the name used to describe the deterioration in strength of a component as a result of the action of variable stresses as a function of time. The stress variation can be deterministic, e.g. a sinusoidal modulation, or more usually a random variable. At the present time the description of fatigue rests almost entirely on a phenomenological basis. This is understandable since any detailed analytical work would require a complete description of the internal microscopic structure of the component and a method for incorporating this into a useful measure of deterioration. Present-day work relies upon Miner's (1945) concept of cumulative damage, which may be described as follows: if a material is subjected to a number of stress reversals (i.e. stress going from $-S$ to $+S$) n_1 at a certain vibration level, and the total number of stress reversals to failure at this level is N_1, then the partial damage D_1 is given by

$$D_1 = \frac{n_1}{N_1}. \tag{9.50}$$

If the vibration level is changed, then the new partial damage, D_2, for n_2 stress reversals with N_2 for failure is given by $D_2 = n_2/N_2$. The total accumulated damage D_T is given by

$$D_T = \sum_{i=1}^{m} \frac{n_i}{N_i}. \tag{9.51}$$

Component failure is said to occur when $D_T = 1$ (Robson, 1963).

When the stresses leading to fatigue are random, the above definition remains the same but must be reinterpreted. In particular, the stress $S(t)$ will be a random function covering a range of values, governed by a statistical law. However, the actual magnitude of the stress is of little relevance without knowledge of the number of cycles that the stress executes in unit time at a specified level. In addition, it will be necessary to have experimental information about $N(S)$, i.e. the number of stress reversals to failure at a given stress level S. This is usually given semi-empirically in the form

$$S^{\alpha} N(S) = S_1^{\alpha} = \text{constant}, \tag{9.52}$$

where S_1 and α are determined by the material and its geometrical shape and associated fixtures. Some useful comments on N–S curves are to be found in Yeh (1973).

To return to the stress probability function, we must calculate the expected number of peaks occurring within a given range during a given time interval. Figure 9.2 shows the situation that must be studied.

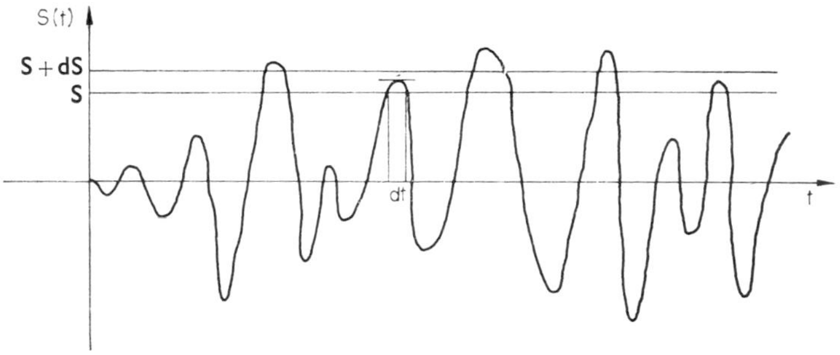

FIG. 9.2. The fluctuations in stress about the mean value. Attention is drawn to the number of peaks in a given stress interval $(S, S+dS)$ and the length of time spent there.

We require knowledge of the probability of a peak in $(S, S+dS)$ in time dt; we call this $p(S)\,dS\,dt$ since we expect it to be a stationary process and therefore dependent only on the time interval dt. Thus in a time T, the total number of peaks in $(S, S+dS)$ will be

$$Tp(S)\,dS \tag{9.53}$$

which in turn will lead to a damage in dS due to peak stresses equal to

$$\frac{Tp(S)\,dS}{N(S)}. \tag{9.54}$$

The total expected damage $\langle D \rangle$ is given by

$$\langle D \rangle = T\int_0^{\infty} \frac{p(S)\,dS}{N(S)} \tag{9.55}$$

and the mean time to failure T_F is when $\langle D \rangle = 1$, i.e.

$$T_F^{-1} = \int_0^{\infty} \frac{p(S)}{N(S)}\,dS. \tag{9.56}$$

The probability of failure during a service life T_S is therefore T_S/T_F.

The calculation of $p(S)$ is not entirely straightforward for the following reason. At first sight it may seem that if the stress amplitudes are distributed in a Gaussian fashion such that $F(S)\,dS$ is the probability that the stress lies between S and $S+dS$, then $p(S)$ is also Gaussian. This, however, is not true as was shown by Rayleigh (1899–1920) for the two-dimensional random walk problem and by Rice (Wax, 1954) for a narrow-band-pass filter. The details of the calculation will not be given here; suffice to say that to obtain $p(S)$ it is necessary to know the joint probability function $p(S, \dot{S}, \ddot{S})$ where dots indicate differentiation with respect to time. More precise details are given by Robson (1963). The result, however, is that $p(S)$ is given by

$$p(S) = \mathcal{N}_p \frac{S}{\sigma^2}\, e^{-S^2/2\sigma^2}, \tag{9.57}$$

where $\mathcal{N}_p$ is the number of peaks per unit time (equal to the half number of zero-crossings) and σ^2 is the mean square stress as obtained from the elastic model employed for the beam (see Section 8.9). In principal, $p(S)$ should depend on position, i.e. where the stress is applied, but we will assume that an appropriate spatial average has been taken (Yeh, 1973).

Inserting eqns. (9.52) and (9.57) into (9.55) and integrating leads to

$$\langle D \rangle = T\mathcal{N}_p \left(\frac{\sqrt{2}.\sigma}{S_1}\right)^{\alpha} \Gamma\left(1+\frac{\alpha}{2}\right) \tag{9.58}$$

and

$$T_F^{-1} = \mathcal{N}_p \left(\frac{\sqrt{2}.\sigma}{S_1}\right)^{\alpha} \Gamma\left(1+\frac{\alpha}{2}\right). \tag{9.59}$$

At this point it is important to note that the above formula is only considered to be valid for narrow-band excitation, i.e. when the dominant frequency which excites the structure lies within the band-pass region of the structure itself. If the exciting frequency forms a broad band then a more satisfactory method of calculating T_F is to divide the stress spectral density into regions $(0, \omega_1)$, (ω_1, ω_2), etc., which cover the broad bandwidth. At the same time, partial mean square stresses and number of peaks per unit time are defined as

$$\sigma_i^2 = \int_{\omega_{i-1}}^{\omega_i} \Phi_{ss}(\omega)\, dw, \tag{9.60}$$

$$\mathcal{N}_p^{(i)} = \frac{1}{2}\left\{\frac{\int_{\omega_1}^{\omega_2} \omega^2 \Phi_{ss}(\omega)\, d\omega}{\int_{\omega_1}^{\omega_2} \Phi_{ss}(\omega)\, d\omega}\right\}^{\frac{1}{2}} \tag{9.61}$$

$$\simeq \frac{1}{2}\left(\frac{\omega_1+\omega_2}{2}\right). \tag{9.62}$$

The total damage $\langle D \rangle_{BB}$ is now

$$\langle D \rangle_{BB} = \sum_i T\mathcal{N}_p^{(i)} \left(\frac{\sqrt{2}\sigma_i}{S_1}\right)^{\alpha} \Gamma\left(1+\frac{\alpha}{2}\right). \tag{9.63}$$

The mean life to failure can be calculated when $\langle D \rangle_{BB} = 1$. Some numerical examples can be found in Yeh (1973) and also a discussion of some other aspects of fatigue calculations and experiment (beware, however, of the numerous typographical errors in the article).

9.5. *Fatigue calculations by the method of zero-crossings*

An alternative approach to damage calculations has been proposed by Crandall *et al.* (1962) which is also convenient for calculating the variance in the expected damage $\langle D^2\rangle - \langle D\rangle^2$ as well as $\langle D\rangle$ itself. This is clearly of practical importance.

The total time interval T over which the damage is to be calculated is divided into M equal sub-intervals and within each sub-interval a partial damage d_i is associated. The total damage D is therefore

$$D = \sum_{i=0}^{M-1} d_i, \tag{9.64}$$

where the d_i are correlated random variables. The mean and variance of D are therefore given by

$$\langle D\rangle = \sum_{i=0}^{M-1} \langle d_i\rangle \tag{9.65}$$

and

$$\text{var}(D) = \langle D^2\rangle - \langle D\rangle^2 = \sum_i \langle d_i d_j\rangle - \langle D\rangle^2. \tag{9.66}$$

If the stress history is a stationary random process, it is legitimate to assume that the statistical averages needed are invariant with respect to a translation of the time axis. Therefore

$$\langle d_i\rangle = \langle d_j\rangle = \langle d_0\rangle \tag{9.67}$$

and

$$\langle d_i d_j\rangle = \langle d_0 d_{j-i}\rangle \tag{9.68}$$

for any i and j. We can now rewrite the sums as follows:

$$\langle D\rangle = M\langle d_0\rangle, \tag{9.69}$$

$$\text{var}(D) = M\{\langle d_0^2\rangle - \langle d_0\rangle^2\} + 2\sum_{k=1}^{M-1} (M-k)\{\langle d_0 d_k\rangle - \langle d_0\rangle^2\}. \tag{9.70}$$

We note that for $k \to M$ (M large), d_0 and d_k become less strongly correlated and $\text{var}(D) \to M\{\langle d_0^2\rangle - \langle d_0\rangle^2\}$.

It remains to define d_i. As we have seen in Section 9.4 the damage for *one* cycle (or stress reversal) is given by N_i^{-1}, where N_i is the number of cycles until failure at maximum stress S_i. To avoid the subtleties of calculating the probability distributions of peak amplitudes, Crandall *et al.* hypothesize that the damage can be written in terms of a zero-crossing of the stress process rather than a peak value. In that case the damage is defined as

$$d_i = \frac{1}{2N_i} \tag{9.71}$$

since there are twice as many zero-crossings as peak values. The zero-crossing is related to the stress amplitude through the slope $\dot{\mathscr{S}} \equiv d\mathscr{S}/dt$ at the time where $\mathscr{S} = 0$. Thus the greater $\dot{\mathscr{S}}$ at $\mathscr{S} = 0$, the greater will be the resulting stress peak; this certainly appears to be a sensible assumption. The maximum stress S resulting from a zero-crossing with slope $\dot{\mathscr{S}}$ is written

$$S = \frac{|\dot{\mathscr{S}}|}{\pi\mathscr{N}}, \tag{9.72}$$

where $\mathscr{N}$ is the number of zero-crossings per unit time. More precisely we can write

$$\mathscr{N} = 2\left[\frac{\int_{-\infty}^{\infty} \omega^2 \Phi(\omega)\, d\omega}{\int_{-\infty}^{\infty} \Phi(\omega)\, d\omega}\right]^{\frac{1}{2}} = \frac{1}{\pi}\left[-\frac{\phi''(0)}{\phi(0)}\right], \tag{9.73}$$

$$\equiv \frac{1}{\pi}\frac{\sigma_{\dot{\mathscr{S}}}}{\sigma_{\mathscr{S}}}, \tag{9.74}$$

where $\sigma_{\mathscr{S}}^2$ is the mean square stress and $\sigma_{\dot{\mathscr{S}}}^2$ the mean square of the rate of change of the stress.

Equation (9.72) would be exact for simple harmonic motion of frequency $\mathscr{N}$, therefore in a narrow-band process it is expected to be a good approximation to the amplitude of the peak stress following the associated zero-crossing. It is believed that the statistics of the zero-crossing damage process, so defined, will not be significantly different from the peak associated damage process.

To proceed we divide T into sub-intervals Δt such that $M\Delta t = T$, where Δt is small enough such that the stress history $\mathscr{S}(t)$ is linear over it. If we now define the joint probability function $p(\mathscr{S}(t), \dot{\mathscr{S}}(t))$ where t is the time at the beginning of the interval Δt, we can calculate the fraction of samples which possess a zero-crossing in the sub-interval Δt. First, however, we note that a zero-crossing will arise only for those $\mathscr{S}$ and $\dot{\mathscr{S}}$ that satisfy the conditions

$$-\dot{\mathscr{S}}\Delta t < \mathscr{S} < 0 \quad (\dot{\mathscr{S}} > 0), \tag{9.75a}$$

$$0 < \mathscr{S} < -\dot{\mathscr{S}}\Delta t \quad (\dot{\mathscr{S}} < 0). \tag{9.75b}$$

We can now calculate the value of $\langle d_0 \rangle$ as follows:

$$\langle d_0 \rangle = \int_{-\infty}^{0} \frac{1}{2N_0(S)}\, d\dot{\mathscr{S}} \int_{0}^{-\dot{\mathscr{S}}\Delta t} p(\mathscr{S},\dot{\mathscr{S}})\, d\mathscr{S} + \int_{0}^{\infty} \frac{1}{2N_0(S)}\, d\dot{\mathscr{S}} \int_{-\dot{\mathscr{S}}\Delta t}^{0} p(\mathscr{S},\dot{\mathscr{S}})\, d\mathscr{S} \tag{9.76}$$

which for small $|\dot{\mathscr{S}}\Delta t|$ becomes

$$\langle d_0 \rangle = \Delta t \int_{-\infty}^{\infty} \frac{d\dot{\mathscr{S}}}{2N_0(S)}\, |\dot{\mathscr{S}}|\, p(0,\dot{\mathscr{S}}). \tag{9.77}$$

On the assumption that $\mathscr{S}(t)$ and $\dot{\mathscr{S}}(t)$ form a Gaussian process we may write (Robson, 1963):

$$p(0,\dot{\mathscr{S}}) = \frac{1}{2\pi\sigma_{\mathscr{S}}\sigma_{\dot{\mathscr{S}}}} \exp\left\{-\frac{\dot{\mathscr{S}}^2}{2\sigma_{\dot{\mathscr{S}}}^2}\right\}. \tag{9.78}$$

Inserting (9.78), (9.72) and (9.52) into (9.77) and using $\sigma_{\dot{\mathscr{S}}} = \pi\mathscr{N}\sigma_{\mathscr{S}}$, we obtain

$$\langle d_0 \rangle = \Delta t \frac{\mathscr{N}}{2}\left(\frac{\sqrt{2}.\sigma_{\mathscr{S}}}{S_1}\right)^{\alpha} \Gamma\left(1+\frac{\alpha}{2}\right), \tag{9.79}$$

hence

$$\langle D \rangle = \frac{\mathscr{N}T}{2}\left(\frac{\sqrt{2}.\sigma_{\mathscr{S}}}{S_1}\right)^{\alpha} \Gamma\left(1+\frac{\alpha}{2}\right) \tag{9.80}$$

which, if we recall that $\mathscr{N}_p = \frac{1}{2}\mathscr{N}$, is identical with the result obtained from the Rayleigh distribution in the previous section. It should be noted that the derivation in this section accounts for the random variation in cycles in the period T as well as the randomness in the amplitudes.

Using the same technique, we can now calculate the variance in the damage as defined by eqn. (9.70). Firstly, we require $\langle d_0^2\rangle$ which is given by

$$\langle d_0^2\rangle = \Delta t\int_{-\infty}^{\infty}\frac{d\dot{\mathscr{S}}}{4N_0^2(S)}\,|\dot{\mathscr{S}}|\,p(0,\dot{\mathscr{S}}), \tag{9.81}$$

$$= \frac{\mathscr{N}\Delta t}{4}\left(\frac{2\sigma_{\mathscr{S}}^2}{S_1^2}\right)^{\alpha}\Gamma(1+\alpha). \tag{9.82}$$

The calculation of $\langle d_0 d_k\rangle$ requires the joint probability function $p(\mathscr{S}(t),\dot{\mathscr{S}}(t);\ \mathscr{S}(t+\tau),\dot{\mathscr{S}}(t+\tau))$, where $\tau = k\Delta t$. By integrating over the sub-region where both sub-intervals have zero-crossing we find by arguments similar to those used above that, for $k \neq 0$,

$$\langle d_0 d_k\rangle = (\Delta t)^2\int_{-\infty}^{\infty} d\dot{\mathscr{S}}_0\int_{-\infty}^{\infty} d\dot{\mathscr{S}}_k\,\frac{1}{2N_0(S)}\,\frac{1}{2N_k(S)}\,|\dot{\mathscr{S}}_0|\,|\dot{\mathscr{S}}_k|\,p(0,\dot{\mathscr{S}}_0;\,0,\dot{\mathscr{S}}_k). \tag{9.83}$$

Using a four-dimensional Gaussian for $p(\ldots)$ (Lawson and Uhlenbeck, 1953; Middleton, 1960), viz.

$$p(0,\dot{\mathscr{S}}_0;\,0,\dot{\mathscr{S}}_k) = \frac{1}{4\pi^2\Lambda^{\frac{1}{2}}}\exp\left\{-\frac{1}{2\Lambda}\left(\Lambda_{22}\dot{\mathscr{S}}_0^2+2\Lambda_{24}\dot{\mathscr{S}}_0\dot{\mathscr{S}}_k+\Lambda_{44}\dot{\mathscr{S}}_k^2\right)\right\}, \tag{9.84}$$

where Λ and Λ_{ij} depend on the auto-correlation function of the stress process $\langle\mathscr{S}(t)\mathscr{S}(t+\tau)\rangle = R(\tau)$ and are given by

$$\Lambda = [R(0)^2-R(\tau)^2][R''(0)-R''(\tau)^2]+2R'(\tau)^2[R(0)R''(0)-R(\tau)R''(\tau)]+R'(\tau)^4, \tag{9.85}$$

$$\Lambda_{22} = \Lambda_{44} = -R''(0)[R(0)^2-R(\tau)^2]-R(0)R'(\tau)^2, \tag{9.86}$$

$$\Lambda_{24} = R''(\tau)[R(0)^2-R(\tau)^2]+R(\tau)R'(\tau)^2, \tag{9.87}$$

$R(\tau)$ will be given by the model used to represent the rod vibrations and is in general a function of position. However, in the present limited description of fatigue it would correspond to the position of maximum stress although, in a more detailed study, the fatigue damage itself might well be position dependent.

Inserting (9.84) into (9.83) and taking α to be an odd positive integer (for mathematical convenience) we obtain

$$\langle d_0 d_k\rangle = \left(\frac{\Delta t}{2\pi}\right)^2\left(\frac{2}{S_1^2\pi\mathscr{N}^2}\right)^{\alpha}\Gamma\left(1+\frac{\alpha}{2}\right)^2\frac{\Lambda^{\alpha+\frac{3}{2}}}{\Lambda_{22}^{\alpha+2}}F\left(1+\frac{\alpha}{2},\,1+\frac{\alpha}{2};\,\frac{1}{2};\,\frac{\Lambda_{24}^2}{\Lambda_{22}^2}\right), \tag{9.88}$$

where $F(\ldots)$ is the hypergeometric function.

Var(D) can now be written in limit as $\Delta t\to 0$ as

$$\operatorname{var}(D) = \frac{\mathscr{N}T}{4}\left(\frac{2\sigma_{\mathscr{S}}^2}{S_1^2}\right)^{\alpha}\Gamma(1+\alpha)+\frac{\mathscr{N}^2}{2}\left(\frac{2\sigma_{\mathscr{S}}^2}{S_1^2}\right)^{\alpha}\Gamma\left(1+\frac{\alpha}{2}\right)^2\int_0^T d\tau(T-\tau)\{g(\tau)-1\}, \tag{9.89}$$

where

$$g(\tau) = \frac{\Lambda^{\alpha+\frac{3}{2}}}{\sigma_{\mathscr{S}}^{2\alpha}\Lambda_{22}^{\alpha+2}(\sqrt{\pi}.\mathscr{N})^{2\alpha+2}}F\left(1+\frac{\alpha}{2},\,1+\frac{\alpha}{2};\,\frac{1}{2};\,\frac{\Lambda_{24}^2}{\Lambda_{22}^2}\right). \tag{9.90}$$

Further progress requires a specific model for $R(\tau)$.

If we take the average fundamental p.s.d. of the stress, as defined by eqn. (9.38), then using the Bernoulli–Euler beam model we can write for the corresponding correlation function

$$R(\tau) = \frac{1}{2\pi}\int_{-\infty}^{\infty} e^{i\omega t}\,\frac{C(\omega)\,d\omega}{(\omega^2-\omega_0^2)^2+4\mu_0^2\omega^2}, \tag{9.91}$$

where $C(\omega)$ is proportional to the average p.s.d. of the random force statistics.

Inverting the transform, we have

$$R(\tau) = \frac{1}{4\mu_0\omega_0^2}\int_{-\infty}^{\infty} dt'\,R_c(\tau-t')e^{-\mu_0 t'}\left\{\cos pt'+\frac{\mu_0}{p}\sin pt'\right\}, \tag{9.92}$$

where $p^2 = \omega_0^2-\mu_0^2$.

For white noise input such that $R_c(\tau) = 4\mu_0\omega_0^2\langle\sigma^2\rangle\delta(\tau)$ we obtain

$$R(\tau) = \langle\sigma^2\rangle e^{-\mu_0\tau}\left\{\cos p\tau+\frac{\mu_0}{p}\sin p\tau\right\} \quad (\tau \geqslant 0) \tag{9.93}$$

when $\langle\sigma^2\rangle$ is the mean square stress.

Using this model it is noted that the hypergeometric function oscillates with period π/ω_0. The integrand in eqn. (9.89) has peaks at $n\pi/\omega_0$ and zeros at $(2n+1)\pi/2\omega_0$ ($n = 1,2,3,\ldots$). Moreover, the integrand is negligible in the range $0 < \tau < \pi/2\omega_0$ and therefore a good approximation to the integral in eqn. (9.89) can be obtained by replacing the quantity in curly brackets by

$$F\left(-\frac{\alpha}{2}, -\frac{\alpha}{2}; 1; e^{-2\mu_0\tau}\right)-1$$

with the limits changed to $(\pi/2\omega_0, \infty)$. This approximation is good for $\mu_0 \ll \omega_0$ and $\mu_0 T \gg 1$. The expression for var(D) then becomes

$$\operatorname{var}(D) = \frac{\mathcal{N}T\omega_0}{2\mu_0}\left(\frac{2\sigma_{\mathcal{G}}^2}{S_1^2}\right)^{\alpha}\Gamma\left(1+\frac{\alpha}{2}\right)\left\{f_1(\alpha)-\frac{2\omega_0 f_2(\alpha)}{\mathcal{N}\mu_0 T}+\frac{2\mu_0 f_3(\alpha)}{\omega_0\mathcal{N}T}\right\}, \tag{9.94}$$

where

$$f_1(\alpha) = \frac{\mu_0}{\pi\omega_0}\int_0^{\infty}\left\{F\left(-\frac{\alpha}{2}, -\frac{\alpha}{2}; 1; e^{-2\mu_0\tau}\right)-1\right\}dt, \tag{9.95}$$

$$f_2(\alpha) = \frac{\mu_0^2}{2\pi^2}\int_0^{\infty}\tau\left\{F\left(-\frac{\alpha}{2}, -\frac{\alpha}{2}; 1; e^{-2\mu_0\tau}\right)-1\right\}d\tau, \tag{9.96}$$

$$f_3(\alpha) = \frac{\Gamma(1+\alpha)}{16\Gamma(1+\alpha/2)^2}. \tag{9.97}$$

For $\mu_0 < 10\omega_0$ and $\alpha < 15$ the above formula is considered by Crandall *et al.* to be adequate for engineering purposes. Moreover, very little extra error is incurred in this range of μ_0 and α if we neglect the terms involving f_2 and f_3 and simply write

$$\operatorname{var}(D) \simeq \frac{\mathcal{N}T\omega_0}{2\mu_0}\left(\frac{2\sigma_{\mathcal{G}}^2}{S_1^2}\right)^{\alpha}\Gamma\left(1+\frac{\alpha}{2}\right)f_1(\alpha) \tag{9.98}$$

which indicates a linear growth with T and an inverse dependence on the damping factor μ_0.

The standard deviation $\sigma_D = (\operatorname{var}(D))^{\frac{1}{2}}$, and hence we write the relative variance for a given material as

$$\frac{\sigma_D}{\langle D\rangle} = \left(\frac{2\omega_0 f_1(\alpha)}{\mu_0\mathcal{N}T}\right)^{\frac{1}{2}} \tag{9.99}$$

which indicates that it decreases inversely as the square root of T, μ_0 and $\mathcal{N}$.

A numerical example is given by Crandall *et al.* to illustrate the above work using a beam clamped at the lower end and with a mass attached to the free end. The beam is set into motion with a dashpot damping device. Treating it as a system with one degree of freedom the correlation function is evaluated and with the appropriate physical data the damage $\langle D \rangle$ and its variance σ_D at the root of the beam are calculated. The interesting comment is made that the definition of failure $\langle D \rangle = 1$ is not precise in that, because of the variance σ_D, there is actually a distribution of failure times T_F. This distribution is unknown but the central limit theorem may be invoked to show that for large T it is Gaussian with a variance about $\langle T_F \rangle$ (given by eqn. (9.59)) obtainable from the solutions, $T_F^{\pm}$, of the equations

$$\langle D \rangle \pm \sigma_D = 1 \tag{9.100}$$

or

$$2\mathcal{N}F_F^{\pm} = \frac{1}{\Gamma\left(1+\frac{\alpha}{2}\right)}\left\{\left[\frac{\omega_0}{\mu_0} f_1(\alpha)+4\left(\frac{S_1}{\sqrt{2}.\sigma_{\mathcal{S}}}\right)^{\alpha}\right] \pm \left(\frac{\omega_0}{\mu_0} f_1(\alpha)\right)^{\frac{1}{2}}\right\}^2. \tag{9.101}$$

Quite clearly there are some important advances to be made in the field of fatigue failure which will involve the use of better probability distributions for the occurrence of stress reversals, a more adequate description of the spatial variation of the fatigue and a calculation of the failure time probability distribution.

Further reading relevant to stress calculation and fatigue prediction may be found in the following papers: Liepmann (1952), Kraichnan (1956*a*, *b*), Lyon (1956*a*, *b*, 1960*a*, *b*, 1961), Maidanik (1961) and Smits *et al.* (1962).

9.6. *Random heat transmission and generation*

In most of our work on random heat generation as used in previous chapters we have assumed that the temperature fluctuations in a component were spatially averaged or, in some unexplained way, space independent. In this section we shall examine this problem in more detail. Two specific situations which should be fully understood are the temperature distributions arising from volume and surface random heat sources. The former can arise from non-uniform fuel enrichment in a particular fuel element whilst the second is connected with fluctuations in the coolant-fuel heat-transfer coefficient, which is a surface effect. A further problem that we will examine in less detail is the effect of a random thermal conductivity or a random density on the resulting temperature distribution. In essence, then, we shall study the spatial variation of temperature noise in typical reactor components and discuss how the conclusions may affect our interpretation of experiments. Some pioneering work in this field is due to Samuels (1966) and we shall base our discussion on his approach to the problem. It should also be borne in mind that non-uniform temperature distributions both in space and time lead to thermal stresses which, as we have seen, lead to fatigue failure after a certain lifetime.

9.7. *Random heat generation in slabs, spheres and cylinders*

In this section we will consider a very simple random heat-generation problem. Consider a slab of material bounded by the planes $x = 0$ and $x = l$ and containing a random heat source $Q(x,t)$. The temperatures, T, at the slab surfaces are fixed such that $T(0,t) = T(l,t) = T_0$, where T_0 is time independent. The equation to be solved, therefore, is

$$\rho C \frac{\partial T}{\partial t} = k \frac{\partial^2 T}{\partial x^2} + Q \tag{9.102}$$

subject to the above boundary conditions.

We will assume that $Q(x,t) = Q_0 + \delta Q(x,t)$ where Q_0 is constant in space and time, and δQ is a stationary random Gaussian process. From linearity we can then infer that corresponding fluctuations in $T(x,t)$ will also be random. We further assume that $\langle \delta Q \rangle = 0$ and write $T(x,t) = \langle T(x) \rangle + \delta T(x,t)$, where $\langle T(x) \rangle$ is the average time-independent value of T. Inserting T and Q into the equation we find

$$0 = k \frac{d^2 \langle T \rangle}{dx^2} + Q_0 \tag{9.103}$$

and

$$\rho C \frac{\partial\, \delta T}{\partial t} = k \frac{\partial^2\, \delta T}{\partial x^2} + \delta Q, \tag{9.104}$$

where $\langle T/(0) \rangle = \langle T(l) \rangle = T_0$ and $\delta T(0,t) = \delta T(l,t) = 0$.

Solving (9.103) leads to

$$\langle T(x) \rangle = T_0 - \frac{Q_0}{2k} x(l-x). \tag{9.105}$$

To obtain the solution for δT we use a finite sine transform (Sneddon, 1951) such that

$$\delta T(x,t) = \frac{2}{l} \sum_{n=1}^{\infty} v_n(t) \sin\left(\frac{n\pi x}{l}\right), \tag{9.106}$$

$$\delta Q(x,t) = \frac{2}{l} \sum_{n=1}^{\infty} q_n(t) \sin\left(\frac{n\pi x}{l}\right), \tag{9.107}$$

where

$$v_n(t) = \int_0^l \delta T(x,t) \sin\left(\frac{n\pi x}{l}\right) dx \tag{9.108}$$

and

$$q_n(t) = \int_0^l \delta Q(x,t) \sin\left(\frac{n\pi x}{l}\right) dx. \tag{9.109}$$

Inserting these expansions into (9.104) we find

$$\rho C \frac{dv_n(t)}{dt} + k \left(\frac{n\pi}{l}\right)^2 v_n(t) = q_n(t). \tag{9.110}$$

Solving this equation, and assuming that the source was "switched-on" in the far distant past, we can write

$$v_n(t) = \frac{1}{\rho C} \int_0^{\infty} dt_0 e^{-\lambda_n t_0} q_n(t-t_0), \tag{9.111}$$

where

$$\lambda_n = \frac{k}{\rho C} \left(\frac{n\pi}{l}\right)^2. \tag{9.112}$$

Inserting into the equation for δT we find

$$\delta T(x,t) = \frac{2}{l\rho C} \sum_{n=1}^{\infty} \sin\left(\frac{n\pi x}{l}\right) \int_0^{\infty} dt_0 e^{-\lambda_n t_0} q_n(t-t_0) \tag{9.113}$$

or using (9.109)

$$\delta T(x,t) = \frac{2}{\rho Cl}\sum_{n=1}^{\infty}\sin\left(\frac{n\pi x}{l}\right)\int_0^l dx'\sin\left(\frac{n\pi x'}{l}\right)\int_0^{\infty} dt_0 e^{-\lambda_n t_0}\delta Q(x',t-t_0). \quad (9.114)$$

Thus we have the temperature fluctuations in terms of the heat source fluctuations. We can now calculate the cross-correlation function $\phi_T(x_1,x_2;\tau)$ as follows:

$$\phi_T(x_1,x_2;t_2-t_1) = \langle \delta T(x_1,t_1)\delta T(x_2,t_2)\rangle, \quad (9.115)$$

$$= \frac{4}{(\rho Cl)^2}\sum_{n=1}^{\infty}\sum_{m=1}^{\infty}\sin\left(\frac{n\pi x_1}{l}\right)\sin\left(\frac{m\pi x_2}{l}\right)\int_0^l dx'\int_0^l dx''\sin\left(\frac{n\pi x'}{l}\right)$$

$$\times\sin\left(\frac{m\pi x''}{l}\right)\int_0^{\infty} dt_0\int_0^{\infty} dt_0' e^{-\lambda_n t_0-\lambda_m t_0'}\langle \delta Q(x',t_1-t_0)\delta Q(x'',t_2-t_0')\rangle, \quad (9.116)$$

where we can write $\quad \langle \delta Q\,\delta Q\rangle \equiv \phi_Q(x',x'';t_1-t_2-t_0'+t_0). \quad (9.117)$

ϕ_Q is, hopefully, prescribed and from the resulting integrations we obtain a measure of the temperature noise. To illustrate the problem consider the case when the source is purely random and hence

$$\phi_Q(x',x'';\tau) = \sigma_Q^2\,\delta(x'-x'')\,\delta(\tau). \quad (9.118)$$

We then obtain

$$\phi_T(x_1,x_2;\tau) = \frac{\sigma_Q^2}{l(\rho C)^2}\sum_{n=1}^{\infty}\frac{e^{-\lambda_n|\tau|}}{\lambda_n}\sin\left(\frac{n\pi x_1}{l}\right)\sin\left(\frac{n\pi x_2}{l}\right) \quad (9.119)$$

which indicates that the correlation time is of order $\lambda_1^{-1} = (\rho C/k)(l/\pi)^2$.

The mean square temperature fluctuation is obtained by setting $x_1 = x_2 = x$ and $\tau = 0$, whence

$$\langle \delta T(x)^2\rangle = \frac{\sigma_Q^2}{l(\rho C)^2}\sum_{n=1}^{\infty}\frac{1}{\lambda_n}\sin^2\left(\frac{n\pi x}{l}\right) \quad (9.120)$$

which can be summed exactly to give

$$\langle \delta T(x)^2\rangle = \frac{\sigma_Q^2 l}{\rho Ck}\left(\frac{x}{l}\right)\left(1-\frac{x}{l}\right). \quad (9.121)$$

The r.m.s. temperature therefore $\sqrt{\langle(\delta T)^2\rangle}$ does not have the same shape as the average temperature; indeed we may write the relative variance as

$$\frac{\sqrt{(\langle \delta T)^2\rangle}}{\langle T\rangle - T_0} = \frac{\sigma_Q}{Q}\left(\frac{4k}{\rho Cl}\right)^{\frac{1}{2}}\frac{1}{\sqrt{[x(l-x)]}}. \quad (9.122)$$

Relative fluctuations are therefore largest at the edges although absolute values of the r.m.s. temperature, which are most important from the practical point of view, are largest in the centre of the slab.

The probability density temperature fluctuation is Gaussian and can be written as

$$p(T) = \frac{1}{(2\pi\langle(\delta T)^2\rangle)^{\frac{1}{2}}}\exp\left\{-\frac{(T-\langle T\rangle)^2}{2\langle(\delta T)^2\rangle}\right\}, \quad (9.123)$$

hence we may obtain the probability that T will exceed some prescribed level T_0 by

$$p(T > T_0) = \int_{T_0}^{\infty} dT\,p(T). \quad (9.124)$$

The extension of this procedure to a cylindrical rod is straightforward and involves the use of a Hankel transform (Sneddon, 1951).

A more common boundary condition at the surface of a heated element in a reactor is the "radiation" condition, viz.

$$k\nabla T(\mathbf{r},t)|_s = -h(T(\mathbf{r}_s,t)-T_c), \tag{9.125}$$

where the subscript s indicates surface value. T_c is a mixed mean coolant temperature and h the heat-transfer coefficient. In general, both h and T_c can be random functions of axial position and also time. Thus in addition to any random heat generation there is also a boundary condition which contains random parameters. If, as is usual, we are concerned with relatively small deviations from the mean, then with $h = h_0+\delta h$ and $T_c = T_{c0}+\delta T_c$ we can rewrite the boundary condition (9.125) as follows:

$$-k\nabla\langle T\rangle = h_0(\langle T\rangle - T_{c0}) \tag{9.126}$$

and

$$-k\nabla\,\delta T = h_0(\delta T-\delta T_c)+\delta h(\langle T\rangle - T_{c0}), \tag{9.127}$$

where we have neglected products of second-order terms. These boundary conditions will account therefore for fluctuations in inlet coolant temperature and fluctuations in coolant velocity through their effect on the heat-transfer coefficient.

A number of examples of random surface conditions for various geometrical configurations have been given by Samuels (1966). We shall consider one such case which illustrates the effect of a random boundary condition. It does not correspond exactly to the boundary condition (9.125), but is sufficient to illustrate an important point regarding the limitations of point model noise analysis.

We consider a sphere of radius a with a random surface temperature but with no heat generation within it. In that case the equation of conduction reduces to

$$k\frac{\partial^2}{\partial r^2}\{rT(r,t)\} = \rho C\frac{\partial}{\partial t}\{rT(r,t)\} \tag{9.128}$$

subject to $T(a,t) = T_s(t)$ (T_s random) and $T(0,t)$ bounded. Taking the Laplace transform of eqn. (9.128) and solving the resulting second-order differential equation, we obtain for the Laplace transformed temperature (s is the transform variable):

$$\overline{T}(r,s) = \frac{a\overline{T}_s(s)\sinh\{(s/K)^{\frac{1}{2}}r\}}{r\sinh\{(s/K)^{\frac{1}{2}}a\}}, \tag{9.129}$$

where $K = k/\rho C$.

Inverting the transform we find

$$T(r,t) = \frac{2K}{r}\sum_{n=1}^{\infty}(-)^{n+1}\left(\frac{n\pi}{a}\right)\sin\left(\frac{n\pi r}{a}\right)\int_0^{\infty}dt_0 e^{-\lambda_n t_0}T_s(t-t_0) \tag{9.130}$$

from which the cross-correlation function can be obtained. However, a quantity of some physical interest is the mean square temperature as a function of position across the sphere. We seek therefore

$$\langle T(r,t_1)T(r,t_2)\rangle = \phi(r;\,t_2-t_1) \tag{9.131}$$

and calculate $\phi(r; 0) = \langle T^2(r)\rangle$, viz.

$$\langle T^2(r)\rangle = \frac{4K^2}{r^2} \sum_{n,m=1}^{\infty} (-)^{n+m} \left(\frac{n\pi}{a}\right)\left(\frac{m\pi}{a}\right) \sin\left(\frac{n\pi r}{a}\right) \sin\left(\frac{m\pi r}{a}\right)$$

$$\times \int_0^{\infty} dt_0 \int_0^{\infty} dt_0' \exp(-\lambda_n t_0 - \lambda_m t_0') \phi_s(t_0' - t_0), \quad (9.132)$$

where

$$\phi_s(t_2 - t_1) = \langle T_s(t_1) T_s(t_2)\rangle \quad (9.133)$$

is the correlation function of the random surface temperature.

For white noise input $\phi_s(\tau) = \sigma_s^2 \delta(\tau)$ and $\langle T^2(r)\rangle$ reduces to

$$\langle T^2(r)\rangle = \frac{2\pi K^2 \sigma_s^2}{r^2} \sum_{m=1}^{\infty} (-)^{m+1} m \frac{\exp[m\pi r/a] - \exp[-m\pi r/a]}{\exp[m\pi] - \exp[-m\pi]} \sin\left(\frac{m\pi r}{a}\right). \quad (9.134)$$

In Fig. 9.3 we plot the quantity

$$\frac{\langle T^2(r)\rangle}{2\pi K^2 \sigma_s^2 / a^2} \quad (9.135)$$

and note that the fluctuations are confined to a thin surface region $r \gtrsim 0{\cdot}9a$. This fact is important since it indicates that in any point model approximation where a weighted average value of $\langle T^2(r)\rangle$ is needed, the value will be very much less than a straight volume weighting. In addition it indicates that noise sources give rise to effects which are highly localized in space and consequently in any measurement it should be ensured that the detector (e.g. thermocouple) is as close to the noise source as possible for maximum efficiency.

We shall consider no further problems in detail, but since the cylinder is an important geometry in nuclear reactor cores we will show how the corresponding conduction equation can be solved for a boundary condition of the form (9.125) in which h and T_c are not random but for which there is random heat generation.

The conduction equation for no axial conduction and radial symmetry may be written

$$\rho C \frac{\partial T}{\partial t} = k\left(\frac{\partial^2 T}{\partial r^2} + \frac{1}{r}\frac{\partial T}{\partial r}\right) + Q(r,t) \quad (9.136)$$

plus the boundary condition $T(0, t)$ finite and

$$\left.\frac{\partial T(r,t)}{\partial r}\right|_a = -h(T(a,t) - T_c). \quad (9.137)$$

Writing $\theta(r,t) = T(r,t) - T_c$, we have

$$\rho C \frac{\partial \theta}{\partial t} = k\left(\frac{\partial^2 \theta}{\partial r^2} + \frac{1}{r}\frac{\partial \theta}{\partial r}\right) + Q(r,t) \quad (9.138)$$

and

$$\left.\frac{\partial \theta}{\partial r}\right|_a = -h\theta(a,t). \quad (9.139)$$

A convenient solution can be obtained using Sneddon's finite Hankel transform (Sneddon, 1951), viz. for any function $f(r)$, we have

$$\bar{f}(m) = \int_0^a r f(r) J_0(\alpha_m r)\, dr \quad (9.140)$$

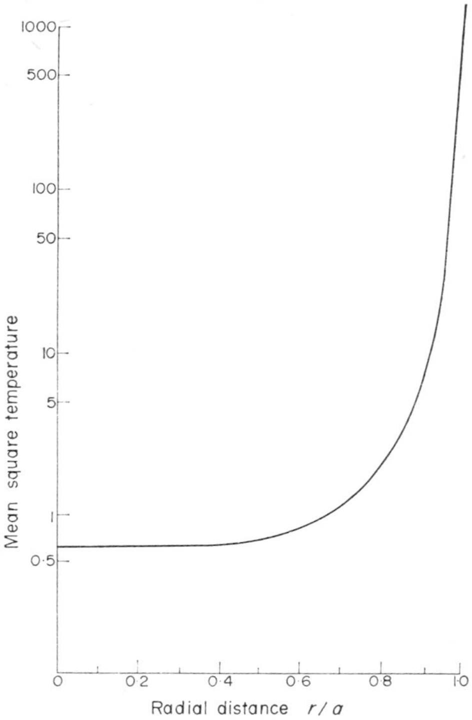

FIG. 9.3. Mean square temperature in a sphere with white noise random surface temperature. (From Samuels (1966) *J. Heat Mass Transfer*, **9**, 301.)

with an inversion theorem

$$f(r) = \frac{2}{a^2}\sum_{m=1}^{\infty}\frac{\alpha_m^2 \bar{f}(m)}{h^2+\alpha_m^2}\frac{J_0(\alpha_m r)}{J_0^2(\alpha_m a)}, \tag{9.141}$$

where α_m is defined by

$$hJ_0(\alpha_m a) - \alpha_m J_1(\alpha_m a) = 0 \quad (m = 1,2,3,\ldots). \tag{9.142}$$

Applying the transform to the equation for $\theta(r,t)$ we get

$$\frac{d\bar{\theta}(m,t)}{dt} + K\alpha_m^2\bar{\theta}(m,t) = \frac{1}{\rho C}\bar{Q}(m,t), \tag{9.143}$$

where

$$\bar{Q}(m,t) = \int_0^a rQ(r,t)J_0(\alpha_m r)\,dr. \tag{9.144}$$

Solving for initial conditions at $t = -\infty$, we get

$$\bar{\theta}(m,t) = \frac{1}{\rho C}\int_0^\infty dt_0 \exp(-K\alpha_m^2 t_0)\bar{Q}(m,t-t_0). \tag{9.145}$$

Using the inversion theorem, we find

$$\theta(r,t) = \frac{2}{a^2\rho C}\sum_{m=1}^{\infty}\frac{\alpha_m^2}{h^2+\alpha_m^2}\frac{J_0(\alpha_m r)}{J_m^2(\alpha_m a)}\int_0^\infty dt_0 \exp(-K\alpha_m^2 t_0) \times \int_0^a r'Q(r',t-t_0)J_0(\alpha_m r')\,dr' \tag{9.146}$$

from which the various correlations and mean square values can be obtained in terms of the cross-correlation function of Q.

9.8. *Burn-out and heat-flux perturbations*

The calculations described above are of interest not just from the point of view of normal operating temperature noise, but also in connection with burn-out predictions. We have already seen from eqn. (9.123) that the temperature exhibits Gaussian fluctuations: thus it is essential to know whether the probability that the temperature will exceed a prescribed level will become appreciable.

Burn-out problems are particularly important in the design of boiling-water reactors (although not exclusively so) since it is essential to maintain the heat flux below the critical value at which film boiling sets in. On the other hand, from the economic viewpoint, it is important to operate at as high a rating as possible. However, because of statistical fluctuations it cannot be guaranteed that a prescribed "safe" heat flux will remain constant. The statistical aspect enters in a number of ways, but in essentially two specific groupings: physical characteristics and process variables. Typical of the physical characteristics variables for a heterogeneous reactor would be the fabrication tolerances, e.g. outside diameter of the fuel element cladding, the concentration of the fissile isotope per unit length, the local equivalent flow diameter of the fuel element assembly in a coolant passage and the mass flow of a coolant through its channel. Process variables are heat-transfer coefficients (intimately connected with mass-flow rate), radial and axial neutron fluxes and the macroscopic fission cross-sections after irradiation. The latter must be considered a random variable, due to uncertainties in the methods employed to calculate burn-up in the fuel elements.

We shall discuss below a general procedure for dealing simultaneously with the random variables listed above. For the moment, however, it will be useful to concentrate on one specific item, namely non-uniform concentration of fissile material, and see what methods can be used to assess the local fluctuations in temperature of the fuel element surface.

9.9. *Non-uniform fuel concentration*

We shall consider this problem in terms of spatial fluctuations only. This is a valid assumption in view of the fact that the local inhomogeneities in fissile material remain constant with time (neglecting burn-up) but randomly distributed in the fuel element volume. We will further assume that the thermal conductivity is non-random and spatially constant. Thus the heat conduction equation for a cylindrical fuel element may be written as

$$k_f\left[\frac{\partial^2}{\partial r^2}+\frac{1}{r}\frac{\partial}{\partial r}+\frac{1}{r^2}\frac{\partial^2}{\partial \theta^2}+\frac{\partial^2}{\partial z^2}\right]T(r,\theta,z)=Q(r,\theta,z). \tag{9.147}$$

The heat-source term Q is given by

$$Q(r,\theta,z) = \mu N(r,\theta,z)\bar{\sigma}_f\,\phi(r,\theta,z), \tag{9.148}$$

where μ is the heat released per fission, $\bar{\sigma}_f$ is the spectrum averaged fission cross-section and $N(r,\theta,z)$ and $\phi(r,\theta,z)$ are the fissile atom number density and neutron flux at the point (r,θ,z). Note that it is essential to use a full (r,θ,z) geometry for this problem since the fuel inhomogeneities are randomly dispersed and do not therefore allow radial symmetry or axial independence to be assumed.

We now assume that Q may be written as

$$Q(r,\theta,z) = Q_0(r,z)+Q_p(r,\theta,z), \tag{9.149}$$

where Q_0 is the heat generated by the *uniformly* distributed fuel and Q_p is due to the "inclusions". These inclusions may be regarded as localized in small volumes v_s into which the element has been divided, and can be represented mathematically by delta functions in position with the appropriate heat-generating strength. Thus we write

$$Q_p(r,\theta,z) = \sum_{s=1}^{M} \epsilon_s \zeta_s \frac{\delta(r-r_s)}{r} \delta(\theta-\theta_s)\delta(z-z_s), \tag{9.150}$$

where ζ_s equals unity if an inclusion (i.e. the point r_s, θ_s, z_s) lies in v_s and zero if not. We also define ϵ_s as the amount of heat generated by the sth inclusion which is located at (r_s,θ_s,z_s). ϵ_s, r_s, θ_s and z_s are random variables. If now the heat-source term is inserted into eqn. (9.147), it is clear that the solution consists of $T_0(r,z)$, the temperature distribution arising from the uniformly distributed heat source, plus $T_p(r,\theta,z)$, the random fluctuation due to the inclusions. The calculation of $T_0(r,z)$ is straightforward, that for T_p requires further thought. It may, however, be written in terms of the Green's function of eqn. (9.147), which we call $G(r,\theta,z|r',\theta',z')$, as follows:

$$T_p(r,\theta,z) = \sum_{s=1}^{M} \epsilon_s \zeta_s G(r,\theta,z|r_s,\theta_s,z_s). \tag{9.151}$$

The corresponding heat flux at the surface, $r = R$, of the element is

$$q(R,\theta,z) = -k\frac{\partial}{\partial r} T(r,\theta,z)|_{r=R} \tag{9.152}$$

$$\equiv q_0(R,z)+q_p(R,\theta,z), \tag{9.153}$$

where

$$q_p(r,\theta,z) = -k\left.\frac{\partial T_p(r,\theta,z)}{\partial r}\right|_{r=R} \tag{9.154}$$

$$= \sum_{s=1}^{M} \epsilon_s \zeta_s H(\theta,z|r_s,\theta_s,z_s) \tag{9.155}$$

in an obvious notation.

If we assume independence of strength and location, q_p will depend on two specific probability distributions, viz.

(1) $$P_1(\zeta) = \bar{M}v_s\delta_{\zeta,1}+(1-\bar{M}v_s)\delta_{\zeta,0} \tag{9.156}$$

which is the probability that $\zeta = 1$ or zero where $\bar{M}$ is the mean number per unit volume of inclusions and we have assumed that the inclusions are randomly distributed throughout the fuel element.

(2) $$P_2(\epsilon),$$

where $P_2(\epsilon)d\epsilon$ is the probability that an inclusion will give rise to a heat source of strength $(\epsilon, \epsilon+d\epsilon)$.

Now the characteristic function (Bartlett, 1962) of a random variable is defined as

$$F_\phi(Z) = \langle e^{i\phi z}\rangle, \qquad (9.157)$$

where $\langle \ldots \rangle$ indicates a statistical average. If for ϕ we use q_p as defined by eqn. (9.155) and remember that the terms in the sum are independent variables, we have

$$F_{q_p}(Z) = \prod_{s=1}^{M} \langle \exp\{iZ\epsilon_s \zeta_s H(\theta, z|r_s, \theta_s, z_s)\}\rangle_{12}, \qquad (9.158)$$

where $\langle \ldots \rangle_{12}$ indicates averages over $P_1 P_2$.

With the average over P_1 performed, we get

$$\langle \ldots \rangle_{12} = 1 - \overline{M}v_s + \overline{M}v_s \int_0^\infty d\epsilon\, P_2(\epsilon) \exp\{iZ\epsilon H(\theta, z|r_s, \theta_s, z_s)\}. \qquad (9.159)$$

Inserting (9.159) into (9.158) and taking logarithms leads to

$$\log F_{q_p}(Z) = \sum_{s=1}^{M} \log\langle \ldots \rangle. \qquad (9.160)$$

Allowing $v_s \to r_s dr_s d\theta_s dz_s$, expanding the logarithm and changing the sum over s to a volume integral, we obtain finally

$$\log F_{q_p}(Z) = -\overline{M}\pi L R^2 + \overline{M} \int_0^L dz_s \int_0^R r_s dr_s \int_0^{2\pi} d\theta_s \int_0^\infty d\epsilon P(\epsilon) \times \exp\{iZ\epsilon H(\theta, z|r_s, \theta_s, z_s)\}. \qquad (9.161)$$

The mean value is obtained by differentiating with respect to Z and setting $Z = 0$. Then we find from (9.157)

$$\langle q_p(\theta, z)\rangle = \overline{M}\langle\epsilon\rangle \int_0^L dz_s \int_0^R r_s dr_s \int_0^{2\pi} d\theta_s H(\theta, z|r_s, \theta_s, z_s). \qquad (9.162)$$

Similarly, the variance $\sigma_q^2 = \langle q_p^2\rangle - \langle q_p\rangle^2$ is given by

$$\sigma_q^2 = \overline{M}\langle\epsilon^2\rangle \int_0^L dz_s \int_0^R r_s dr_s \int_0^{2\pi} d\theta_s H^2(\theta, z|r_s, \theta_s, z_s). \qquad (9.163)$$

If we can assume that the heat flux at the surface due to a single inclusion is always small compared with the effect of the others then the probability distribution for q_p is approximately Gaussian, viz.

$$P(q_p) \simeq \frac{1}{\sqrt{2\pi}\,.\,\sigma_q} \exp\{-(q_p - \langle q_q\rangle)^2/2\sigma_q^2\}. \qquad (9.164)$$

The probability that q_p will not exceed some prescribed level q_m is clearly

$$P(q < q_m) = \frac{1}{\sqrt{2\pi}\,.\,\sigma_q} \int_{-\infty}^{q_m} P(q_p)\, dq_p. \qquad (9.165)$$

Further examples of the technique described above may be found in Knowles and Fox (1972) together with some numerical calculations for specific systems.

9.10. *Random heat-transfer coefficient*

A further example of random heat flux which arises from surface imperfections of a heated element can be given if we assume that such imperfections are reflected in a random heat-transfer coefficient. Thus we consider a slab of width $2a$ containing a spatially uniform, non-random heat source Q_0, with the following boundary conditions:

$$-k\frac{dT(x,z)}{dx} = h_1(z)(T(x,z)-T_{c0}) \quad (x = a), \tag{9.166a}$$

$$k\frac{dT(x,z)}{dx} = h_2(z)(T(x,z)-T_{c0}) \quad (x = -a). \tag{9.166b}$$

x is the transverse dimension and z the axial one. $h_1(z)$ and $h_2(z)$ are taken to be random functions of z (due, for example, to surface roughness) and the problem is to find the properties of the corresponding random heat flux at the surface. Assuming that the heat-transfer coefficients can be written $h_1 = h_0+\delta h_1$ and $h_2 = h_0+\delta h_2$ where $|\delta h| \ll h_0$, we can linearize the boundary condition by setting $T = \langle T\rangle+\delta T$ and neglecting $\delta h\,\delta T$. We then find the following equations must be solved:

$$k\frac{d^2\langle T\rangle}{dx^2}+Q_0 = 0 \tag{9.167}$$

plus the boundary conditions $d\langle T\rangle/dx = 0$ at $x = 0$ and

$$-k\frac{d\langle T\rangle}{dx} = h_0(\langle T\rangle-T_{c0}) \quad (x = a). \tag{9.168}$$

Also we have

$$\frac{d^2\delta T}{dx^2} = 0 \tag{9.169}$$

together with

$$-k\frac{d}{dx}\,\delta T = h_0\,\delta T+\delta h_1(\langle T\rangle-T_{c0}) \quad (x = a), \tag{9.170a}$$

$$k\frac{d}{dx}\,\delta T = h_0\,\delta T+\delta h_2(\langle T\rangle-T_{c0}) \quad (x = -a). \tag{9.170b}$$

These equations are readily solved and yield

$$\langle T\rangle-T_{c0} = \frac{Q_0 a}{h_0}+\frac{Q_0}{2}(a^2-x^2), \tag{9.171}$$

$$\delta T = \frac{Q_0 a}{2h_0}\left\{\frac{\delta h_1+\delta h_2}{h_0}+\frac{(\delta h_2-\delta h_1)}{(k+ah_0)}\,x\right\}. \tag{9.172}$$

Clearly $\langle\delta T\rangle = 0$. However, on the assumption that $\langle(\delta h_1)^2\rangle = \langle(\delta h_2)^2\rangle = \langle(\delta h)^2\rangle$, we can write the mean square fluctuation of T as

$$\langle(\delta T)^2\rangle = \frac{1}{2}\left(\frac{Q_0 a}{h_0}\right)^2\left\{\frac{1}{h_0^2}+\frac{x^2}{(k+ah_0)^2}\right\}\langle(\delta h)^2\rangle. \tag{9.173}$$

At the surface $x = a$ this becomes

$$\langle(\delta T)^2\rangle = (\langle T\rangle-T_{c0})^2\,\frac{1}{2}\left\{1+\frac{a^2h_0^2}{(k+ah_0)^2}\right\}\frac{\langle(\delta h)^2\rangle}{h_0^2}. \tag{9.174}$$

Similarly, the heat flux
$$q = -k\frac{dT}{dx} \tag{9.175}$$
may be written, at $x = a$,
$$q = \langle q\rangle - k\frac{Q_0 a}{h_0}\frac{(\delta h_2-\delta h_1)}{2(k+ah_0)} \tag{9.176}$$
or
$$\langle q^2\rangle-\langle q\rangle^2 = \tfrac{1}{2}(\langle T\rangle - T_{c0})^2\frac{k^2\langle(\delta h)^2\rangle}{(k+ah_0)^2}. \tag{9.177}$$

From this expression we can obtain an estimate of the fluctuations and note the parameters of importance in promoting them.

9.11. *General formulation for surface temperature fluctuations*

Abernathy (1961) has given a useful general approach which enables the effects of a number of fluctuating variables to be assessed as far as their influence on the fuel element surface temperature is concerned. Thus, if we write the surface temperature $T_s = T_s(u_1, u_2, u_3, \ldots, u_N)$ as a function of the random variables u_i (physical or process variables), then we can expand T_s about the value it takes when $u_i = \bar{u}_i$, their average values, viz.

$$T_s(u_1,u_2,u_3,\ldots,u_N) = T_s(\bar{u}_1+\Delta u_1, \bar{u}_2+\Delta u_2, \bar{u}_3+\Delta u_3, \ldots, \bar{u}_N+\Delta u_N), \tag{9.178}$$
$$\simeq T_0 + \sum_{i=1}^{N}\overline{\frac{\partial T_s}{\partial u_i}}\Delta u_i \tag{9.179}$$
$$\equiv T_0 + \sum_{i=1}^{N} a_i \Delta u_i, \tag{9.180}$$

where $T_0 = T_s(\bar{u}_1, \bar{u}_2, \bar{u}_3, \ldots, \bar{u}_N)$ and we have assumed that $|\Delta u_i| \ll \bar{u}_i$.

The variance in u_i is $\langle(\Delta u_i)^2\rangle = \sigma_i^2$, so that for independent random variables we can write the variance of T_s as follows:

$$\sigma_s^2 = \langle(T_s - T_0)^2\rangle \equiv \langle\theta^2\rangle \tag{9.181}$$
$$= \sum_{i=1}^{N} a_i \sigma_i^2. \tag{9.182}$$

Even if the individual variables u_i are not Gaussian, it may be shown, via the Central Limit Theorem, that for N large the distribution of θ will approach Gaussian. In view of this fact we may write

$$P(\theta) = \frac{1}{\sqrt{2\pi}\,.\,\sigma_s}\exp\left(-\frac{\theta^2}{2\sigma_s^2}\right) \tag{9.183}$$

and hence calculate $P(\theta < \theta_m)$.

To illustrate how this method might be employed in practice, consider the temperature distribution at the surface of the cladding in a gas-cooled reactor with the simple configuration shown in Fig. 9.4. Assuming an axial heat source of cosine shape $Q_0 \cos Bz$, a coolant mass flow rate $\dot{m}$ and a heat-transfer coefficient h, we can easily write down the cladding surface temperature at any height z (El-Wakil, 1971), namely

$$T_{\text{cl}}(z) = T_{\text{ci}} + A_f Q_0\left[\frac{1}{B\dot{m}C_p}\left(\sin Bz + \sin\frac{BH}{2}\right) + \frac{1}{hG}\cos Bz\right], \tag{9.184}$$

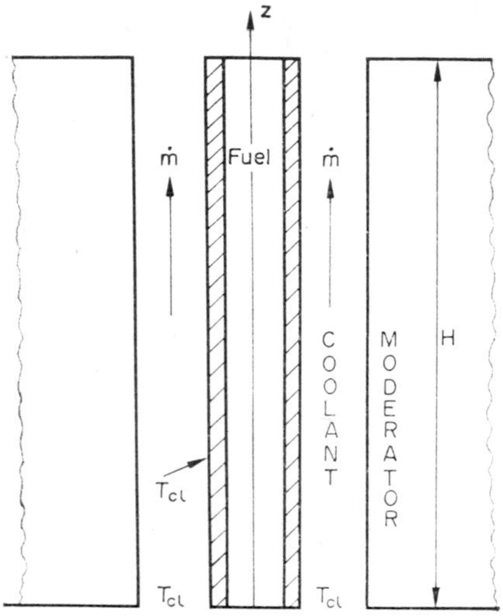

FIG. 9.4. Typical fuel channel in graphite moderated-gas cooled reactor. $T_{cl}(z)$ is the surface temperature of the cladding (shaded area) at height z.

where T_{ci} is the inlet coolant temperature, A_f is the area of the fuel region, C_p is the specific heat of the coolant and C is the circumference of the cladded fuel. In addition, we have a "correlation" for the heat-transfer coefficient which can be typically written (El-Wakil, 1971) as

$$\frac{hD_e}{k} = 0{\cdot}023 \left(\frac{\dot{m}D_e}{A_c \mu}\right)^{0{\cdot}8} \left(\frac{C_p \mu}{k}\right)^{0{\cdot}4}, \tag{9.185}$$

where A_c is the coolant flow area, μ the viscosity of the coolant and D_e the equivalent flow diameter equal to $4A_c$/wetted perimeter. Each of the parameters in $T_{cl}(z)$ has a fluctuation and, by using the formalism given above, the net effect can be obtained. However, it should be noted that separate experiments or calculations may be required to find the individual variances. Abernathy (1961) applies this technique to temperature fluctuations in the Experimental Gas-cooled Reactor at Oak Ridge (ORNL-2500, 1958).

9.12. *Burn-out in boiling-water reactors*

There appears to be little basic theory regarding this topic, with most of the engineering calculations based upon empirical results. Those by L. S. Tong (1968) are considered to be fairly reliable and have been obtained by a series of painstaking experiments to detect the departure from nucleate boiling (DNB) for a given heat flux.

The results are given in the form

$$P(BO) = \frac{1}{\sqrt{2\pi}\,.\,\sigma_1} \int_{-\infty}^{\eta_0} \exp\{-(\eta' - a)^2/2\sigma_1^2\}\, d\eta', \tag{9.186}$$

where a and σ_1 are obtained from experiments for various geometric and surface conditions. $P(BO)$ is the probability of a burn-out event for a prescribed minimum local burn-out heat flux margin, in this case defined as η_0^{-1}.

It is clear that to go beyond this semi-empirical representation will require a detailed statistical knowledge of fuel-element surface conditions and bubble behaviour in the neighbourhood of a heated surface. Such an investigation has yet to be carried out successfully.

9.13. *Thermally induced random stress fluctuations*

It is well known that when materials are heated they change their dimensions unless they are in some way constrained. If the heated element is constrained then stresses will be set up within it: we call these thermal stresses and a full discussion may be found in Timoshenko and Goodier (1951) and also in Smith (1956).

As we have shown in previous sections, a reactor operating at steady state is subject to random temperature fluctuations both in time and space. We may conclude, therefore, that these give rise to random stresses in the various core components. It follows then, by analogy with random vibrations, that these temperature fluctuations can lead to structural fatigue and possible failure according to the Miner criterion. In this section we shall briefly indicate the method for calculating random thermal stress and show how the results can be used to calculate damage.

If it is assumed that there is no axial variation of temperature and that the temperature is radially symmetric, we can write for the three components of stress S_r, S_θ, S_z in a solid cylinder of radius a the following expressions:

$$S_r = \frac{\alpha E}{1-\nu}\left(\frac{1}{a^2}\int_0^a T(r')r'\,dr' - \frac{1}{r^2}\int_0^r T(r')r'\,dr'\right), \tag{9.187}$$

$$S_\theta = \frac{\alpha E}{1-\nu}\left(\frac{1}{a^2}\int_0^a T(r')r'\,dr' + \frac{1}{r^2}\int_0^r T(r')r'\,dr' - T(r)\right), \tag{9.188}$$

$$S_z = \frac{\alpha E}{1-\nu}\left(\frac{2}{a^2}\int_0^a T(r')r'\,dr' - T(r)\right), \tag{9.189}$$

where α is the linear coefficient of expansion and ν is Poisson's ratio.

It was assumed in the derivation of these stresses that the axial displacement of the rod was zero. On the other hand, the radial displacement, u, can be deduced from

$$u = \frac{1+\nu}{1-\nu}\alpha\left[(1-2\nu)\frac{r}{a^2}\int_0^a T(r')r'\,dr' + \frac{1}{r}\int_0^r T(r')r'\,dr'\right]. \tag{9.190}$$

To find the time dependence of the stress corresponding to random temperature fluctuations, let us use eqn. (9.146). Then for $S_z(r,t)$ we find

$$S_z(r,t) = \frac{\alpha E}{1-\nu}\frac{2}{\rho C a^2}\sum_{m=1}^{\infty}\frac{\alpha_m^2}{h^2+\alpha_m^2}\frac{1}{J_0^2(\alpha_m a)}\left\{\frac{2}{a\alpha_m}J_1(\alpha_m a) - J_0(\alpha_m r)\right\}$$
$$\times\int_0^\infty dt_0\exp(-K\alpha_m^2 t_0)\int_0^a r'\,Q(r', t-t_0)J_0(\alpha_m r')\,dr', \tag{9.191}$$

from which the mean square stress $\sigma_z^2(r)$ can be found. Similarly for the other components. Using the maximum values of the stresses the time to failure can be found as outlined in Section 9.4.

Transient thermal stress does not seem to have received a great deal of attention in connection with reactor component failure, but useful pioneering work may be found in the papers of Murray and Young (1944), Murray (1945), Clark (1950), Stein (1957) and Karush (1944). A general discussion on steady-state thermal stresses with passing reference to transient and fatigue problems is given by Thompson and Rodgers (1956).

9.14. *Bubble formation and diffusion*

"...There has certainly been a great deal of effort put into understanding boiling. If you consider noise in boiling and various possible geometries, void distributions can be extraordinarily complicated. I have spent a good deal of thought on it, and people at M.I.T. in their boiling laboratory have done a good deal of work. To correlate the statistics of noise with the statistics of the bubble configuration would be an extraordinarily difficult job...."

This statement was made by Garrett Birkhoff at a conference on "Industrial Needs and Academic Research in Reactor Kinetics" held at Brookhaven National Laboratory in 1968 (BNL-50117-T-497). Faced with this statement by such an eminent applied mathematician one cannot help being pessimistic about the possibility of an easy solution to the problem of bubble behaviour and its influence on reactor dynamics. However, whilst the difficulty of the problem is not in dispute, there are many ingenious approximate methods that have been developed for gaining understanding of boiling. It is, moreover, fortunate that most reactor transient effects are not too sensitive to the detailed bubble behaviour but can be described adequately by a lower level of description. We shall discuss below some of these methods and indicate their relative merit as far as reactor noise studies are concerned.

It is convenient to separate the problem into two parts: (1) examination of the individual bubble dynamics and (2) the behaviour of bubbles *en masse*. In order to understand how these parts combine to form a quantity of relevance to kinetic behaviour, let us consider the density of a liquid containing an ensemble of bubbles. The mixture density of steam and liquid in a boiling system can be written

$$\rho_m(t) = \rho_v\alpha(t)+\rho_l(1-\alpha(t)), \tag{9.192}$$

where ρ_v and ρ_l are the densities of vapour and liquid, respectively, and α is the void fraction, being equal to the ratio of volume of vapour to total volume.

Rewriting (9.192), we have

$$\rho_m(t) = \rho_l+\alpha(t)(\rho_v-\rho_l). \tag{9.193}$$

Under steady boiling conditions the fraction of space occupied by the vapour is

$$\alpha(t) = \tfrac{4}{3}\pi \sum_i N_i(t) R_i^3(t), \tag{9.194}$$

where $N_i(t)$ is the density (number per unit volume) of bubbles of radius $R_i(t)$. In view of the random nature of boiling both N_i and R_i are time-dependent. Indeed there should in principle be a space-dependent coordinate in N_i since the system is not necessarily homogeneous in bubble distribution. From eqn. (9.194) it is clear that both R_i and N_i must be studied. In principle, these two parameters interact; however, for the sake of simplicity we will assume that the time-dependent behaviour of a bubble in a mixture can be obtained by examining an isolated bubble.

9.15. *Dynamics of a single bubble*

We shall consider the rate of growth of a single bubble in a superheated liquid since this is a crucial factor in determining the density changes and heat transfer in a boiling system. This problem has occupied physicists and mathematicians for many years, both theoretically and experimentally. Theoretical studies have dealt with the necessary conditions for the formation of a bubble nucleus and the formulation and solution of equations which describe the time-dependent behaviour of the newly created bubble of finite size. Experimental work has centred around high-speed photographic methods for examining boiling and more recently in the use of laser beams to calculate bubble-size distributions.

It is the view of Frenkel (1946) and others that a bubble originates from a nucleation site within the liquid. These nucleation sites can be very small bubbles of entrained gas or impurities in the form of minute particles. The absence of such centres would make it necessary to overcome very large intermolecular forces to initiate boiling thus permitting a very high degree of superheat to be attained: this situation has been demonstrated in very pure water systems. The fact that, in practice, high superheats are not attained indicates that the processes mentioned above are the most likely causes of bubble formation.

The problem of describing the growth of a bubble, once formed, was first studied in depth by Lord Rayleigh (1917) in the incompressible, inviscid flow approximation. He obtained an equation which bears his name, viz.

$$R\ddot{R}+\tfrac{3}{2}\dot{R}^2 = \frac{P_L(R)-P_\infty}{\rho_l}, \tag{9.195}$$

where P_∞ is the system pressure and $P_L(R)$ is the pressure in the liquid at the bubble wall.

Equation (9.195) involves some incomplete physical assumptions and was extended by Plesset (1949) to include the effect of surface tension by using the relation

$$P_L(R) = P_v - \frac{2\sigma}{R}, \tag{9.196}$$

where P_v is the vapour pressure in the bubble and σ the surface tension. The *asymptotic* solution of (9.196) for constant P_v using this correction leads to

$$R(t) \sim R(0) + \left(\frac{2}{3}\frac{P_v - P_\infty}{\rho_l}\right)^{\frac{1}{2}} t \tag{9.197}$$

which does not accord with experiment (Dergarabedian, 1953).

Further developments in this field recognized that the temperature difference between liquid outside and vapour inside would be an important factor in bubble growth. It was further noted that the temperature at which the process of boiling can commence must be greater than the saturation temperature T_∞ corresponding to the external system pressure P_∞. If we consider an embryonic bubble with radius R_0, and initial superheat temperature T_0, then T_0 must exceed T_∞ by an amount which corresponds to the excess value of P_{v_0} compared with P_∞, where P_{v_0} is the vapour pressure corresponding to T_0. This pressure difference is clearly $2\sigma/R_0$. Because of these factors,

Plesset and Zwick have shown that the temperature at the liquid–vapour interface decreases and thereby reduces bubble growth rate.

The effect of a changing interface temperature means that with the Plesset correction for surface tension the bubble radius is given by

$$R\ddot{R}+\tfrac{3}{2}\dot{R}^2+\frac{2\sigma}{\rho_l R} = \frac{P_v(t)-P_\infty}{\rho_l}, \tag{9.198}$$

where $P_v(t)$ is an unknown functional relationship due to the effect of the moving boundary.

For low superheat and neglecting spatial variation of temperature within the bubble the Clausius–Clapeyron equation can be used, viz.

$$\frac{dP}{dT} = \frac{L}{T(V_v-V_l)}, \tag{9.199}$$

where L is latent heat and V_v and V_l are the specific volumes of vapour and liquid, respectively.

Integrating (9.199) we find

$$P_v(t)-P_\infty = \frac{L}{T(V_v-V_l)}(T_v-T_\infty). \tag{9.200}$$

This equation has to be supplemented by an expression for the interface temperature which Greenfield *et al.* (1954) show to be

$$T_v-T_0 = -\frac{L\rho_v}{C_l\rho_l(\pi a_l)^{\frac{1}{2}}}\int_0^t \frac{R(t')\dot{R}(t')}{R(t)\sqrt{(t-t')}}\,dt' \tag{9.201}$$

(a_l = thermal diffusivity of liquid).

This leads finally to a non-linear integro-differential equation for bubble growth, viz.

$$R\ddot{R}+\tfrac{3}{2}\dot{R}^2+\frac{2\sigma}{\rho_l R} = \frac{L}{\rho_l T(V_v-V_l)}\left[T_0-T_\infty-\frac{L\rho_v}{C_l\rho_l(\pi a_l)^{\frac{1}{2}}}\int_0^t \frac{R(t')\dot{R}(t')}{R(t)\sqrt{(t-t')}}\,dt'\right]. \tag{9.202}$$

For a uniformly distributed heat source we must add to the terms in square brackets the quantity

$$\frac{1}{\rho_l C_l}\int_0^t Q(t')\,dt'. \tag{9.203}$$

It may also be necessary to take into account the generation of new bubbles from surfaces and nucleation sites as the temperature of the system increases.

Numerical solutions of the integro-differential equation have been made and, with experimental distributions for N_i, the calculated density changes are in reasonable agreement with experiment (Greenfield *et al.*, 1954).

We also mention the work of Flatt (1960) who has examined the rate of formation of bubbles in a homogeneous reactor caused by the diffusion of radiolytic gases formed in the reactor. In this work a diffusion equation is set up for the concentration of dissolved gases and this in turn determines the rate at which bubbles grow due to diffusion through the vapour–liquid interface. The question of initial bubble formation is, however, not considered.

9.16. *Diffusion of bubbles in heated channels*

Let us consider now the behaviour of bubbles as a statistical assembly with the assumption that $R(t)$ itself is not of direct interest but rather the fraction of voids at time t that arise from bubbles of radius $R(t)$.

We follow Houghton (1961) in this respect and treat the problem as a special case of Brownian motion. Thus, in normal operation, the liquid in a heated channel is in turbulent motion and we consider therefore a bubble as a free particle maintained in a constant irregular motion by the turbulent eddies in the liquid. Superimposed upon this random agitation is the directional effect of buoyancy and mass notion of the liquid phase through the channel.

Using the methods developed by Langevin and extended by Chandrasekhar (Wax, 1954), we can write down a Langevin equation for $\mathbf{W}_R(t)$, the velocity of a bubble of radius R. Using a simple equation of motion we find

$$m\frac{d\mathbf{W}_R}{dt} = -\beta[\mathbf{W}_R - \mathbf{U}(\mathbf{r}, t)] + m[\mathbf{A}_R(t) + \mathbf{K}_R(\mathbf{r}, t)], \tag{9.204}$$

where (1) $\beta[\mathbf{W}_R - \mathbf{U}]$ is the dynamical friction term according to Stokes's law, $\mathbf{U}$ being the bulk liquid velocity; (2) $\mathbf{K}_R$ is the external force term due to gravity; (3) $\mathbf{A}_R$ is a rapidly fluctuating contribution arising from the effect of the turbulent eddies on bubble motion; it is analogous to the random molecular impacts experienced by a particle undergoing true Brownian motion due to molecular impacts.

Finally, we note that m is not the true mass of the bubble as would be the case for solid particles, but due to the buoyancy effect is given by

$$m = (\rho_v + \chi\rho_l)\tfrac{4}{3}\pi R^3. \tag{9.205}$$

χ is a fraction of the mass of fluid displaced and, for water, lies in the range 0·46–0·59 (Dryden *et al.*, 1956). From this we may conclude that vapour bubbles have a much greater inertia than might be anticipated from considerations of the mass of vapour alone.

We now assume that the motion of the bubbles can be represented as a Markoff process and apply the equation deduced by Chandrasekhar (1943, p. 41) for the probability distribution, corresponding to eqn. (9.204), viz.

$$\frac{\partial C_R(\mathbf{r}, t)}{\partial t} = \nabla . \left\{\frac{k\theta(\mathbf{r}, t)}{\beta}\nabla C_R(\mathbf{r}, t) - \left[\mathbf{U}(\mathbf{r}, t) + \frac{m}{\beta}\mathbf{K}_R(\mathbf{r}, t)\right] C_R(\mathbf{r}, t)\right\}$$

$$\equiv -\nabla . \mathbf{J}_R(\mathbf{r}, t), \tag{9.206}$$

where $C_R(\mathbf{r}, t)d\mathbf{r}$ is the probability of finding a bubble of radius R in the volume element $d\mathbf{r}$ about $\mathbf{r}$ at time t. We assume that $d\mathbf{r}$ is sufficiently large to make this statement meaningful. An alternative interpretation of C_R is that it represents the fraction of voids arising from bubbles of radius R. However, again it is necessary to assume that the bubbles are relatively small for this interpretation to be valid—remember that the Brownian particles are considered as point masses, whereas bubbles can be macroscopically large. This is a defect of the theory but, as we shall see, does not prevent it from yielding some useful conclusions.

The bulk liquid velocity at any point is related to the entrance velocity $\mathbf{u}$ by the relation

$$\mathbf{U} = \frac{\mathbf{u}}{1-\alpha(\mathbf{r},t)}, \tag{9.207}$$

where

$$\alpha(\mathbf{r},t) = \sum_R C_R(\mathbf{r},t) \tag{9.208}$$

is the total void fraction for all bubble radii.

It is also convenient to define the bubble slip velocity $\mathbf{v}_R$ due to buoyancy as $m\mathbf{K}_R/\beta$ and the bubble eddy diffusivity as $E_B = k\theta/\beta$ (where θ is the absolute temperature of the liquid). Equation (9.206) can be interpreted as a mass balance which accounts for the effects of eddy diffusion, convection and gravitation. However, the volume of the vapour phase may be increasing due to nucleation and individual bubble growth, thus a more appropriate balance equation can be written as

$$\frac{\partial C_R}{\partial t} = -\nabla.\mathbf{J}_R + p_R + q_R, \tag{9.209}$$

where $\mathbf{J}_R$ is the flux vector defined above and q_R denotes the fractional volumetric growth or collapse rate of bubbles of radius R after leaving the channel wall. The term p_R is the rate of creation of bubble nuclei of radius R in the bulk superheated liquid.

Since eqn. (9.209) applies to bubbles of any radii, we can sum it directly over R to obtain

$$\frac{\partial \alpha}{\partial t} = \nabla.\left(E_B\nabla\alpha - \frac{\mathbf{u}\alpha}{1-\alpha} - \mathbf{v}\alpha\right) + p + q, \tag{9.210}$$

where $\mathbf{v}\alpha \equiv \Sigma_R \mathbf{v}_R C_R$, $\mathbf{v}$ being the slip velocity and p and q are self-evident.

A complete description of the system requires a further equation representing heat transfer to the liquid phase. We can write this in terms of a heat balance as follows:

$$\rho_l C_p \frac{\partial \theta}{\partial t} = -\nabla.\mathbf{q}_H - \rho_v L(p+q). \tag{9.211}$$

The left-hand side is the rate of change of heat in the liquid phase, the first term on the right-hand side is the net heat flow into the system due to diffusion and the second term arises from heat loss due to the conversion of water into steam, L being the latent heat of vaporization. The heat flux vector $\mathbf{q}_H$ is given in a partially empirical manner as

$$\mathbf{q}_H = -E_W\nabla\theta + \frac{\mathbf{u}}{1-\alpha}(\theta-\theta_s) - \mathbf{v}\alpha(\theta-\theta_s), \tag{9.212}$$

where θ_s is the saturation temperature and E_W is the turbulent diffusivity of water. The first term denotes heat flow from turbulent diffusion, the second from mass motion of the coolant and the third from buoyancy.

Equations (9.210) and (9.211) are coupled non-linear partial differential equations which, given $\mathbf{U}(\mathbf{r},t)$, E_B, E_W, $\mathbf{v}$, p and q, could be solved numerically. These data might be difficult to obtain with sufficient accuracy to warrant such a detailed calculation. In view of this fact the sensible view has been put forward by Houghton that the equations be linearized and the solutions obtained for small void fractions. To fix ideas we consider a steady-state two-dimensional problem of a heated channel, as

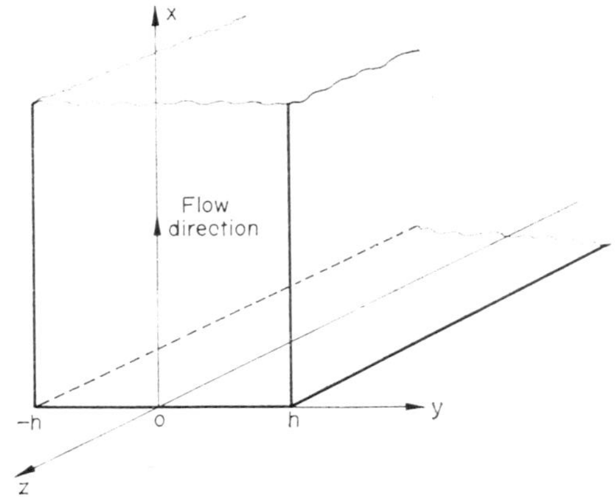

FIG. 9.5. Schematic diagram of a heated channel. Coolant flows in through the base at $x = 0$, with a profile $u(y)$, and is bounded by the heated surfaces $y = \pm h$. The system is considered infinite in the z-direction and semi-infinite in $x \geqslant 0$.

shown in Fig. 9.5. The liquid enters the base in the x-direction with a profile $u_x = u(y)$, $u_y = 0$, $u_z = 0$. The gravitational field is unidirectional and we can set $v_y = v_z = 0$ and

$$v_x = v = \sum_R \frac{(\rho_l - \rho_v) mgE_B}{(\rho_v + \chi\rho_l)k\theta} \tag{9.213}$$

(see Chandrasekhar, 1943, p. 59).

In addition we assume that the liquid phase is incompressible, i.e. $\nabla . \mathbf{u} = 0$, and also that turbulent diffusion in the x-direction is negligible compared with the dispersion produced by mass flow. With these restrictions, eqn. (9.210) becomes

$$\left[\frac{u(y)}{(1-\alpha)^2} + v\right]\frac{\partial\alpha}{\partial x} = \frac{\partial}{\partial y}\left(E_B \frac{\partial\alpha}{\partial y}\right) + p + q. \tag{9.214}$$

Similarly, the equation for heat transport is

$$\frac{\partial}{\partial x}\left[\left(\frac{u(y)}{1-\alpha} - v\alpha\right)T\right] = \frac{\partial}{\partial y}\left(E_W \frac{\partial T}{\partial y}\right) - \frac{\rho_v L}{\rho_l C_p}(p+q), \tag{9.215}$$

where $T = \theta - \theta_s$.

Linearization is accomplished by ensuring that $\alpha \ll 1$, neglecting velocity changes in both the liquid and vapour phases and assuming isotropic turbulence. Then for $v \leqslant u$, eqns. (9.214) and (9.215) become

$$(u+v)\frac{\partial\alpha}{\partial x} = E_B \frac{\partial^2\alpha}{\partial y^2} + p + q, \tag{9.216}$$

$$u\frac{\partial T}{\partial x} = E_W \frac{\partial^2 T}{\partial y^2} - \frac{\rho_v L}{\rho_l C_p}(p+q). \tag{9.217}$$

For initial and boundary conditions we use:

at $\quad x = 0, \quad y \geqslant 0;$

$$\alpha = 0, \quad T = \theta_i - \theta_s \equiv T_i;$$

at
$$y = 0, \quad x \geqslant 0;$$
$$\frac{\partial \alpha}{\partial y} = 0, \quad \frac{\partial T}{\partial y} = 0;$$
at
$$y = h, \quad x \geqslant 0;$$
$$E_B \frac{\partial \alpha}{\partial y} = N(x),$$
$$E_W \rho_l C_p \frac{\partial T}{\partial y} = Q(x) - \rho_v LN(x).$$

$N(x)$ is the rate of bubble production on the wall and $Q(x)$ is the corresponding heat flux due to conduction in the liquid.

On the premise that experimental void fractions cannot be measured to better than $\pm 0{\cdot}05$, Houghton estimates that the above linearization should not lead to serious error for void fractions less than 0·3.

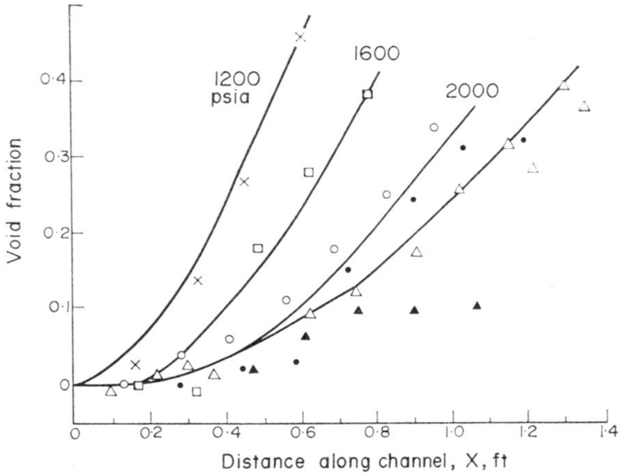

FIG. 9.6. Comparison of theoretical and experimental void fraction distributions in heated channels. The full lines are theoretical results and the circles, crosses, etc., are the experiments: ○, △, □, × are from Maurer (1960); ●, ▲ are from Egen *et al.* (1957). The complete figure is from Houghton (1961), *Nucl. Sci. Engng*, **11**, 121.

We shall not continue Houghton's analysis here since it may be found readily in the reference cited. It is sufficient to say that the above equations have been integrated with approximate, but physically acceptable, forms for Q, N, p and q. The results for the average void fraction across the system

$$\alpha_m(x) = \frac{1}{h} \int_0^h \alpha(x, y)\, dy$$

have been compared with experiment at various heights up the heated channel and very reasonable agreement with experiment is claimed. Some typical theoretical results are shown in Fig. 9.6 compared with the experiments of Egen *et al.* (1957) and Maurer (1960). To obtain this fit, two constants appearing in the theory must be adjusted; nevertheless, the theoretical result is far from a "correlation" since it accounts adequately for the variation of $\alpha_m(x)$ up to $\alpha_m = 0{\cdot}4$ and $x/L < 0{\cdot}7$, L being the length of the test section.

It would seem that further studies using the Fokker–Planck approach described above could prove profitable, particularly if time-dependent studies were included.

9.17. *Phase-space representation of a bubble cloud*

Hulburt and Katz (1964) have applied the statistical mechanical approach to the behaviour of suspended particles and derived an equation analogous to that of Liouville. An extension of this idea to bubbles has been proposed by Sandervag (1971). His argument goes as follows. Let the individual bubble be characterized by a number of coordinates; for a spherical bubble one coordinate would be the radius, R, which is a measure of the linear extension. The external coordinates are those of spatial location, viz. $\mathbf{x} = (x_1, x_2, x_3)$.

In the phase space defined by the four coordinates $(R, \mathbf{x})$ each instantaneous bubble is represented by a point and the *state* of the system evolves in time along a trajectory determined by the velocity in phase space. These velocities are $\dot{X}_i(R, \mathbf{x}, t)$ and $\dot{R}(R, \mathbf{x}, t)$. If $d\tau = dR\,dx_1\,dx_2\,dx_3$ is a volume element in phase space, we define $p(R, \mathbf{x}, t)$ as the number of bubbles with phase-space coordinates $(R, \mathbf{x})$ in $d\tau$ at time t. If $h(R, \mathbf{x}, t)$ is the net number of bubbles introduced into the system per unit time in $d\tau$ with phase-space coordinates $(R, \mathbf{x})$ then the differential equation for $p(R, \mathbf{x}, t)$ is

$$\frac{\partial}{\partial t}p(R,\mathbf{x},t)+\frac{\partial}{\partial R}\{\dot{R}(R,\mathbf{x},t)p(R,\mathbf{x},t)\}+\sum_{i=1}^{3}\frac{\partial}{\partial x_i}\{\dot{X}_i(R,\mathbf{x},t)p(R,\mathbf{x},t)\} = h(R,\mathbf{x},t). \tag{9.218}$$

This equation is basically very formal since most of the parameters in it are unknown. $\dot{R}$ should be available from the Rayleigh equation and its modification, and $\dot{X}_i$ can be calculated from a knowledge of the liquid velocity and properties. Similarly, $h(R, \mathbf{x}, t)$ is dependent on the flow properties of the system, the detachment of bubbles from the wall, the bubble generation in the bulk fluid, the coalescence of bubbles and bubble break-up. A formidable effort is required to calculate any one of these sources and therefore the phase-space approach is to be regarded as an idealized goal rather than a practical approach. Nevertheless, simple model approximations based on the Liouville equation may shed some light on important mechanisms involved in bubble dynamics.

References

ABERNATHY, F. H. (1961) The statistical aspects of nuclear reactor fuel element temperature, *Nucl. Sci. Engng*, **11**, 290.

BAKEWELL, H. P. (1962) Wall pressure correlations in turbulent pipe flow, p. 45, AD-283683, U.S. Naval Underwater Sound Laboratory.

BAKEWELL, H. P. (1964) Narrow band investigations of the longitudinal space-time correlation function in turbulent air flow, *J. Acoust. Soc. Am.* **36**, 146.

BAROUDI, M. Y. (1963) An experimental investigation of turbulence excited panel vibration and noise (boundary-layer noise), p. 19, AGARD-465, Paris.

BARTLETT, M. S. (1962) *An Introduction to Stochastic Processes*, Cambridge University Press.

BOGDANOFF, J. L. and GOLDBERG, J. E. (1960) On the Bernoulli–Euler beam theory with random excitation, *J. Aerospace Sci.* **27**, 371.

BOURGAINE, A. (1970) Methods of computing structural response to a random environment, *La Recherche Aérospatiale*, **6**, 301.
BURGREEN, D. *et al.* (1958) Vibration of rods induced by water in parallel flow, *Trans. ASME*, **80**, 991.
CHANDRASEKHAR, S. (1943) Stochastic problems in physics and astronomy, *Rev. Mod. Phys.* **15**, 1.
CLARK, MELVILLE (1950) Temperature of and stresses in cylindrical fuel elements during pile flashes. Application to Brookhaven reactor, USAEC, BNL-86.
CORCOS, G. M. (1962) Pressure fluctuations in shear flows, pp. 65–66, Univ. of California, Berkeley, Inst. of Engineering Research, Series No. 193, Issue No. 2.
CRANDALL, S. H. and YILDIZ, A. (1962) Random vibration of beams, *J. Appl. Mech.* (June), p. 267.
CRANDALL, S. H. *et al.* (1962) The variance in Palmgren–Miner damage due to random vibration, *Proc. 4th U.S. Nat. Cong. Appl. Math. (ASME)*, p. 119.
DERGARABEDIAN, P. (1953) The rate of growth of vapour bubbles in superheated water, *J. Appl. Mech.* **20**, 537.
DRYDEN, H. L. *et al.* (1956) *Hydrodynamics*, Dover, N.Y.
EGEN, R. A. *et al.* (1957) Vapour formation and behaviour in boiling heat transfer, Battelle Memorial Institute Report, BMI-1163.
EL-WAKIL, M. M. (1971) *Nuclear Heat Transport*, International Textbook Co., Penn.
FLATT, H. P. (1960) Transient bubble growth in a homogeneous reactor, Atomic International Rep. NAA-SR-3923.
FRENKEL, J. (1946) *Kinetic Theory of Liquids*, Oxford University Press.
GORMAN, D. J. (1969*a*) The role of turbulence in the vibration of reactor fuel elements in liquid flow, AECL-3371.
GORMAN, D. J. (1969*b*) An analytical and experimental investigation of reactor fuel element vibration. in two-phase parallel flow, AECL-CRNL-339.
GREENFIELD, M. L. *et al.* (1954) Studies on density transients in volume-heated boiling systems, USAEC, AECU-2950.
HOUGHTON, G. (1961) Some theoretical aspects of vapour void formation in heated vertical channels, *Nucl. Sci. Engng*, **11**, 121.
HULBURT, H. M. and KATZ, S. (1964) Some problems in particle technology, *Chem. Eng. Sci.* **19**, 555.
KARUSH, W. (1944) Temperature rise in a pitted surface, USAEC, AECD-2961.
KNOWLES, J. B. and FOX, P. F. (1972) Local heat flux variations and burn-out with heterogeneous nuclear reactor fuel, UKAEA, AEEW-R748.
KRAICHNAN, R. H. (1956*a*) Pressure field within homogeneous anisotropic turbulence, *J. Acoust. Soc. Am.* **28**, 64.
KRAICHNAN, R. H. (1956*b*) Pressure fluctuations in turbulent flow over a flat plate, *J. Acoust. Soc. Am.* **28**, 378.
LAWSON, J. D. and UHLENBECK, G. E. (1953) *Threshold Signals*, McGraw-Hill, N.Y.
LIEPMANN, H. W. (1952) On the application of statistical concepts to the buffeting problem, *J. Aero. Sciences*, **19**, 793.
LYON, R. H. (1956*a*) Propagation of correlation functions in continuous media, *J. Acoust. Soc. Am.* **28**, 76.
LYON, R. H. (1956*b*) Response of strings to random noise fields, *J. Acoust. Soc. Am.* **28**, 391.
LYON, R. H. (1960*a*) On the vibration statistics of a randomly excited hard-spring oscillator, *J. Acoust. Soc. Am.* **32**, 716.
LYON, R. H. (1960*b*) Response of a non-linear string to random excitation, *J. Acoust. Soc. Am.* **32**, 953.
LYON, R. H. (1961) On the vibration statistics of a randomly excited hard-spring oscillator, II, *J. Acoust. Soc. Am.* **33**, 1395.
MAIDANIK, G. (1961) Use of delta function for the correlations of pressure fields, *J. Acoust. Soc. Am.* **33**, 1598.
MAURER, G. W. (1960) A method of predicting steady state boiling vapour fractions in reactor coolant channels, *Bettis Technical Review*, USAEC, Report WAPD-BT-19, pp. 59–70.
MIDDLETON, D. (1960) *An Introduction to Statistical Communication Theory*, McGraw-Hill, N.Y.
MILES, J. W. (1954) On structural fatigue under random loading, *J. Aero. Sciences*, **21**, 753.
MINER, M. A. (1945) Cumulative damage in fatigue, *J. Appl. Mech.* **12**, A159.
MURRAY, F. H. (1945) Thermal stresses and strains in a finite cylinder with no surface forces, USAEC, AECD-2966.
MURRAY, F. H. and YOUNG, G. (1944) On thermal stresses, strains and warping, USAEC, AECD-2960.
NICHOLSON, R. B. (1958) A high frequency reactor instability mechanism, *Nucl. Sci. Engng*, **3**, 620.

PAIDOUSSIS, M. P. (1965) The amplitude of fluid induced vibration of cylinders in axial flow, AECL-2225.

PAIDOUSSIS, M. P. (1966) Dynamics of flexible slender cylinders in axial flow, Part I. Theory, *J. Fluid Mech.* **26**, 717.

PAIDOUSSIS, M. P. (1969) An experimental study of vibration of flexible cylinders induced by nominally axial flow, *Nucl. Sci. Engng*, **35**, 127.

PAIDOUSSIS, M. P. (1974) The dynamical behaviour of cylindrical structures in axial flow, *Ann. Nucl. Sci. Engng.* (formerly *J. Nucl. Energy*), **1**, No. 2.

PLESSET, M. S. (1949) The dynamics of cavitation bubbles, *J. Appl. Mech.* **71**, 277.

RAYLEIGH, LORD (1899–1920) *Collected Scientific Papers*, Cambridge University Press, vols. 1–6.

RAYLEIGH, LORD (1917) Pressure developed in a liquid during the collapse of a spherical cavity, *Phil. Mag.* **34**, 94.

REAVIS, J. R. (1969) Vibration correlation for maximum fuel element displacement in parallel turbulent flow, *Nucl. Sci. Engng*, **38**, 63.

ROBSON, J. D. (1963) *Random Vibrations*, Edinburgh University Press.

SAMUELS, J. C. (1966) Heat conduction in solids with random external temperatures and/or random internal heat generation, *J. Heat Mass Transfer*, **9**, 301.

SAMUELS, J. C. and ERINGEN, A. C. (1958) Response of a simply supported Timoshenko beam to a purely random Gaussian process, *Trans. ASME* (Dec.), p. 496.

SANDERVAG, O. (1971) Thermal non-equilibrium and bubble size distributions in an upward steam water flow, Institute for Atom Energy, Kjeller Research Establishment Report, KR-144.

SMITH, C. O. (1956) Stress analysis for Orsort students, USAEC, CF-56-10-139 (Oak Ridge School of Reactor Technology).

SMITS, T. I. *et al.* (1962) Crest and extremal statistics of a square-law derived random process, *J. Acoust. Soc. Am.* **34**, 1859.

SNEDDON, I. N. (1951) *Fourier Transforms*, McGraw-Hill, N.Y.

STEIN, R. P. (1957) Transient heat transfer in reactor coolant channels, USAEC, AECU-3600.

THOMPSON, A. S. and RODGERS, O. E. (1956) *Thermal Power from Nuclear Reactors*, John Wiley, N.Y.

THOMSON, W. T. (1965) *Vibration Theory and Applications*, Prentice-Hall, New Jersey.

THOMSON, W. T. and BARTON, M. V. (1957) The response of mechanical systems to random excitation, *J. Appl. Mech.* **24**, 248.

TIMOSHENKO, S. (1957, 1966) *Strength of Materials*, vols. I and II, 3rd ed., Van Nostrand.

TIMOSHENKO, S. and GOODIER, J. N. (1951) *Theory of Elasticity*, McGraw-Hill, N.Y.

TONG, L. S. (1968) An evaluation of the departure from nucleate boiling in bundles of reactor fuel rods, *Nucl. Sci. Engng*, **33**, 7.

WAX, N. (1954) *Selected Papers on Noise and Stochastic Processes*, Dover, N.Y.

YEH, L. (1973) A broad band acoustic fatigue theory as applied to multi-model structural vibrations, *J. Brit. Nucl. Energy Soc.* **12**, 107.

APPENDIX

Noise Equivalent Sources

IT IS important to note that the autocorrelation function or p.s.d. obtained from artificially generated noise has an amplitude proportional to the square of the average neutron density. On the other hand, examination of eqn. (3.29) or (3.46) shows that the correlation function or p.s.d. due to inherent fission noise is directly proportional to the first power of the average neutron density. Also, whilst the functional dependence on ω and τ are the same for the two cases, the amplitudes are certainly not. It is not possible, therefore, without some additional assumption, to employ the Langevin technique to analyse the noise that arises directly from the inherent fluctuations due to fission. This can be seen more clearly if we compare eqn. (5.41) using (5.60), which makes no assumptions regarding the spectral shape of the artificial noise source. The one-delayed-neutron group form of eqn. (3.46) is:

$$\Phi_{NN}(\omega) = \epsilon\lambda_f \langle N\rangle\left\{1+\frac{\epsilon}{\Lambda^2}\frac{\overline{\nu(\nu-1)}}{(\bar{\nu})^2}\frac{(\omega^2+\lambda^2)}{(\omega^2+\alpha_1^2)(\omega^2+\alpha_2^2)}\right\} \tag{A. 1}$$

and from eqn. (5.60)

$$\Phi_{nn}(\omega) = \frac{\langle N\rangle^2}{\Lambda^2}\Phi_{\Delta\Delta}(\omega)\frac{(\omega^2+\lambda^2)}{(\omega^2+\omega_1^2)(\omega^2+\omega_2^2)}. \tag{A. 2}$$

Since $\alpha_i = \omega_i$ we note a perfect functional match between the correlated parts of these expressions provided the noise source is white. This result is not surprising since both methods are essentially measuring the transfer function of the system. The problem remains, however, as to what noise source must be added to the reactor kinetics equations to yield the correct form of $\Phi_{NN}(\omega)$ for inherent fission fluctuations from the Langevin technique (as was done for Brownian motion).

Let us return to the set of equations (5.24) and (5.25) and assume that ρ is constant but that $S(t)$ is a "noise equivalent source" which in some way simulates the effect of the inherent fission process, such that the correct correlation function is predicted. We will further assume that the mean value of $S(t)$ is zero, thereby ensuring that the equations for the averages $\langle N\rangle$ and $\langle C\rangle$ are faithfully reproduced.

The correlation function for $N(t)$ can be immediately written down from (5.39), viz.:

$$\Phi_{NN}(\tau) = \int_0^\infty dt_0 \int_0^\infty dt_0'\, G(t_0)\, G(t_0')\, \Phi_{SS}(\tau+t_0-t_0') \tag{A. 3}$$

or

$$\Phi_{NN}(\omega) = \Phi_{SS}(\omega)\frac{(\omega^2+\lambda^2)}{(\omega^2+\omega_1^2)(\omega^2+\omega_2^2)}. \tag{A. 4}$$

We concentrate on the correlated events only, since it is clear that uncorrelated processes are equal to the mean detection rate. Thus for $\Phi_{SS}(\omega)$ to reproduce the exact result we must have

$$\Phi_{SS}(\omega) = \epsilon\lambda_f \langle N \rangle \frac{\epsilon}{\Lambda^2} \frac{\overline{\nu(\nu-1)}}{(\overline{\nu})^2} \tag{A. 5}$$

$$\equiv (\epsilon\lambda_f)^2 \frac{\langle N \rangle}{\Lambda} \frac{\overline{\nu(\nu-1)}}{\overline{\nu}}, \tag{A. 6}$$

i.e. white noise.

The further assumption needed to make the Langevin method applicable is the Schottky formula (Rice, see Wax, 1954). This is a noise equivalent source and is particularly useful for representing discrete random events. The formula reads

$$A = \sum_i q_i^2 \overline{m}_i, \tag{A. 7}$$

where q_i is the net number of events (in this case neutrons) produced as a result of the occurrence of one nuclear reaction of type i, and $\overline{m}_i$ is the average number of reactions of type i occurring per second per unit volume. It is assumed that each reaction, i, is independent.

We consider two situations to obtain A: (1) non-fission capture in which $q_1 = -1$ and $\overline{m}_1 = \overline{v\Sigma_a} \langle N \rangle \Sigma_c/\Sigma_a$ and (2) a prompt fission yielding ν neutrons, such that $q_2 = (\nu-1)$ and $\overline{m}_2 = \overline{v\Sigma_a} \langle N \rangle p(\nu) \Sigma_f/\Sigma_a$, where $p(\nu)$ is the probability that ν neutrons are released in a single event. The Schottky formula now gives

$$A = \frac{\overline{v\Sigma_a}\langle N \rangle}{\Sigma_a} [\Sigma_a - \overline{\nu}\epsilon_f + \overline{\nu(\nu-1)}\, \Sigma_f] \tag{A. 8}$$

or

$$A = \frac{\langle N \rangle}{\Lambda} \left[\rho + \frac{\overline{\nu(\nu-1)}}{\overline{\nu}}\right], \tag{A. 9}$$

which, since $\rho \ll \overline{\nu(\nu-1)}/\overline{\nu}$, we can write as

$$A = \frac{\langle N \rangle}{\Lambda} \frac{\overline{\nu(\nu-1)}}{\overline{\nu}}. \tag{A. 10}$$

Now since the process of correlation considered above includes two detections we have

$$\Phi_{SS} = (\epsilon\lambda_f)^2 A, \tag{A. 11}$$

which coincides precisely with the desired result.

A number of comments are needed here concerning delayed neutrons. These were neglected in the derivation of A and indeed their inclusion would have precluded us from considering the source as white since they have a delay time λ_i. Nevertheless, when due allowance is made (Sheff and Albrecht, 1966) it is found that the correction is of order β compared with unity and hence the white noise assumption is a good one. The correction arises if ν_{di}, the number of delayed neutrons per ith precursor, is considered as a random variable. This effect was not included in our original probability balance equations (2.25) or (2.46) and is the reason for its effect not being shown in eqn. (3.46).

Some further comments in connection with the Langevin technique are useful. Thus, for example, it should be noted that eqn. (5.151) differs from eqn. (5.108) (neglecting delayed neutrons) because it is based upon an approximation of the exact equation of probability balance rather than synthesized from the Langevin equations. Thus, whilst both equations are of the Fokker–Planck type, they have very different physical origins. One is for artificially generated noise and the other for inherent fluctuations. This distinction is important since it shows yet again that the Langevin technique in its most basic form cannot predict, correctly, discrete fission events unless additional noise equivalent sources are added via the Schottky formula. A detailed derivation of the appropriate Langevin equations from which the fission fluctuation processes may be derived has been given by Dalfes (1963).

Saito (1974) also describes a complete theory of noise based upon "noise equivalent sources" and he particularly stresses the different approaches used by various authors to introduce the external noise source. In particular, it is noted that sometimes the noise is introduced in the form of random reactivity modulations, even the fission process being treated in this way. On other occasions a hybrid approach is used with the fission noise introduced by white Schottky noise and mechanical effects by reactivity modulations. Quite clearly such an approach is a sensible one since there are many fluctuations arising from "engineering" processes which are virtually impossible to describe in the microscopic manner necessary for employment of the Schottky formula. Saito accepts this fact and acknowledges that in many cases the appropriate noise source must be determined by engineering judgement.

References

DALFES, A. (1963) The Fokker–Planck and Langevin equations of a nuclear reactor, *Nukleonik*, **5**, 348.

SAITO, K. (1974) On the theory of power reactor noise, Parts I, II, III, *Ann. Nucl. Sci. Engng* (formerly *J. Nucl. Energy*), **1**, Nos 1, 2, 3.

SHEFF, J. R. and ALBRECHT, R. W. (1966) The space dependence of reactor noise, parts I and II, *Nucl. Sci. Engng*, **24**, 246 and **26**, 207.

WAX, N. (1954) *Selected Papers on Noise and Stochastic Processes*, Dover Press.

Index